S0-DUS-462
Fuel for Thought
BUILDING ENERGY AWARENESS IN GRADES 9-12

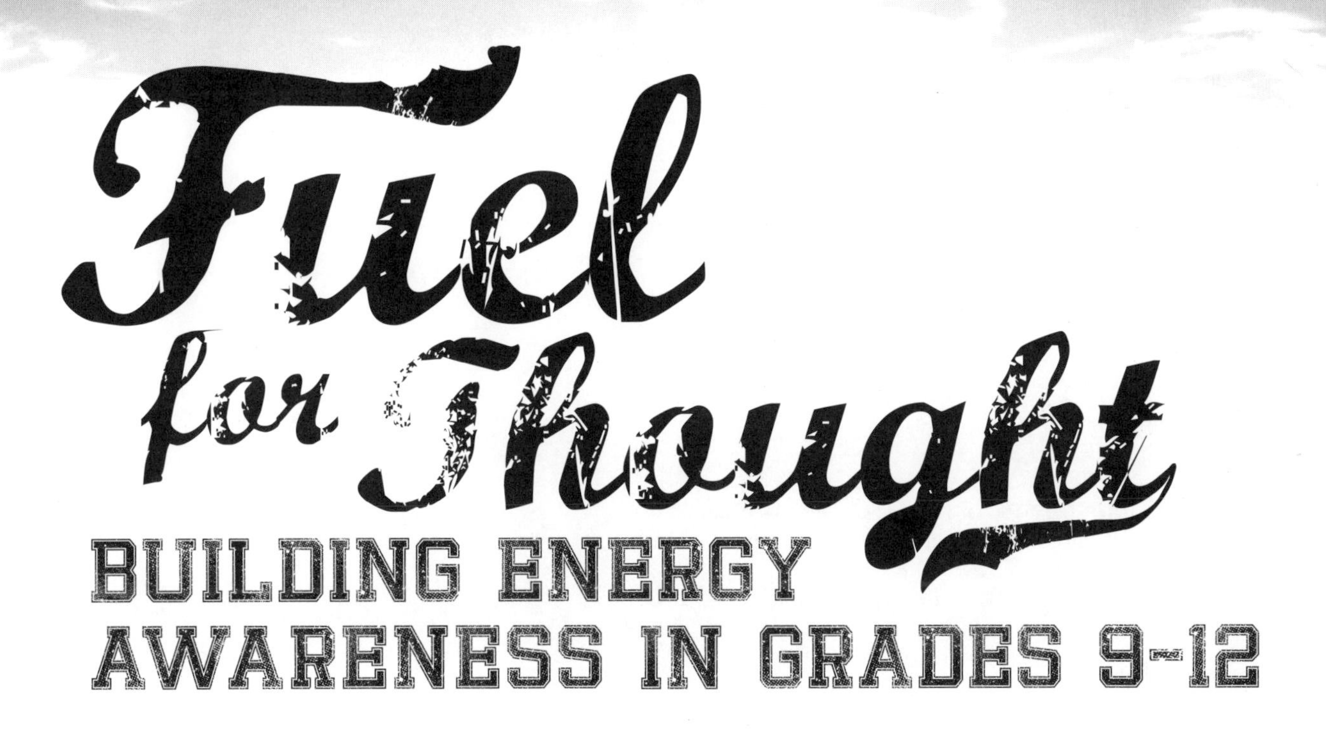

Edited by
Steve Metz

Arlington, Virginia

15 14 13 12 4 3 2 1
1840 Wilson Blvd., Arlington, VA 22201
www.nsta.org/store
For customer service inquiries, please call 800-277-5300

Library of Congress Cataloging-in-Publication Data
Fuel for thought : building energy awareness in grades 9-12 / edited by Steve Metz.
p. cm.
Includes bibliographical references and index.
ISBN 978-1-936137-20-6 (print) -- ISBN 978-1-936959-93-8 (electronic) 1. Force and energy--Study and teaching (Secondary) 2. Power resources--Study and teaching (Secondary) I. Metz, Steve. II. National Science Teachers Association.
QC73.6.F84 2011
530.071'2--dc23

2011036819

Contents

Contents

Contents

ENERGY ON THE MOVE—VEHICLES AND FUELS

LIVING CONNECTIONS

Contents

CARBON, CLIMATE, AND THE ENVIRONMENT

ALTERNATIVE ENERGY SOURCES

Contents

Contents

Introduction

By Steve Metz

As you cross the bridge over the Merrimac River into my hometown, the skyline view is a spectacular image of New England churches and historical buildings. A recent addition seems disconcerting to some longtime residents. Standing as tall as the Statue of Liberty alongside colonial era structures, this interloper seems to welcome visitors as she waves her arms in the wind. Our new high-tech wind turbine adds a modern touch to this historical town. It is perhaps a metaphor of our times, a bold visual statement that says "the times they are a-changin'."

The new energy reality is really no more than a heightened realization of the critical importance of energy in our world. Access to available energy sources fueled the Industrial Revolution and continues to be the lifeblood of modern economies. Energy heats and lights our schools, cooks our food, plays songs on the radio, runs our computers, and warms our planet. As global population and energy needs increase hand in hand, it is critically important that students learn about energy in their science classes.

The concept of energy is central to all the science disciplines. Life as we know it is impossible without energy—in fact the right *sort* of energy—so a clear understanding of energy is essential for life science students. Energy-related topics run throughout the biology curriculum: photosynthesis, cellular respiration, ecosystems, energy pyramids, and cellular transport. Energy transformations also are fundamental to understanding basic processes in chemistry and physics, from rusting cars and exploding dynamite to electric motors, wind turbines, and the photoelectric effect. In Earth and space science classes, energy conversions affect climate, tectonic plate movements, earthquakes, and wind and ocean currents. More than any other single topic, energy connects science, technology, and mathematics in an authentic way.

If energy is the most important concept we teach, it can also be the most difficult. You can't hold it in your hand. Energy involves the very small atomic-level processes (such as electrons passing from molecule to molecule, or protons shuttling across a cell membrane) and the very large (such as the gigatons of carbon dioxide we emit into the atmosphere each year, or the vast period of time it took to create fossil fuels). Learning about energy requires understanding fundamental physics and chemistry; strange-sounding units like joules, BTUs, and kilowatt-hours; and often misunderstood processes like conduction, convection, and radiation.

For this generation of students, there can be no more pressing issue for our nation and planet. The world is poised on the brink of a major shift in energy policy, as finite petroleum supplies are replaced with cleaner, renewable energy sources. The most significant environmental and social issues facing our planet are mainly issues of energy management. Food and water supplies, environmental degradation, habitat loss, the needs of an increasing population, and even disease all have some connection to energy availability and consumption. Practical alternative energy sources must be developed, and soon.

With great challenges comes great opportunity. The development of environmentally friendly energy technologies will improve economic security, mitigate global damage from climate change, and lessen the world's reliance on unstable international sources of petroleum. The new energy reality will provide employment opportunities for our students, especially those with the necessary education and training. Jobs will be created in solar, wind, bioenergy, geothermal, and hydrothermal renewable energy industries. This emerging clean energy economy will require a new generation of engineers, scientists, electricians, machinists, and skilled laborers. Science teachers can play a central role in making this vision a reality by involving students in interesting, meaningful science activities such as those compiled in this book.

In this volume you will find materials closely focused on physics topics involving energy, but we also take you on a wide-ranging interdisciplinary tour of energy topics and student investigations. Subjects include biofuels, nanosunscreens, penguins and sea ice, solar panels, convection currents, wind turbines, and climate and coal. We look at the ecological role of wildfires, the health effects of ionizing radiation, and the appearance of fall colors in our forests. You will learn how to build a lemon battery and a digital thermometer and see how one school uses renewable energy sources to defrost its sidewalks. Chapters also include suggestions for further reading and links to additional resources.

Most of the materials included in this book were designed and tested in high school science classes. A few of the activities were developed for upper–middle school classes, but these were intentionally included because they easily can be modified for use at the high school level. Similarly, you will find a wide range of inquiry levels, from activities that are more teacher-directed to investigations that involve a greater degree of student-centered, open inquiry. This too can easily be modified to provide more or less inquiry, depending on the requirements of a particular class and setting.

A WORD ABOUT ORGANIZATION

This book is organized into three basic parts or overarching domains for learning and teaching about energy. Part 1, *Student Activities and Investigations*, includes student activities and is subdivided into subcategories such as "Energy, Heat, and Temperature," "Energy on the Move—Vehicles and Fuels," and "Living Connections." Part 2, *Student Projects and Case Studies*, presents examples of more extensive, question-driven student projects. Finally, Part 3, provides background content knowledge suitable either for developing deeper teacher understanding or for use with students.

Because of the interdisciplinary nature of energy, many of these chapters can fit in more than one unit. For example, the chapters on biofuels fall easily under "Vehicles and Fuels", but might also have been placed in the "Living Connections" or "Alternative Energy Sources" categories. I hope that teachers will use these diverse materials to create interdisciplinary connections, perhaps bringing physics concepts into biology class or discussing Earth science applications in the context of a chemistry investigation.

SAFETY NOTE

These activities include safety precautions. However, they should only be attempted by a trained teacher in a proper setting where appropriate safety and personal protective equipment are present. All procedures should follow local safety standards, hazardous materials management protocols, and proper disposal procedures for chemical and biological waste. Before attempting any activity with students, work through it step-by-step on your own to know what to expect, and add whatever supplemental safety instructions or warnings you feel are necessary and appropriate.

Part 1

Student Activities and Investigations

Chapter 1

WHAT EXACTLY IS ENERGY?

By William Robertson

Seems like a simple question, so you might expect a simple answer. For that simple answer, let's head to your average science textbook and retrieve the following: "Energy is the ability to do work." Okay, but what does that mean? What do we mean by *work*? Even if we know what work is, does that give us a good feel for what *energy* is? Is energy something tangible, like a rock? Can you hold an energy in your hand? Is energy something we can model in the same way we draw models of atoms or representations of magnetic field lines?

IT'S KINETIC OR POTENTIAL

Before we answer those questions, let's back up a bit and discuss a few basics. First, it's relatively easy to recognize when something has energy. Anything that's moving has energy, and the faster the object is moving, the more energy it has. Also, the more massive a moving object, the more energy it has. A semi-truck moving at 60 miles per hour has more energy than a Yugo moving at 60 miles per hour. For the record, we call the energy an object has because of its motion *kinetic energy* (so named because the Greek word *kinesis* means motion).

Things that aren't moving also have energy. A charged-up battery has energy. A boulder poised at the edge of a cliff has energy. A stretched rubber band or spring has energy. Two magnets that have been pulled apart have energy (watch them snap back together when you let go). In these latter examples, the objects have energy because of their relative shape or position. The battery has energy because of the arrangement of dissimilar substances such as carbon and zinc; the rubber band and spring have energy because of their shape; and the magnets and the boulder (include the Earth here because gravity is important) have energy because of their position. Energy that is a result of relative shape or position is called *potential energy.*

A CONSERVED QUANTITY

It turns out that all kinds of energy—sound energy, light energy, chemical energy, nuclear energy, and on and on—fit into the two categories of kinetic energy and potential energy. That simplifies things a bit, but you might be aware that even with these two categories, there are lots of formulas one can use to calculate energy. Why calculate amounts of energy? Because by using formulas to keep track of energy as it moves from one place to another, we can better understand how various living and nonliving things interact with one another. If we understand the energy transformations throughout an ecosystem, then we know more about the organisms living there. If we understand how potential energy transforms to kinetic energy and back, then we know how to design a roller coaster.

Okay, great. We have this abstract thing called energy, and we have formulas for calculating the energy of objects as energy takes on different forms. One more thing will help us understand what energy is, and that thing is that energy (as represented by all these formulas) is a *conserved quantity.* Before explaining what that means, I need to explain what a *closed system* is. A closed system is a collection of objects surrounded by an imaginary boundary. If the system is closed, then no energy crosses the boundary in either direction, in or out. Imagine a perfect thermos that doesn't let heat in or out. This would be a closed system. A perfectly oiled machine that loses absolutely no energy due to friction would also be a closed system. Of course, it's practically impossible to create a truly closed system, but the concept is useful. If we were able to create a closed system, with no energy in or out, the total energy possessed by all the objects in the system would be *constant*—always the same total number. Objects transfer energy to other objects, and the energy changes form continuously (from one kind of kinetic energy to another, from kinetic energy of one object to potential energy of other objects, etc.). During these transformations, the total energy is constant, and that is a very useful thing for figuring what objects are doing now and what they will be doing in the future. And in case you didn't know, figuring out what objects are doing and what they will do in the future is what much of science is about. This applies to satellites in orbit, electrons hanging around atoms, the operation of ecosystems, the operation of animal body systems, and roller coasters negotiating a loop successfully.

So what is energy? It's this abstract, conserved quantity that helps us solve a multitude of science problems. No, you can't hold an energy in your hand, and you can't draw a picture of an energy, but you can use the idealized concept of a closed system, plus the fact that energy is conserved in a closed system, to analyze many situations.

THE WORK-ENERGY THEOREM

To finish, I really should explain how one calculates formulas for different types of energy. There is something called the work-energy theorem, which states that the total work done on an object is numerically equal to the change in kinetic energy of the object. We write that as follows:

Total work done = change in kinetic energy

We use a special definition of work for this theorem. Work is defined as the net (total) force applied to an object multiplied by the distance the object moves in the direction of the force. I don't expect to be able to explain fully the concept of work in this brief chapter, so we'll focus on the results of the work-energy theorem. If you apply a single force to an object for a short length of time, calculating the work done on the object (using a little bit of calculus) results in the expression $\frac{1}{2} mv^2$, which is a formula for kinetic energy that you might recognize. By carefully considering the work done by certain kinds of forces, we can also end up with formulas that represent different kinds of potential energy. The most important result, though, is that the work-energy theorem leads directly to the realization that energy is a conserved quantity. In this sense, the work-energy theorem is what really tells us what energy is.

We can also analyze the work-energy theorem in reverse. If you do work on something to give it energy, it sort of makes sense that an object that has energy can do work for you. Just imagine a bowling ball that has lots of kinetic energy. It has the ability to do work (exert a force on something in the direction of the thing's motion) on the bowling pins. So we have the expression "energy is the ability to do work." Now, doesn't that ease your mind?

Chapter 2

BURNING A CANDLE AT BOTH ENDS

CLASSROOMS AS COMPLEX SYSTEMS

By Thomas O'Brien

A long candlewick is ignited at both ends and bent into a C-shape with the two burning ends placed 2–4 vertical inches apart. When the lower end of the wick is blown out, the flame from the upper end leaps downward to reignite it. (Instead of a candlewick, you can also use two candles. See "Procedure," step 1, p. 9.)

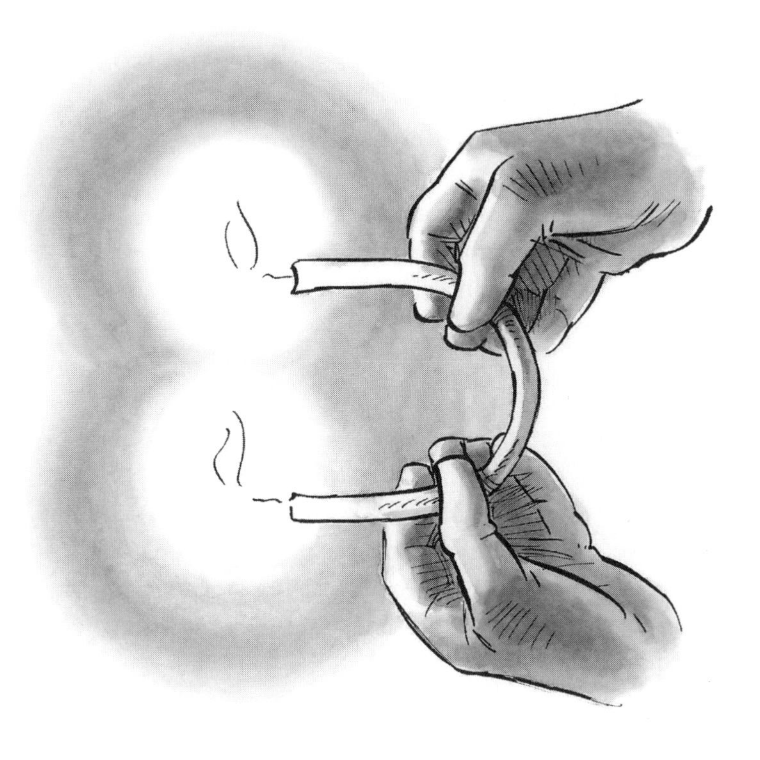

SCIENCE CONCEPTS

For a fire to ignite, it is necessary to have fuel, an oxidizing agent (e.g., oxygen), and sufficient heat to reach the fuel's kindling temperature. A seemingly simple candle is actually a surprisingly complex, dynamic system that allows paraffin (a linear, alkane type of hydrocarbon) to burn after undergoing two phase changes. The room-temperature solid melts to form a liquid that rises up the wick via capillary action (from a pool that gathers at the base of the wick). The liquid on the wick then evaporates to form a combustible gas. Convection currents then cause the burning gas (and its carbon dioxide and water by-products) to rise while fresh, more oxygenated, cooler, denser air displaces the burning gas by-products mixture upward as they in turn feed the flame from below. When the bottom flame is extinguished, a trail of invisible, hot, unburned paraffin vapor (plus some visible, partially combusted carbon particles) continues to rise (for a brief time) and ignites if it encounters the top flame and then leaps downward to reignite the extinguished lower wick.

This long explanation is provided to emphasize that the age or developmental level of students (and the course content focus) should determine how deeply the teacher explores any scientific phenomenon. This demonstration can also be used to highlight the way that we can allow memorized answers to mask conceptual holes and misconceptions. For instance, the commonly used phrase *heat rises* causes many students to envision heat as a fluid (analogous to the incorrect caloric theory of the 1700s and early 1800s). It is actually more scientifically correct to say that, under the influence of gravity, cooler denser fluids fall and displace (upward) warmer, less-dense fluids. But even this wording begs the question of how heat causes density differences. A given fluid is less dense when it is at a higher temperature because heat energy causes molecules to move faster and spread farther apart in space (thereby decreasing the heated fluid's mass/volume ratio). Without this kinetic molecular theory perspective, convection may appear to happen magically by virtue of the assumption that "heat rises."

SCIENCE EDUCATION CONCEPTS

This introductory activity models how simple it is to prepare and execute interactive, discrepant-event demonstration-experiments. They can be used on a daily basis to activate students' perceptual attention, catalyze cognitive processing, and energize interest in science phenomena. The predict-observe-explain (POE) instructional sequence elicits learners' prior knowledge, facilitates dialogue and active meaning-making, and develops science-mystery-solving skills that can uncover the scientific explanation behind various simple magic tricks. **Note:** American author Edgar Allen POE helped create the genre of mystery, science fiction, and horror short stories that included discrepant events and anomalies that are resolved by the end of the story, often with an unexpected twist—much like good science teaching but without the horror element!

This activity also serves as a *visual participatory analogy* (science education analogy) for classrooms as complex systems where teachers and students interact with one another and phenomena. Teachers can "light the fire" of interest in their sometimes undermotivated students; likewise, students who actively participate in learning can "relight" their occasionally burnt-out teachers. Similarly, teachers-helping-teachers forms of professional development enable peers to reignite each other's fire and passion for learning and for teaching.

MATERIALS

- 1, 10–14 in. waxed wick (available from candle making websites)
- OR 2, 6–12 in. candles
- OR make a wick by dipping a multistrand cotton string into hot wax

Points to ponder

As a little wood can set light to a great tree, so young pupils sharpen the wits of great scholars: Hence much wisdom have I learned from my teachers, more from my colleagues, but from my students most of all.
—Rabbi Hanina Ben Dosa (1st century AD) in the Talmud

The true aim of everyone who aspires to be a teacher should be, not to impart his own opinion, but to kindle minds.
—Frederick W. Robertson, English clergyman (1816–1853)

Education is not filling a pail but the lighting of a fire.
—William Butler Yeats, Irish writer (1865–1939)

PROCEDURE

(see "Answers to questions in procedure, steps 3 and 5" at the end of this chapter.)

1. Introduce the discrepant-event demonstration with a statement that includes a widely used verbal analogy: "People who are overworked [including teachers and students!] often feel that they are 'burning the candle at both ends.' Today we're going to study the science behind that phenomenon." For a more humorous effect, you might display this written message: "Today's lesson will focus

on the physics and chemistry that occur when a slender illumination device is subjected to rapid carbonization on its antipodal points" and then ask, "What do you think we are going to be exploring?" In either case, light both ends of the long wick (or the exposed wick ends of two separate candles). For a more dramatic visual effect, turn the lights in the room off or down.

2. Bend the long wick into a C-shape (or place the flame of one candle just below the flame of a second candle) so that the two wick ends are 2–4 vertical inches apart.
3. Ask the learners to *predict* what will happen if the lower flame is blown out. What will happen if the upper flame is blown out? Do you expect the results to be about the same or different and why? What scientific principles support these predictions? Receive the learners' ideas without commentary.
4. Start by blowing out the upper flame. After nothing unexpected happens, reignite the extinguished upper wick by touching it to the lower flame. Then repeat the process, except blow out the lower flame. In this case, the learners will *observe* that the flame on the upper wick will leap downward to reignite the extinguished lower wick. (**Note:** Be sure to adjust the position of the upper burning wick to align with the rising vapors from the extinguished lower flame. This will only work if the airflow in the room is not too turbulent, so stand away from ventilation ducts and fans, which will disperse and dilute the rising unburned hydrocarbon vapors.)
5. Elicit a variety of initial, possible, plausible *explanations* of this phenomenon. Do not prematurely focus on the one "right" answer; let learners wrestle with the questions posed by the phenomenon. Explaining how a candle "works" (including this surprising effect) involves a number of science concepts that might cut across several instructional units.

DEBRIEFING

When working with teachers

This demonstration can be used as a prop at the start of a professional development program or course to express empathy for the demands of teaching that cause teachers to sometimes experience burnout. (If you use the alternative scientific-sounding description in "Procedure," step 1, point out that facility with science terminology does not necessarily mean that someone understands the science behind a particular phenomenon.)

Use the quotes (all of which use the analogy of learning and fire) to discuss the value of daily use of interactive demonstrations or experiments to reignite both students' and teachers' interest in and understanding of science. Also, discuss the analogical connection of both informal and formal teacher-to-teacher collaborations. If time permits, briefly discuss the not-so-obvious science behind a burning candle. You can also have teachers read the detailed explanation under "Science concepts" or explore the resources cited in "On the web" as their homework.

The use of discrepant-event demonstrations for teaching can be traced at least as far back in history to Michael Faraday's famous public Christmas Lectures for England's Royal Society in the mid-1880s. In fact, Faraday presented a series on the chemistry and physics of flames predicated on his belief that "There is not a law under which any part of this universe is governed which does not come into play and is touched upon in these phenomena. There is no better, there is no more open door by which you can enter into the study of natural philosophy than by considering the physical phenomena of a candle." His book *The Chemical History of a Candle* (1860), from which the quote in the previous sentence was taken, is still in print and offers many other intriguing activities (see "On the web"). One is the use of a glass tube to channel off some of the unburned hydrocarbon vapors from the dark portion of a burning candle flame. These vapors can be ignited at the other end of the tube, thus getting two flames from one end of a single candle! Call it natural philosophy or science—this is still a magical experience!

When working with students

This demonstration could be used as either an Engage- or Elaborate-phase discrepant-event activity in 5E-based units that deal with the physics of convection or the chemistry of phase changes and combustion (e.g., oxidation of hydrocarbons). If used as an Engage-phase activity, don't use the Explain-phase teacher "answers" until after the mysterious phenomena has been further studied in the Explore phase. If the demonstration is used during the Elaboration phase, students should be able to explain the reigniting phenomenon as dependent on convection currents that push hot unburned paraffin gas upward. Ask students to predict whether convection experiments would work in a zero-gravity or microgravity environment and how they might test that (see the NASA site listed in "On the web" to study the effect of "turning off" gravity and convection currents during the free fall of a burning candle).

A variety of other real-world application questions can be used as follow-ups. For example, challenge students to consider why the direction of spin of room ceiling fans is different for summer and winter. (In summer the fan is set on "forward"; in winter the fan is set on "reverse" to push warm air down to the floor where people are.) Ask why heaters are better placed on the floor than the ceiling (and just the opposite for air conditioners).

EXTENSIONS

1. *"Magic" Birthday Candles.* These candles, which self-reignite when blown out, are fun to use with students as a related discrepant event. (They also serve as a visual analogical model for the "inextinguishable" energy of dedicated teachers….) (see "On the web.")
2. *Science Mentor.* Michael Faraday's mentor, Sir Humphrey Davy, also studied the physics and chemistry of flames and invented (and deliberately did not patent) the coal miner's safety lamp (see "On the web": Whelmer #42).
3. *Misconceptions Matter.* Teachers can use any of the following references to read about the nature of student misconceptions and their implications for integrated curriculum, instruction, and assessment: Driver, Guesne, and Tiberghein 1985; Driver et al. 1994 (especially useful); Duit 2009; Fensham, Gunstone, and White 1994; Harvard-Smithsonian Center for Astrophysics (MOSART); Keeley, Eberle, and Farrin 2005; Keeley, Eberle, and Tugel 2007; Kind 2004; Meaningful Learning Research Group; Olenick 2008; Operation Physics; Osborne and Freyberg 1985; Science Hobbyist; Treagust, Duit, and Fraser 1996; and White and Gunstone 1992.

Some authors use related terms that are somewhat less judgmental than *misconceptions* (e.g., *preconceptions, alternative conceptions, naive conceptions*, or *children's science*) to refer to the ideas that students have about a concept before a teacher formally teaches it. These student-generated ideas usually reflect a creative (though typically *not* completely correct) blending of "lessons learned" from informal life experiences and from formal schooling. Students' misconceptions are commonly activated when discrepant event activities are used.

On the web

How does a candle work? *http://home.howstuffworks.com/question267.htm*

How do trick birthday candles work? *http://science.howstuffworks.com/question420.htm*

Candle flame in microgravity: *http://quest.nasa.gov/space/teachers/microgravity/9flame.html*

The chemical history of a candle by Michael Faraday: *www.gutenberg.org/etext/14474*

Modern history sourcebook; The chemical history of a candle, 1860: *www.fordham.edu/halsall/mod/1860Faraday-candle.html*

University of Iowa Physics and Astronomy Lecture Demonstrations (five demonstrations on convection under Heat and Fluids): *http://faraday.physics.uiowa.edu*

Whelmer #42; Fire sandwich: *www.mcrel.org/whelmers/whelm42.asp*

ANSWERS TO QUESTIONS IN PROCEDURE, STEPS 3 AND 5

The detail and depth of explanation should, of course, be appropriate for your students and their prior understandings. A short version of the explanation under "Science concepts" is that paraffin actually burns in the gaseous state, and it is the trail of the partially unburned, hot paraffin vapors rising via convection currents that reignites when the lower candle (or wick end) is extinguished.

Safety note:

1. Make sure the work area is clear of combustible materials (e.g., paper) and flammable (e.g., oil, alcohol) materials.
2. Students and teacher should wear indirectly vented chemical-splash goggles.
3. An ABC-type extinguisher should be available in case of emergency.
4. Caution students about working with active flames and high heat.

Chapter 3

MAKE YOUR OWN DIGITAL THERMOMETER!

USING THE 5E INSTRUCTIONAL MODEL TO DESIGN AND CALIBRATE A SCIENTIFIC INSTRUMENT

By Timothy Sorey, Teri Willard, and Bom Kim

More and more, our world depends on digital measuring devices. They tell us whether we should wear a heavy or light jacket (digital thermometer), if we are running late (digital watch), and even at what speed we should pedal our bicycles to get to school on time (digital speedometer). Digital probeware has become commonplace in high school science laboratories. However, too often students use these devices without understanding how they work.

In the hands-on, guided-inquiry lesson presented in this chapter, high school students create, calibrate, and apply an affordable scientific-grade instrument (Lapp and Cyrus 2000). In just four class periods, they build a homemade integrated circuit (IC) digital thermometer, apply a math model to calibrate their instrument, and ask a researchable question that can be answered using the thermometer they create. This activity uses the 5E Instructional Model—engage, explore, explain, elaborate, and evaluate—to help physical science students discover the many connections between math and science (Karplus 1979).

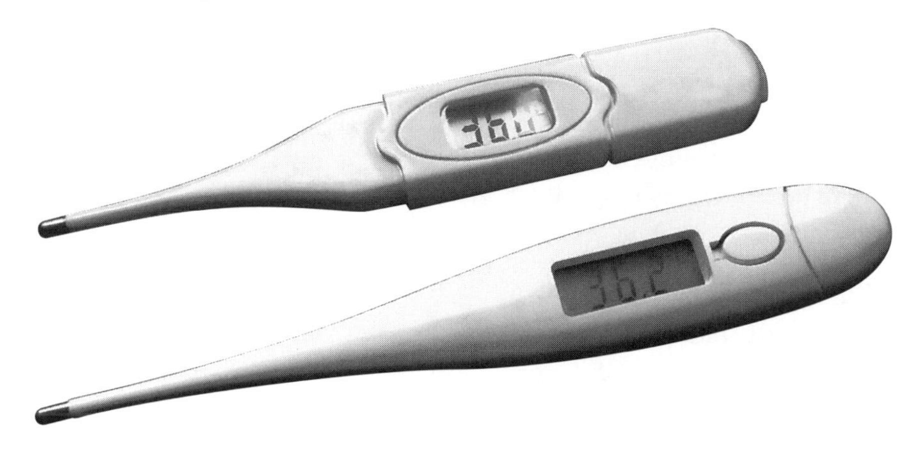

PERIOD 1: ENGAGE

The first class period begins as two glass beakers filled with 500 ml of water are passed around the classroom. (**Safety note:** Students and instructor must wear safety glasses or goggles for this demo.) Students handle beaker A, which has a water temperature of 5–10°C, and beaker B, which has a water temperature of 40–50°C. They then make qualitative observations to determine which beaker is hotter and which is cooler.

When asked to estimate the temperature of beaker B, students may disagree. A student will often suggest that we use a glass thermometer to determine its temperature. Instead, I offer them a store-bought digital thermometer.

As a class, we then discuss how instruments, such as thermometers, impact our lives and the various biological, chemical, and physical quantities they allow us to measure. We also investigate the many uses of digital thermometers—from personal health to cooking meat. After this engaging discussion, I suggest that we build our own digital thermometers.

Cost and materials

The thermometers in this activity are constructed from basic, commercially available materials, including

- 1 m of standard telephone cable
- 8–12 cm of 0.5 cm inner-diameter glass tubing (fired on both ends for safety and smoothness)
- 1 piece of 7.5 × 1.0 cm shrink tubing
- 2 pieces of 5 × 0.33 cm shrink tubing
- a 9 V battery lead
- a 9 V battery
- an LM35 sensor

These materials can be purchased for just under $10. An LM35 sensor, which records temperature in volts, accounts for about $1.20 of the total. When completed, students have a device that ranges from -18° to 150°C and is more durable than a typical lab glass thermometer. Detailed instructions for making the digital thermometer are available online (see "On the web").

Building digital thermometers

Initially, most students do not know that the LM35 sensor is a Celsius-based sensor, with an output of 0.01 volts (V) per 1°C, but they discover this later in the experiment. Figure 1 provides a circuit diagram displaying the orientation

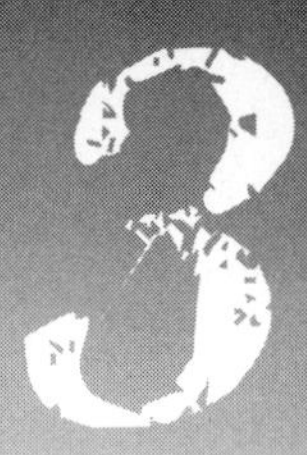

of the LM35 temperature sensor, the three colored wires, the battery poles (positive and negative), and the digital voltmeter (DVM).

Working in pairs, students trim the main plastic shielding of the telephone cable and remove 0.5 cm of the inner plastic insulation from three of the four colored wires. (**Safety note:** Use caution when dealing with pointed objects.) With the yellow, red, and green wires pushed through the glass tubing, each wire is carefully soldered to the LM35 sensor's power (V_s), output (V_{out}), and ground (GND) leads, respectively (Figure 1).

Figure 1

Circuit diagram

This diagram shows the LM35 sensor, battery, digital voltmeter, and three (colored) telephone wires

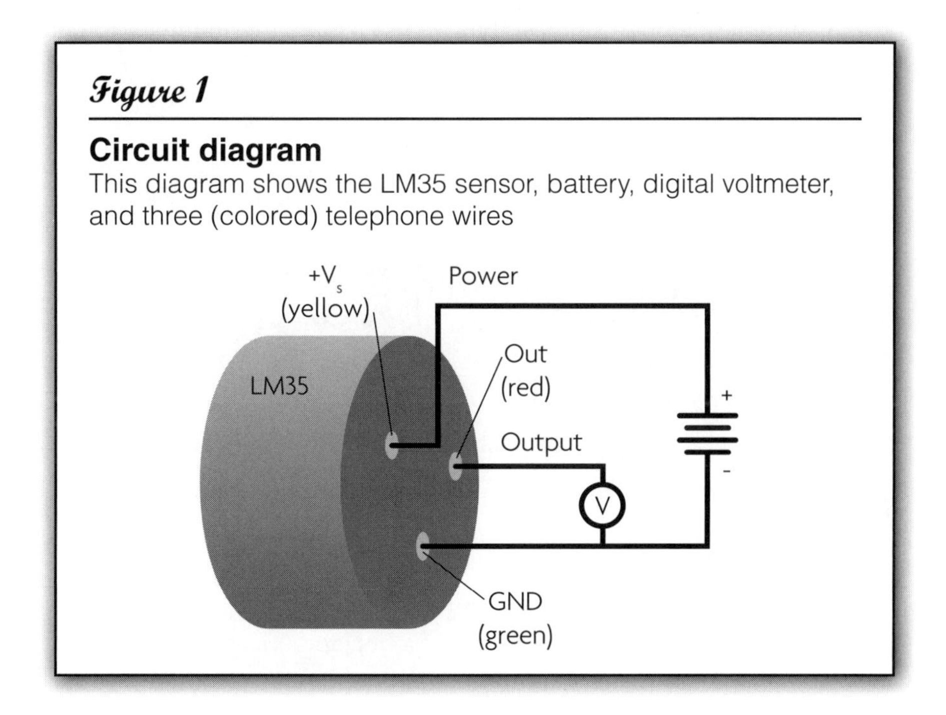

After shrink tubing is placed over the solder joints, wires are caulked with bathtub caulking and pulled back into the glass tube until the LM35 sensor is seated at its tip. The device should then be left to dry for 24 hours.

PERIOD 2: EXPLORE

The next class period, students complete the construction of their instruments by using a 7.5 cm length of 1 cm diameter shrink tubing to enclose and fortify the telephone cable–glass tube connection. They then solder the correct colored wires to the 9 V battery connector.

Before receiving a 9 V battery and DVM, each student pair predicts its instrument's output. To test their predictions, students plug in the 9 V battery and use alligator clips to attach the red positive lead of the DVM to the LM35

Figure 2

Digital thermometer with power supply (9 V battery) and readout (digital voltmeter)

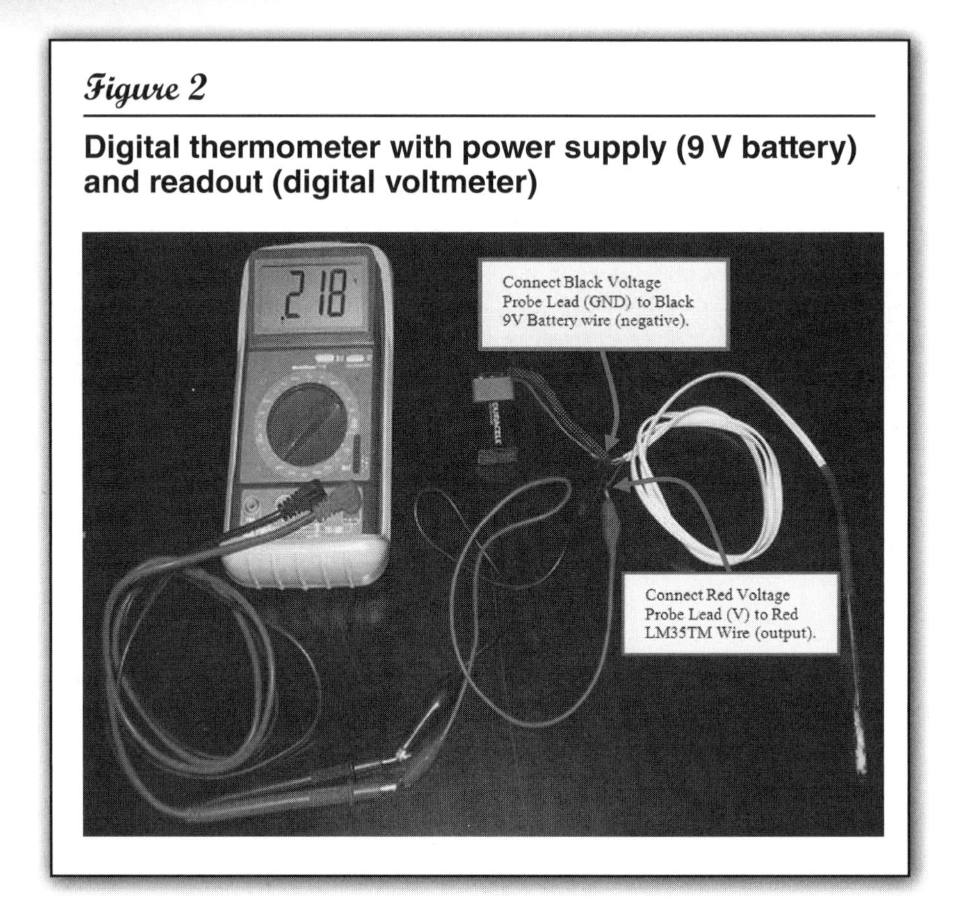

sensor output, and the black DVM ground lead to the black 9 V battery wire (Figure 2; Hill and Horowitz 1989; Skoog, Holler, and Nieman 1998).

Once the apparatus is completed and students begin to use their thermometers, many are surprised to find that their numbers do not match the values they have predicted. At this point, a historical discussion of thermometry—from Aristotle and Galileo to Daniel Fahrenheit and Anders Celsius—is introduced. Although I use a class discussion, students can also do library or internet research on the history of temperature measurement.

Students are then asked a guiding question: How did Daniel Fahrenheit and Anders Celsius convert the linear expansion of mercury in glass thermometers into degrees Fahrenheit and degrees Celsius? Students brainstorm and design calibration experiments to collect the data needed to answer this question. Some decide to calibrate with freezing or boiling water; others decide to measure water baths of various temperatures with both their digital thermometer and a glass thermometer (Figure 3). I encourage students to collect at least three ordered pairs of data (x = V, y = °C). Table 1 shows six ordered data pairs.

Figure 3

Measuring water bath temperature

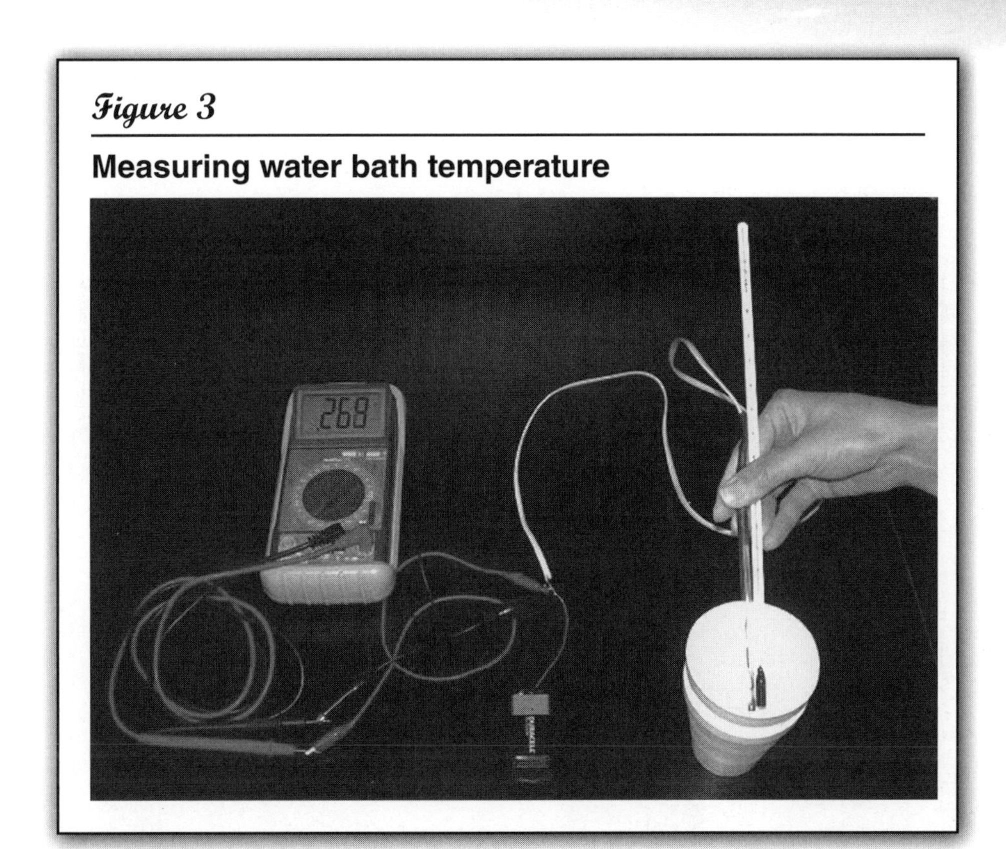

Table 1

Temperature data obtained using LM35 sensor

Water bath descriptions	Reading from LM35 sensor in volts (*x*)	Reading from thermometer in °C (*y*)
Ice water slurry	0.035	1.0
Below room temperature	0.123	11.0
Room temperature	0.224	22.0
Slightly above room temperature	0.321	32.0
Above room temperature	0.417	41.0
Well above room temperature	0.476	50.0

PERIOD 3: EXPLAIN

Many of my students have previously studied scatter plots and linear regressions in math class. These students typically see that it is logical to plot their data on graph paper, observe a trend, and draw a best-fit line between the points.

Most students choose two points on their best-fit line and write the equation that models related volts to degrees Celsius. They use graphing calculators and computer programs to graph their results, although the data can also be graphed using a pencil and paper. For the data in Table 1, an equation, °C = 107.0 V – 2.79, was determined using the graph in Figure 4.

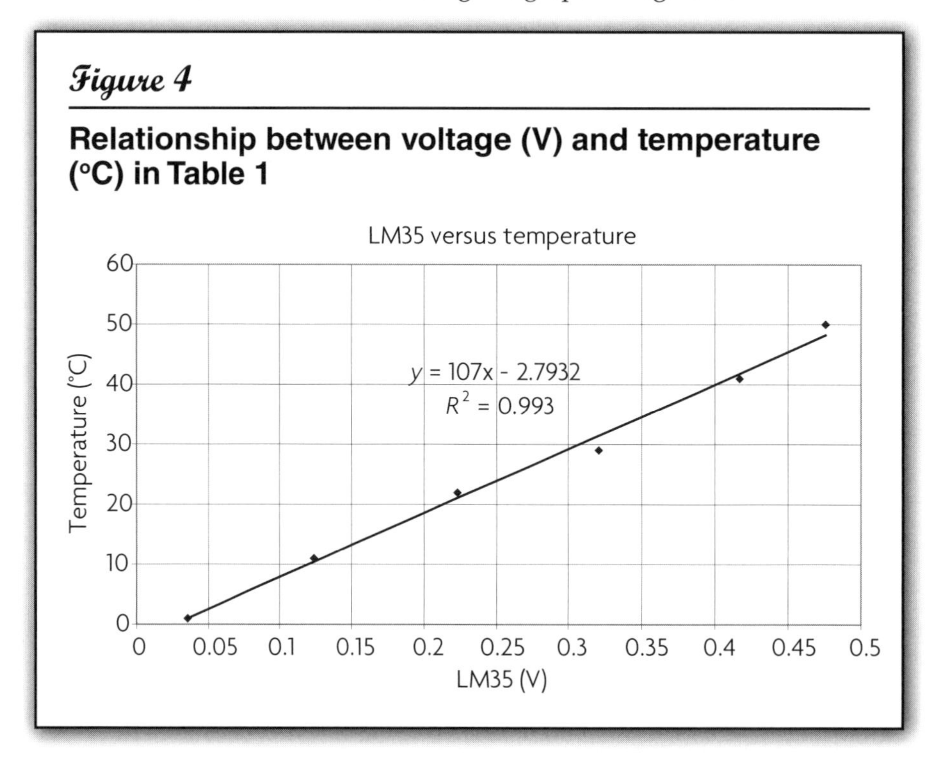

Figure 4

Relationship between voltage (V) and temperature (°C) in Table 1

Some students do not understand that, on the graph, *y* represents degrees Celsius and *x* represents voltage. Students are used to plotting *x* and *y* values, but often fail to recognize that these variables can represent physical quantities such as volts and degrees. At this point, students compare graphs and calibration equations. In my class, students recognized that everyone's data appeared to be linear but contained slightly different values for slope and *y*-intercept. Each group obtained an equation for the relationship between degrees Celsius and volts that was close to the expected value: °C = 100.0 V – 0.

As a final check, students measure their body temperatures at the armpit using their digital thermometers and apply their calibration equations to convert from volts to degrees Celsius. (**Safety note:** Use caution in placing the glass tube under the armpit so as not to break the instrument.) In our class, students' body temperatures ranged from 0.34–0.38 V, which corresponds to a range of 33.6–37.9°C. One student commented, "If I simply multipl[y] these voltage values by 100, that is about the same as the [degrees Celsius] values." This revelation was crucial for students in realizing that their digital thermometers did, in fact, work. Through the application of mathematics, they discovered that the LM35 sensor is a Celsius-based sensor.

The digital thermometer that students create contains almost all of the basic components found in an electronic sensor. Unlike a standard digital thermometer, this student-constructed version does not contain a "data processor" to convert volts to degrees Celsius. Instead, students apply mathematics to find a calibration equation they can use to convert data.

PERIOD 4: ELABORATE AND EVALUATE

Equipped with their calibration equations, lab pairs then formulate an experiment that both answers a researchable question and validates the calibration of their digital thermometers. Some students need guidance in devising experiments. In a ninth-grade physical science class, typical research questions might range from "How fast does water cool down in a beaker compared to a foam cup?" to "When shaking different types of metal shot in an enclosed plastic bottle, does the type of metal affect the temperature increase?" In an advanced chemistry or physics class, student questions might include "What is the freezing point and boiling point of glacial acetic acid?" and "What is the equilibrium temperature of 100 ml of 20°C water mixed with 50 ml of 50°C water?" Guiding students toward a researchable question requires patience so that students can clearly identify manipulated and responding variables for their experiment.

SAFETY

Students must be trained for safety in the use of all manufacturing devices prior to creating this sensor and wear indirectly vented chemical-splash goggles at all times—especially during the cutting and firing of the glass tube and the soldering of electronic joints. Students should use the soldering station one at a time. Hazardous fumes from silicone caulking and soldering should be avoided by use of a fume hood or direct ventilating exhaust system. Make sure that students wash their hands with soap and water after using the solder—as it contains lead, a poisonous metal. An ABC-type fire extinguisher must be located in close proximity so that it can be immediately available should the need arise. Floor areas under and within 11 m (35 ft.) of the soldering operation must be swept clean of combustible and flammable materials.

ASSESSMENT

Over the course of four periods, students are assessed with a rubric (see "On the web"). For further evaluation, students can calibrate a new sensor using the same methods used to create their original sensor. The LM34 (Fahrenheit-based) and LM335 (Kelvin-based) sensors can be used as additional tools. Students can also be given data from these sensors and asked to find a new calibration equation.

CONCLUSION

The purpose of this hands-on activity is to have students learn the skills and knowledge needed to create a reliable scientific-grade instrument. Today, many educational, computer-based labs have been reduced to "plug-and-play" devices that acquire data, but are often implemented with little or no thought about calibration and precision.

This activity demonstrates a real-world connection to science—and encourages students to search for more connections between science and mathematics in the classroom and beyond. Together, science and math teachers must strive to provide students with hands-on, cross-curricular experiences that enhance their understanding and empower their explorations. As a result, students discover just how interconnected and inseparable the fields of science and mathematics truly are.

On the web

Constructing your digital thermometer: *www.nsta.org/highschool/connections/201003Constructing.pdf*

Rubric for guided-inquiry activities: *www.nsta.org/highschool/connections/201003Rubric.pdf*

References

Hill, W., and P. Horowitz. 1989. *The art of electronics.* 2nd ed. Cambridge: University of Cambridge.

Karplus, R. 1979. Science teaching and the development of reasoning. *Journal of Research in Science Teaching* 14: 69–175.

Lapp, P., and U. Cyrus. 2000. Using data-collection devices to enhance students' understanding. *Mathematics Teacher* 93: 504–511.

National Research Council (NRC). 1996. *National science education standards.* Washington, DC: National Academies Press.

Skoog, D., F. Holler, and T. Nieman. 1998. *Principles of instrumental analysis.* 5th ed. Philadelphia: Saunders College.

Chapter 4

EVAPORATING IS COOL

By Richard Hand

Many students hold misconceptions about evaporation. This is a short exercise I use to get them to apply the kinetic molecular theory to explain how cold water can evaporate and to observe the cooling effect of evaporation (see "Exploring evaporation," p. 25). The activity asks students to develop their own evaporation experiments. My students' experiments have included putting water in a dark place to see if it evaporates, measuring the temperature of water as it evaporates, and even placing a small, uncovered container of water in the refrigerator to see if evaporation occurs in a dark, cold place.

TEACHER BACKGROUND

So how is it possible for cold water to evaporate? The key idea to understanding the evaporation of water is in the motion of the water molecules. Water evaporates when fast-moving water molecules escape from the liquid. Because water molecules are always moving and bumping each other, the speed of individual molecules is constantly changing. After molecules bump into each other, one may bump away with greater speed and the other with less speed. In hot water many molecules are moving fast enough to escape and the water quickly evaporates. However, even in cold water, some of the water molecules are moving fast enough and have enough energy to escape

from the liquid and go into the air. (Water may also absorb heat from warmer surroundings, but it is not necessary for evaporation. The energy for evaporation is extracted from the water itself, which leaves the remaining liquid water even colder.)

What evidence supports this explanation? If the faster molecules are escaping, the slower ones will be left behind in the liquid. If the faster molecules are being removed, the average speed of the remaining molecules will decrease. These molecules have less energy and the temperature of the water decreases. We feel cooler when we come out of a swimming pool on a hot day because the faster-moving water molecules escape into the air and this makes the water remaining on the skin cooler. After emerging from the pool, evaporation cools the water on the skin. The escaping water molecules carry off energy. This loss of energy cools the water. The cooler water then absorbs heat from the skin and thus cools the skin.

If this is true, some students want to know why sweating makes them feel hot. Sweating is the body's way of trying to keep cool, and it works well during dry weather when water can evaporate. As the water evaporates from the skin, the remaining water becomes cooler. The cooler water absorbs heat from the skin and cools the body. When all the water has evaporated, the cooling effect will cease. Water molecules move randomly and at varying speeds. As the faster molecules escape from the water, they remove energy from the water. When the water gets colder, it begins to absorb heat from the skin.

How else is the cooling effect of evaporation used? In dry climates, homes can be cooled by hanging wet mats in the entrances and allowing water to evaporate. Water can be stored in porous earthenware vessels. Water evaporating from the outside of the vessel keeps the water inside cool. Some water canteens are covered with felt. Keeping the felt wet cools the water in the canteen. Parks with many trees are cooler than treeless city streets because water evaporating from the plants lowers the temperature.

Water, alcohol, and acetone have different effects on the temperature. Of the three liquids, water has the strongest attraction between molecules, and therefore only the very fast molecules can escape from the water. In "Exploring evaporation," the thermometer with water on the tissue verifies this by showing the smallest decrease in temperature. The attraction is less between the alcohol molecules and even weaker between the acetone molecules. Even the slow acetone molecules can escape from their liquid, and therefore only the very slow acetone molecules are left on the tissue. Slower moving molecules means lower temperature. The thermometer with acetone verifies this by showing the greatest decrease in temperature. Squirting a dropper full of each liquid on the chalkboard and watching how much quicker the acetone

Exploring evaporation

Can cold water evaporate or does only boiling water evaporate? How does a water puddle dry up? Does the water soak into the ground, is it blown away by the wind, or does it evaporate? Must water be heated by the Sun in order to evaporate? What experiments can be done to answer these questions?

Part A

1. Develop a hypothesis to answer one of the above questions.
2. Briefly describe an experiment that will provide evidence for your hypothesis.
3. After receiving teacher approval, carry out your experiment and summarize the results.
4. State the kinetic molecular theory of matter and explain how this theory helps explain how cold water can evaporate.

Part B

How do you think evaporation affects the temperature of the surface from which it evaporates? Give the reasoning behind your answer. Carry out the following procedure to get an answer to this question.

1. Put on safety glasses.
2. Record the starting temperature of a dry thermometer.
3. Wrap a 3 × 4 cm piece of tissue around the bulb of the thermometer.
4. Slowly soak the tissue with 2 ml of water. Record the temperature every 30 seconds for 5 minutes.
5. Remove the tissue and repeat the procedure using alcohol.
6. Your teacher will carry out the procedure for acetone and provide you with the readings to record.

Safety Note: Alcohol and acetone are both flammable and should only be used with caution and proper ventilation.

Analysis

1. How were your results similar for all three liquids? How did they differ? Give your reasoning for each of these answers.
2. How does the human body make use of the cooling effect of evaporation?
3. How else are the cooling effects of evaporation used?

evaporates can show the difference in attraction between molecules in these three liquids.

I assess student understanding by showing them the toy "drinking bird." The bird does not drink, but by dipping its beak in water, it keeps its head wet. Students must explain where evaporation and cooling are taking place. Water evaporating from the head cools it and causes some of the gas in the head of the bird to condense. As the volume of gas decreases, liquid rises in the tube until the bird tips forward and "drinks." Then I ask students to think of ways to increase or decrease the rate of drinking without altering the bird. They must then explain how this change in drinking rate is produced. For example, placing the bird in a warm or sunny place will increase the rate of evaporation and cooling, and thus increase the drinking rate. A fan placed at a right angle to the bird will also increase evaporation and drinking. Placing the bird in an enclosed space, such as an empty aquarium, will prevent evaporation and stop the bird's movement.

Use students' own observations—water evaporates from lakes and oceans without boiling; clothes, dishes, and wet heads will all dry without heating them to the boiling point—to reinforce the idea that heat makes water evaporate faster, but water will evaporate even without being heated or brought to a boil. Even in a freezer, water molecules slowly escape from ice cubes, making them smaller. The random motion and random speeds of the molecules means that even some molecules in cold water have enough speed to escape from the water. As the faster-moving molecules escape from water, they remove energy from the water and cool it. As the water becomes cooler than its surroundings, the water begins to absorb energy from it. Then more water molecules escape into the air and continue to cool the water until the water evaporates.

Chapter 5

SAVE THE PENGUINS

TEACHING THE SCIENCE OF HEAT TRANSFER THROUGH ENGINEERING DESIGN

By Christine Schnittka, Randy Bell, and Larry Richards

Many scientists agree that the Earth is warming, and that human activities have exacerbated the problem (NRC 2001; 2002). Engineers, scientists, and environmental groups around the globe are hard at work finding solutions to mitigate or halt global warming. One major goal of the curriculum described here, Save the Penguins, is to help students recognize that what we do at home can affect how penguins fare in the Southern Hemisphere. The energy we use to heat and cool our houses comes from power plants, most of which use fossil fuels. The burning of fossil fuels has been linked to increased levels of carbon dioxide in the atmosphere, which in turn has been linked to increases in global temperature. This change in temperature has widespread effects on Earth, including effects on the life of penguins. If homes were better insulated, they would require less energy for heating and cooling, reducing fossil fuel use and carbon dioxide emissions. This is the problem presented to students in the beginning of the Save the Penguins curriculum.

In the Save the Penguins curriculum, students learn how engineers are addressing global warming by designing energy-efficient building materials. Students learn the science of heat transfer, design experiments to test materials, and then assume the role of engineer to design, create, and test their own energy-efficient dwellings.

PENGUINS

Penguins are in peril. As the Earth warms, the oceans warm, pack ice melts, and penguins lose habitat. They also lose food sources such as krill, which rely on the protection of pack ice and feed on the algae that grow underneath the ice (Gross 2005). The emperor penguins in Antarctica are in severe decline due to climate change (Jenouvrier et al. 2008). South African penguins are actually leaving their nests to cool off in the water, placing their eggs at risk to attacks by gulls. Park rangers at Boulders Beach in Cape Town, South Africa, have created little semienclosed "huts" for penguins to nest in, which keep them cooler and protect their eggs from predation (Nullis 2009). Several short videos found on YouTube help engage students in this issue and address the impacts of global warming (see "On the web"). If your school blocks access to YouTube, download and save the videos at home using an online tool such as the one provided free at Downloader9.com, then save them to a disk or flash drive for use at school.

SETTING THE STAGE FOR ENGINEERING DESIGN

The problem presented to students in this curriculum is an analogy with symbolic meaning—to build dwellings that protect penguin-shaped ice cubes from increasingly warming temperatures. Through their work on the project, students learn about thermal energy transfer by radiation, convection, and conduction. Students test materials for their ability to slow thermal energy transfer in order to keep the ice penguins cool. After testing materials, students build their penguin homes, and then see how well the dwellings keep the penguin-shaped ice cubes from melting in a test oven.

STUDENTS' ALTERNATIVE CONCEPTIONS

Involving students in a series of discrepant-event demonstrations helps them form more scientific understandings of heat transfer. The five demonstrations that follow can be used to target common student misconceptions. Other demonstrations are possible, but the following were used in research and empirically shown to be effective (Schnittka 2009).

THE CANS: UNDERSTANDING INSULATION AND CONDUCTION

Materials

- 6 cans of soda refrigerated overnight
- paper towel
- aluminum foil
- plastic wrap
- wool sock
- cotton sock
- 6 thermometers

If you ask students what they would wrap around a can of soda to keep it cold on a field trip, most will suggest wrapping it in aluminum foil. They might explain that aluminum foil "keeps the cold in." They would not dream of wrapping the can in a wooly sock because of the prevalent conception that wool makes things warm. Demonstrate this fallacy by wrapping cold cans in different materials and taking their internal temperatures several hours later. Be sure to include a control (no wrapping) for fair comparison (Figure 1). Have students make predictions about which can is the coldest. They are usually quite surprised to find that the wool sock keeps the soda coldest. Use this demonstration to help your students understand what insulators and conductors are. This demonstration and the interpretive discussion that follows usually take 20 minutes.

Figure 1

The cans demonstration

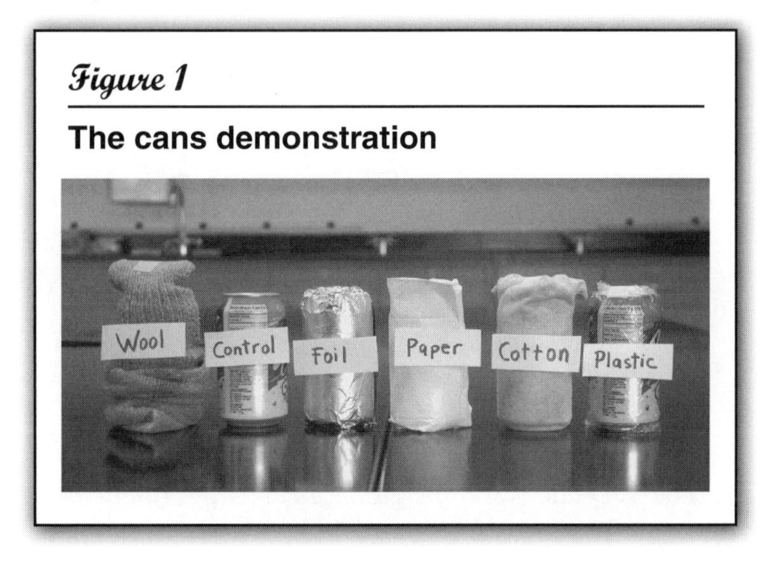

THE TRAYS AND THE SPOONS: UNDERSTANDING WHY METALS FEEL COLD

Materials

- plastic tray
- silver (or silver plate) tray
- aquarium thermometer strips
- silver (or silver plate) spoons
- plastic spoons
- penguin-shaped ice cubes

Figure 2

Plastic and silver-plate trays

Figure 3

Taking the temperature of the trays

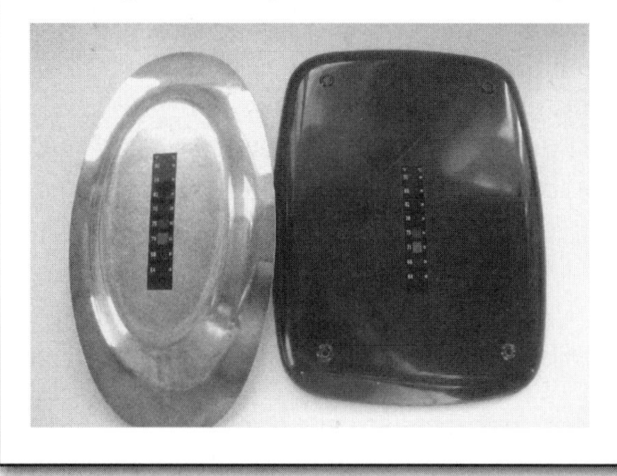

To prepare for the following demonstrations, borrow or bring from home two trays—one metal and one plastic (Figure 2). Tape an aquarium thermometer strip to the underside of each tray. Each thermometer should display room temperature (Figure 3). Flat LCD thermometers in a large, easy-to-read size can be purchased from PetSmart. Alternatively, Ideal brand #61-310 multimeters with thermocouple probes can be used. They can be purchased from a number of online sources.

For the first part of the demonstration, walk around with the trays and allow students to touch them and describe which tray feels colder. While both trays are at room temperature, students will insist the silver tray is colder. After the demonstration with spoons (description follows), reveal the actual temperatures of the trays to students. Students should understand that both of the trays are the same temperature, but that heat transfers faster from a warm hand to the silver tray than it does from a warm hand to the plastic tray.

The second part of this demonstration involves spoons and penguin-shaped ice cubes. Lékué brand penguin ice cube trays (model # 39004) can be purchased online from a number of sources (Figure 4). Give each student group a plastic and a silver spoon to hold. Silver plate or stainless steel will suffice, but silver is a better conductor of heat. Students may comment that the silver spoon feels colder and infer that metals are naturally colder than plastics, just as they may have done with the two trays. Have students predict which spoon works best to keep an ice cube from melting. Most students predict the silver spoon because it feels colder, as did the silver tray. Place penguin-shaped ice cubes on the spoons and have students take turns holding the spoons (Figure 5). Ask students to explain why the penguin in the plastic spoon does not melt significantly, while the penguin in the silver spoon quickly turns to water.

Figure 4

Lékué brand penguin ice cube tray

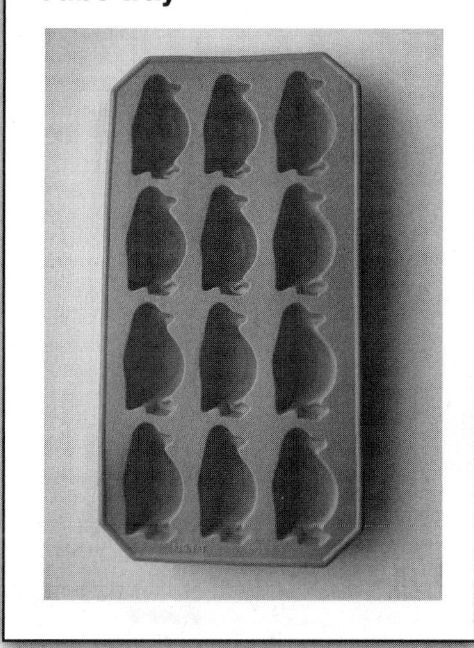

Figure 5

Silver and plastic spoons with penguin-shaped ice cubes

Through group and class discussions, bring students to an understanding that heat transfers from where the temperature is higher to where the temperature is lower, and that heat transfers faster through the silver spoon than the plastic spoon because silver is a better conductor than plastic. The silver spoon feels colder because the hand, which is warmer, is losing thermal energy in the heat transfer. While students may try and insist that cold transfers because they feel their hands get cold, remind students that there is no such thing as cold transfer; only thermal energy, or heat, can transfer. The feeling that the hand is cold when touching the metal is evidence that the hand is a very good detector of rapid heat loss.

Next, reveal the equal temperatures of the silver and plastic trays. This discrepant event promotes a lively discussion and helps students come to a deeper conceptual understanding about the conduction of heat.

THE HOUSE: UNDERSTANDING CONVECTION IN AIR AND RADIATION FROM LIGHT

Materials

- cardboard house made from a box
- digital thermometers or thermocouple probes
- stopwatch or watch with second hand
- computer with spreadsheet software (optional)

The following demonstrations help students understand that convection occurs when less-dense fluids rise and more-dense fluids fall. It can also be used to demonstrate how black surfaces absorb radiation and become quite hot, while reflective surfaces do not. These demonstrations and the interpretive discussions that follow usually take 20 to 30 minutes.

Make a hollow cardboard house and paint the roof black with acrylic paint. Shine a shop light onto the house, and place temperature probes in the attic and the first floor. The attic gets quite hot inside, approaching 38°C (100.4°F). (**Safety note:** The cardboard black roof gets very hot and can start to smoke if the shop light is placed too closely. Be sure to monitor this, and have a fire extinguisher nearby just in case.)

After you turn off the light, ask students why the attic remains hotter than the first floor. The likely answer is "Heat rises!" The scientifically correct explanation would be that cooler air falls because it is more dense (particles are closer together), and this forces hot air, which is less dense, to rise. There are several popular demonstrations of convection involving water and food coloring that can be used, but this demonstration helps students understand that convection is the movement of all warmer and cooler fluids (liquids and gases), not just water.

After the light has been off for a moment, flip the house upside down. The air masses change places due to density differences, and the thermometers or temperature probes reveal this change. Have students record the temperature of the attic and first floor as the air masses change places. They can graph the data by hand or enter the data into a spreadsheet to graph the changes over time. Without access to computer-based probes, we usually have one pair of volunteers, wearing safety goggles, call out the temperatures

while one volunteer calls out the time every five seconds and another pair of volunteers record the temperatures. See Table 1 for sample data collected from a demonstration.

Make sure students come to understand that the air masses are changing places as cooler, more-dense air falls and warmer, less-dense air rises. You might also want to repeat the demonstration with several layers of aluminized Mylar film protecting the house, as seen in Figure 6, in order to demonstrate the difference between a black roof and a reflective roof (optional).

Table 1

Data and graph from house demonstration

Time (sec.)	Attic temperature (°C)	First floor temperature (°C)
5	37	22
10	36	23
15	35	24
20	34	25
25	33	26
30	32	27
35	31	28
40	30	29
45	29	28
50	28	27
55	27	26
60	26	25

THE SHINY MYLAR: UNDERSTANDING RADIATION AND HOW TO REFLECT IT

Materials

- shop lights (1 per group with 150 W lightbulb)
- ring stands (1 per group)
- several sheets of aluminized Mylar (per group)

The following demonstration helps students understand that light can be reflected, preventing heat transfer from radiation. This demonstration and the interpretive discussion that follows usually take 15 minutes.

Mylar is a polyester film that can be coated with aluminum to be very shiny. It is usually thin and can be somewhat translucent, but when layered, it is excellent at reflecting light and preventing heat transfer. It is commonly used in snack-food packaging, helium-filled balloons, and emergency blankets. We have found that the easiest and

Figure 6

Mylar on a hot black roof

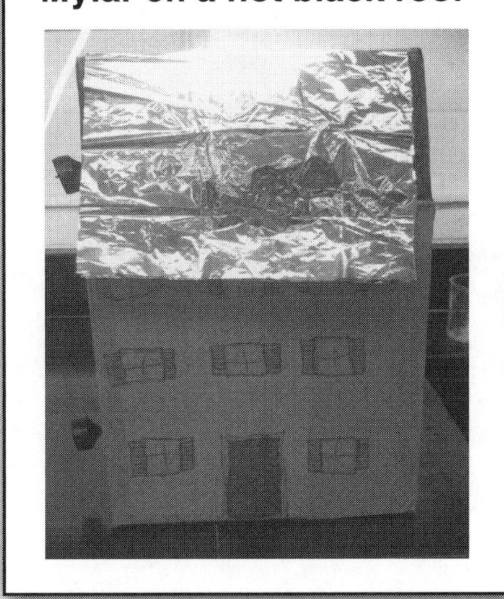

least expensive way to purchase aluminized Mylar is at a craft or stationary store as metallic tissue wrapping paper.

Have students feel the radiation from the lights with their hands. (**Safety note:** Make sure students keep a safe distance from the shop light. To do this safely, have the shop light mounted on a ring stand at a safe distance pointing toward the counter.) Have students working in groups take turns placing their hands on the counter under the lamp as their partners slip the shiny Mylar layers in between the light and their hands. Students will immediately feel the reduction of heat transfer, especially with multiple layers. Use this demonstration to help students understand that radiation is a type of heat transfer that occurs nearly instantaneously, at the speed of light, and that the light is easily reflected by shiny materials. Students are amazed at how suddenly the heat transfer is blocked the moment the Mylar is placed between their hands and the light.

CREATING A DWELLING FOR ICE CUBE PENGUINS

Testing building materials

After students have a good, basic understanding of how heat transfers, introduce the challenge: to build a dwelling for a penguin-shaped ice cube in order to keep the penguin from melting. First, students must decide which materials to build with.

Figure 7

Samples of materials for building dwelling

Students engage in scientific inquiry as they test materials and eventually design and test dwellings. Student groups are given materials such as felt, foam, cotton balls, paper, shiny Mylar, and aluminum foil to test for their effectiveness at preventing some form of heat transfer (Figure 7). It is best for the teacher to prepare all of the samples ahead of time. We find that cutting the fabrics, foils, papers, and foams into 7.6 × 7.6 cm (3 × 3 in.) squares with a 7.6 × 45.7 cm (3 × 18 in.) clear acrylic quilting ruler on a cutting mat with a rotary cutter works best. These items can be purchased from sewing stores. (**Safety note:** All safety protocols need to be followed with building and cutting—students should wear safety glasses.) Decide as a class which materials or combinations of materials can be compared and tested, then divide the work up among the different lab groups. It is reasonable for each group to run three or four tests. For example, one group might compare aluminum foil to shiny Mylar, while another group compares white felt to white foam. Some groups could test bubble wrap with different-color papers on top. Students can compare materials under a shop light mounted to a ring stand shining on a black surface such as a black countertop or black plastic tray. Give students access to thermometers and timers to fairly test samples under the light or on the hot black surface. As students explore the materials, they begin to formulate ideas about how to build a dwelling for a penguin-shaped ice cube so that the least amount of ice is melted. Allow students to procure additional materials after they decide which ones are better building materials. We usually price the materials and give each student group a budget to work with.

Encourage open inquiry and allow students to come up with their own testing ideas. Be sure student groups share the results of their experiments so that the knowledge gained is communal, encouraging a more collaborative and less competitive environment. During sharing time, it is helpful to encourage students to explain to each other why they tested certain materials, what their results were, and why they think the results turned out as they did. This is the perfect opportunity to help students understand heat transfer as it applies to each experiment performed. Expect students to spend at least one class period testing materials and half of a class period sharing their results with each other.

Building the dwelling

After students have tested different building materials and shared their results with the class, they will be eager to start using the materials to build little houses to keep the penguin-shaped ice cubes from melting. Students will finally be able to apply what they have learned from the demonstrations, discussions, and materials testing. As students take on the role of "engineer" to keep ice from melting, remind them that engineers are designing innovative materials for houses, schools, and other buildings to prevent heat transfer. Preventing heat transfer is energy efficient, and efficient buildings use less energy. The less energy needed to heat and cool a building, the less negative impact it has on the environment.

Provide students with tape and glue, and scissors to cut paper and fabrics. Most students will use the materials as given, but some will want to modify them by cutting, folding, crumpling, or layering. Students need to be sure to create an opening so that the penguin can be easily placed in the dwelling and easily removed after testing. As the designs evolve, students may need to conduct further testing, discuss results with other groups, and receive support for their ideas from the teacher. A class period should be sufficient for this phase of the curriculum. Engineers work with constraints of time, space, and money, so give your students a time deadline, too.

Testing the design

Eventually, students are ready to evaluate their dwellings in the oven. The oven can be a large plastic storage bin lined with aluminum foil on four sides and spray-painted black on the bottom, with three 150 W shop lights shining inside so that all three forms of heat transfer can occur. If you use a black plastic storage bin, simply line the sides with foil and attach the shop lights. See Figure 8 for a suggested test oven design. Houses placed in this preheated oven experience conduction with and radiation from the black floor, radiation from all sides, and convection as hot air rises off the black bottom.

Figure 8

A suggested test oven design

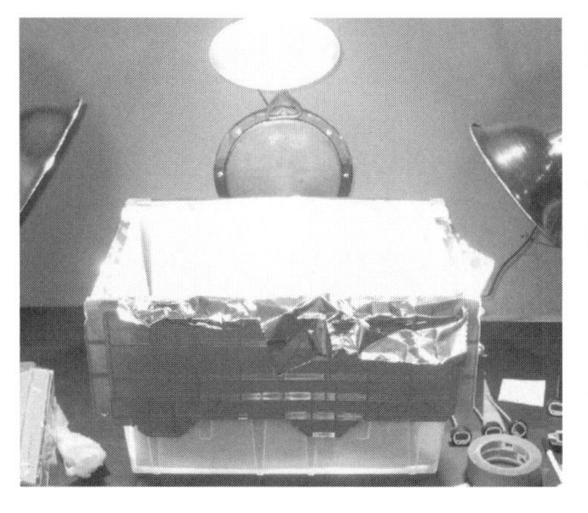

Fill the ice cube trays the night before with 10 ml of water for each penguin. We use a medical syringe to be accurate. The carefully created and frozen 10 g ice penguins are simultaneously placed inside the dwellings and then placed in the oven and subjected to 20 minutes of intense heat. See Figure 9 to see how dwellings are placed in the oven. The oven can accommodate more dwellings depending on its size, but students need to be told about the space limitation. Typically, we have eight teams in each class, so we tell students that their dwelling cannot be larger than one-eighth of the floor space in the oven.

Figure 9

Placement of dwellings in the oven

During the 20-minute wait, we sometimes show students a slide show on energy-efficient building materials such as smart windows, aerogels, radiant barriers, and solar panel roofing tiles. Sometimes we show an excerpt from the movie *March of the Penguins* or an excerpt from *An Inconvenient Truth* to spark discussion. After 20 minutes, the dwellings are finally removed and the remaining solid portions of the ice penguins are placed in little plastic cups for mass measuring.

(Determine the mass of the plastic cups ahead of time and write the mass on the cups.) Students always enjoy the thrill of rescuing their dwelling from the oven and seeing how much of the ice cube penguin they were able to save. After the solid penguin remains are placed in plastic cups, it will not matter if some melting occurs prior to measuring the mass because the water stays in the cups. This provides a teachable moment to discuss melting and the conservation of mass. It usually takes a few minutes for each group to take a turn determining the mass of their cup and penguin on a digital mass scale, then subtracting the mass of the cup. (**Safety note:** Mount the test-oven shop lights on ring stands so that they can be turned off and moved away when students place their penguin dwellings in the oven or retrieve them.)

The results are shared and discussed after all penguin-ice-cube masses are measured. Some student groups are able to save at least half their penguin, but some only have a few grams remaining.

It is always interesting to discuss which design features were best at preventing conduction with the black oven bottom. Which design features were best at preventing radiation from the heat lamp from penetrating the dwellings? Which design features were best at preventing the convection of hot air moving? Students discuss and decide. They analyze the results, and then, if time is available, students go back to the drawing board to make revisions and improvements for a second round of testing.

When time is available, students are able to use their shared results and ideas to make revisions that help save even more ice penguins. This step helps to mitigate competition, because each group of students is a winner if their revised design is better than their first one. The process mimics how engineers continually work together in an iterative process to make things better. It usually takes students an entire class period to redesign their dwellings, bake them in the oven, measure the mass of the penguins, and discuss the final results.

See Figure 10 for an initial and revised design after students were able to learn from their first experience and share results with each other. Notice that students decided to plug up the door to prevent

Figure 10

Initial and revised designs of one student group

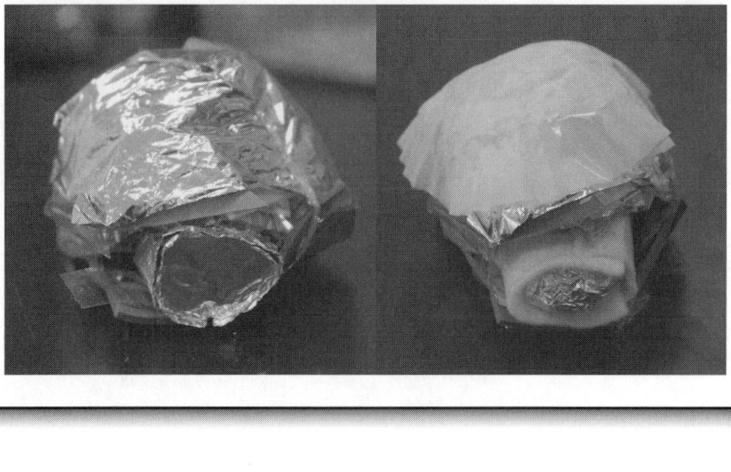

convection and add insulation in the "tube" to prevent the ice from directly contacting the metal. Many students decide to put their dwelling up on stilts of some sort, and many provide extra protection from radiation on all sides.

When the second trial is conducted, or when results are compared among classes, make sure the ice is at the same starting temperature each time. Ice will be the temperature of the freezer when it is removed, and it warms until it becomes 0°C (32°F) and melts. The standard freezer temperature is -17°C (1.4°F), but each freezer can vary. Therefore, the results will be the most accurate if penguins are removed from the same freezer each time they are used, and used right away.

CONCLUSION

This unique engineering, design-based approach to learning science in a context relevant to students' lives has been shown to be effective in (1) improving students' conceptual understandings about science; (2) increasing their knowledge about and attitudes toward engineering; and (3) increasing their motivation to learn science (Schnittka 2009; Schnittka et al. 2010). Research using the Save the Penguins curriculum has shown that with scaffolding, students can connect the heating and cooling of their homes to the burning of fossil fuels for energy production, global warming, melting sea ice, and penguins in peril. In the meantime, students are sharpening their inquiry and process skills, working in teams, and thinking creatively. Students learn the science best and engineer better dwellings when the demonstrations are used early on to target their alternative conceptions about heat. Engineering design activities can be used effectively in science classrooms, but careful attention should be paid to pedagogically sound science teaching. Students' alternative conceptions must be acknowledged, addressed, and targeted in science lessons. Otherwise, there is no expectation that students will implicitly learn scientific concepts just because they participate in an engineering activity. The engineering design activity gives students a tangible application for their science conceptions, insight into the world of engineering, and practice with 21st-century skills: innovation, problem solving, critical thinking, communication, and collaboration.

The complete Save the Penguins curriculum with daily lesson plans is available online at *www.uky.edu/~csc222/ETK/SaveThePenguinsETK.pdf.*

On the web

Climate change likely to devastate emperor penguin populations in Antarctica: *www.youtube.com/watch?v=RqNJ6B1CSss*

Penguin in a pickle: *www.youtube.com/watch?v=Jz-5Y7WgVEE*

Penguins are melting: *www.youtube.com/watch?v=rqUvf9Rxxj4*

References

Gross, L. 2005. As the Antarctic ice pack recedes, a fragile ecosystem hangs in the balance. *PLOS Biology* 3 (4): 557–61. *http://www.plosbiology.org/article/info%3Adoi%2F10.1371%2Fjournal.pbio.0030127.*

Jenouvrier, S., H. Caswell, C. Barbraud, M. Holland, J. Stroeve, and H. Weimerskirch. 2008. Demographic models and IPCC climate projections predict the decline of an emperor penguin population. *Proceedings of the National Academy of Sciences* 106 (6): 1844–47. *www.pnas.org/content/106/6/1844.full.pdf+html.*

National Research Council (NRC). 2001. *Climate change science: An analysis of some key questions.* Washington, DC: National Academies Press.

National Research Council (NRC). 2002. *Abrupt climate change: Inevitable surprises.* Washington, DC: National Academies Press.

Nullis, C. 2009. Save South Africa's penguins: Give them a home. Associated Press. March 29. *http://addistimes.com/africa-news/1873-save-south-africas-penguins-give-them-a-home.html*

Schnittka, C. G. 2009. Engineering design activities and conceptual change in middle school science. PhD diss., University of Virginia.

Schnittka, C. G., M. A. Evans, B. Jones, and C. Brandt. 2010. Studio STEM: Networked engineering projects in energy for middle school girls and boys. Proceedings of the American Society of Engineering Education, Louisville, Kentucky. *http://soa.asee.org/paper/conference/paper-view.cfm?id=23015*

Chapter 6

WARMING TO GLOBAL WARMING

SUNSPOTS AND SEA SURFACE TEMPERATURE

By Erich Landstrom

Solar radiation—or radiant energy emitted by the Sun—is the dominant, direct energy input into Earth's atmosphere and climate. In my astronomy class, I assign students (in grades 11 and 12) a problem-based laboratory activity in which they evaluate the causality of changes on the solar surface in regard to climate change and warming in Earth's environment. Students use graphing calculators and real-time data from the internet to research the possible effects of sunspot activity on ocean temperatures in the Atlantic. Sunspots are relatively cool areas of magnetic disturbance on the Sun's surface that appear as dark blotches contrasted against the rest of the photosphere (Figure 1). For this activity, I use the BSCS 5E Instructional Model—Engage, Explore, Explain, Elaborate, and Evaluate (Bybee 1997)—to analyze a false hypothesis linking sea surface temperature to the Sun.

Figure 1

Image of a large sunspot group compared to Earth

On September 23, 2000, the sunspot area within active region 9169 spanned an area a dozen times larger than the entire surface of the Earth.

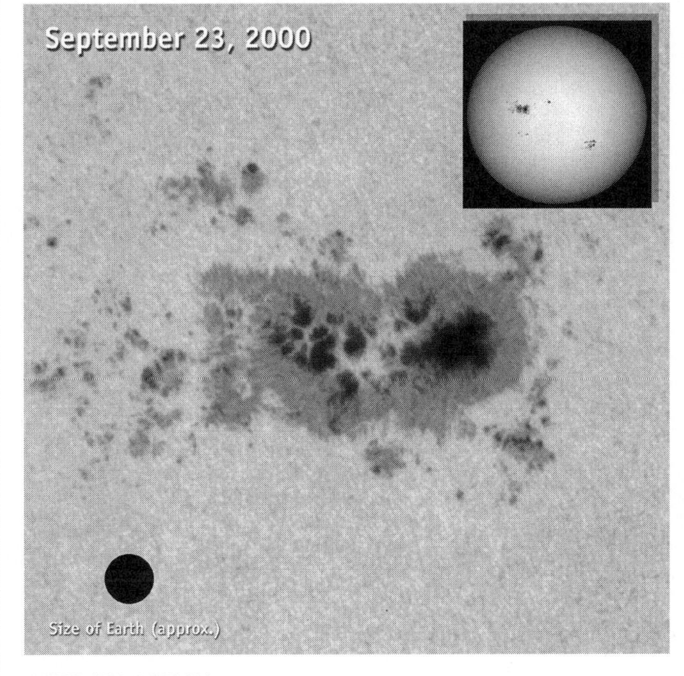

SOHO (ESA & NASA)

ENGAGE

To generate interest in the global warming issue, students are engaged

with the ideas of British economist William Jevons (1835–1882), who hypothesized that sunspots directly affect economic prosperity. Jevons reasoned that sunspots might have an effect on solar energy, in turn affecting the weather on Earth, which in turn, affects crops. These crop changes would then result in economic changes because agricultural production contributes significantly to the national economy—fluctuations in farm output, prices, and incomes could cause instability in overall business activity. Global warming due to sunspot activity could also lead to drought, which may lead to crop failure, which might depress the economy by causing famine and starvation.

Although Jevons' thinking was incorrect, I deliberately omit alternative explanations. I suggest to students that astronomers could count the number of sunspots as a possible proxy for measuring the magnitude of solar energy output. More sunspots might mean more radiation was being made at the solar interior by fusion—resulting in more magnetic disturbances, or sunspots.

Continuing with this (flawed) logic, I propose that when the sunspot count—and consequently solar energy—rises, Earth's ocean temperature should also rise. If the hydrosphere covers approximately 70% of Earth with water, then the majority of the surface should respond with a temperature change. Conversely, as the sunspot count falls, so too should the water temperature.

After students are presented with this (faulty) logic, they are asked to provide proof of Jevons's ideas by plotting monthly sunspot numbers versus time over a 16-year period (or 192 months) on a graphing calculator, and, in a second graph, sea surface temperature versus time over the same period. Students compare the two graphs to determine whether their data analysis identifies a relationship between the Sun and Earth's temperature.

EXPLORE

Students explore sunspot data directly on the National Oceanic and Atmospheric Administration (NOAA) website (see "On the web"). Monthly sunspot numbers are prepared by NOAA's Space Weather Prediction Center (SWPC), the nation's official source of space weather alerts and warnings. The SWPC continually monitors and forecasts Earth's space environment; provides accurate, reliable, and useful solar-terrestrial information; conducts and leads research and development programs to understand the environment and to improve services; advises policy makers and planners; plays a leadership role in the space weather community; and fosters a space weather services industry. The recent solar indices (preliminary) of observed monthly mean values provide recent solar indices running from January 1991 to the present. Table 1 displays the figures from 1991. (**Note:** An instructor's guide for using graphing calculators to analyze sunspot numbers can be downloaded in PDF format

from NASA's Imager for Magnetopause-to-Aurora Global Exploration (IMAGE) website [see "On the web"].)

Students explore ocean-temperature data provided by NOAA's Climate Prediction Center (CPC) (see Table 2 for 1991 data). The CPC collects and produces monthly data and maps for various climate parameters, such as temperature for the United States, Pacific Islands, and other parts of the world. The CPC also compiles data on historic and current atmospheric and oceanic conditions, El Niño, and other climate patterns, such as the North Atlantic and Pacific Decadal Oscillations. The sea surface temperature from 1950 to present is available for the North Atlantic (5–20°N, 60–30°W), South Atlantic (0–20°S, 30°W–10°E), and the global tropics (10° S–10°N, 0–360°).

Table 1

Recent solar indices (preliminary) of observed monthly mean values: 1991

Sunspot numbers (#) observed		
YR	MO	#
1991	01	213.5
1991	02	270.2
1991	03	227.9
1991	04	215.9
1991	05	182.5
1991	06	231.8
1991	07	245.7
1991	08	251.5
1991	09	185.8
1991	10	220.1
1991	11	169.0
1991	12	217.7

Table 2

Sea surface temperature data: 1991

YR	MO	N. ATL	S. ATL	TROP
1991	01	25.80	25.48	27.73
1991	02	25.41	26.37	27.88
1991	03	25.21	27.04	28.29
1991	04	25.49	27.09	28.70
1991	05	25.96	26.51	28.70
1991	06	26.23	25.04	28.20
1991	07	26.66	23.62	27.63
1991	08	27.16	22.89	27.23
1991	09	27.71	22.65	27.21
1991	10	27.65	23.00	27.50
1991	11	27.22	23.44	27.71
1991	12	26.56	24.56	27.79

Although the NOAA data is available online (see "On the web"), and this activity could be completed successfully using a computer spreadsheet, I prefer that students enter data into a graphing calculator rather than into

Figure 2

Sunspot versus sea surface temperature graph: January 1991–April 2007

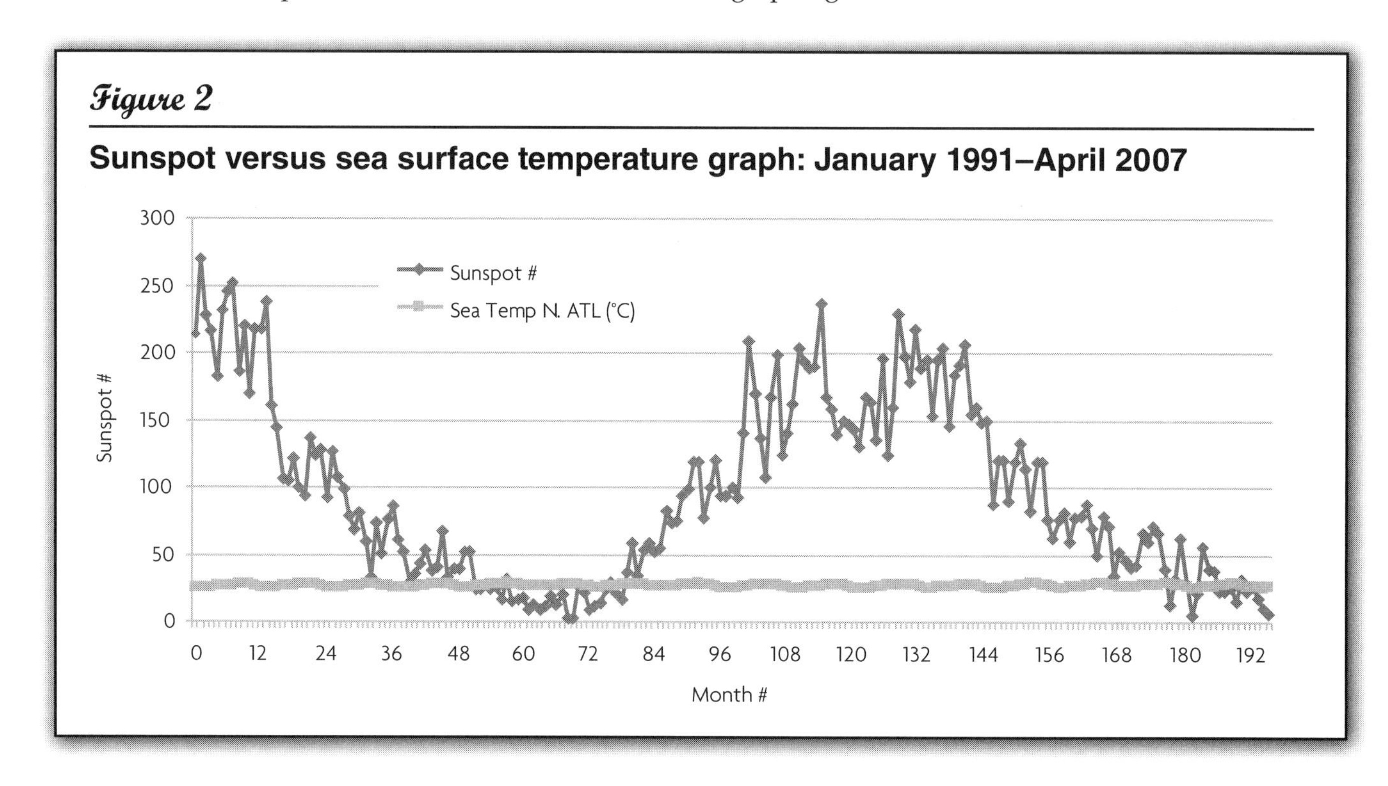

spreadsheet software. This is to discourage students from just cutting and pasting information without obtaining a contextual sense of what the numbers actually mean. Also, I have found that for the results to graph properly, it is necessary to strip out the information for the year and month and relabel it sequentially as "month #." Figure 2 provides an example of a sunspot versus sea surface temperature graph.

EXPLAIN

Teacher explanation of the time-plot should focus on cycles that emerge. In the first cycle, students will notice sea surface temperatures rising and falling like a sine wave, in small amounts, over the course of a given year. This is due to the change of seasons, and it is helpful to elaborate on how climatic patterns on Earth result from an interplay of many factors (e.g., Earth's topography, its rotation on its axis, solar radiation, the transfer of heat energy where the atmosphere interfaces with lands and oceans, and wind and ocean currents). The rise and fall of sunspot numbers over decades is the second cycle. The Sun goes through an approximately 11-year cycle, from a solar max of peak

sunspot numbers to the relative quiet of solar minimum, and back again. This solar variation is known as the Schwabe cycle.

ELABORATE

During elaboration, we examine the applications of the data. First, students are asked about how their initial expectations of the amount of ocean temperature change—and, by implication, global warming—contrasts with their postlab concepts. During prelab discussion, a majority of students accept Jevons's logic.

Second, students are asked to examine the chain of logic and find the weakest link for factors that can influence the absorption of radiation. If students think that water-based measurements are suspect and temperature should be taken aerially or on land, this can lead to a teachable moment of physical science with a review of calorimetry and thermal equilibrium concepts—particularly specific heat and how much energy that water, air, and soil can absorb before 1 g is raised by 1°C.

EVALUATE

To evaluate comprehension after the lab, students are asked if, based on their sunspot versus sea surface temperature graph, a change in sunspot activity affects ocean temperature in a direct relationship or an indirect relationship, or if there is no relationship at all. As part of their written evaluation, students must specifically address whether their results support or oppose Jevons's sunspot theory of economics, and support answers with information from their graphs. Sample responses I have received include:

- *The data opposes Jevons's theory because my data shows that the sunspots have no effect on Earth's temperature.*
- *It has no relationship because whether the sunspot number increases or decreases, the ocean temperature is staying the same all throughout the decade.*
- *No relation. The results oppose Jevons's theory. The graph shows the ocean data going across, while the sunspot data starts high, drops, and rises again, but is unrelated to the ocean temperatures.*
- *The sunspot goes from high to low to high, and the sea level temperature does not change at all. I do not support Jevons's theory because it does not match up.*
- *In my data, there is no relationship. My results would oppose Jevons's theory. The graph shows that the sunspot numbers and ocean temperature are not affected by each other. So in conclusion, sunspots do not affect the economy.*

CONCLUSION

Student achievement with this project-based science exercise on global warming has been very high, with greater than 90% earning passing grades. The exercise is challenging across several domains of science, mathematics, computer programming, and economics. It functions at the highest levels of Bloom's Taxonomy for analysis and evaluation, despite the disparate information that students must test in relation to their hypotheses. Students also respond positively to using the graphing calculators for lab work, with or without probeware attachments.

In investigating a possible correlation between sunspot activity and ocean temperature, students master three of Florida's Sunshine State Standards in Science (see "On the web"). Students use the scientific processes and habits of mind to solve problems; understand that most natural events occur in comprehensible, consistent patterns; and understand that science, technology, and society are interwoven and interdependent. When students engage in problem-based lab activities such as this one, their interest in and understanding of the subject matter can be greatly enhanced.

On the web

NOAA's Space Weather Prediction Center (SWPC): *www.swpc.noaa.gov*

NOAA's Climate Prediction Center (CPC): *www.cpc.noaa.gov/index.php*

Florida's Sunshine State Standards in Science: *www.fldoe.org/bii/curriculum/sss*

The sunspot cycle (teachers guide): *http://image.gsfc.nasa.gov/poetry/activity/s1.pdf*

References

Bybee, R. W. 1997. *Achieving scientific literacy.* Portsmouth, NH: Heinemann.

National Oceanic and Atmospheric Administration Climate Prediction Center (CPC). 2011. Sea Surface Temperature. *www.cpc.noaa.gov/data/indices/sstoi.atl.indices.*

National Oceanic and Atmospheric Administration Space Weather Prediction Center (SWPC). 2011. Recent Solar Indices (Preliminary) of Observed Monthly Mean Values. *www.swpc.noaa.gov/ftpdir/weekly/RecentIndices.txt.*

Chapter 7

MODELING CONVECTION

A SIMPLE APPARATUS FOR DYNAMIC MODELING OF PAIRED CONVECTION CELLS TEACHES STUDENTS ABOUT EARTH'S PROCESSES

By James R. Ebert, Nancy A. Elliott, Laura Hurteau, and Amanda Schulz

Students must understand the fundamental process of convection before they can grasp a wide variety of Earth processes, many of which may seem abstract because of the scales on which they operate. Presentation of a very visual, concrete model prior to instruction on these topics may facilitate students' understanding of processes that are largely invisible (e.g., atmospheric convection) or that operate on large spatial and temporal scales (e.g., mantle convection and plate tectonics).

Typically, teachers use simple models that employ differences in temperature and density to help students visualize convection. However, most of these models are incomplete or merely hint at (instead of model) convective circulation. For example, teachers commonly model convection by introducing heated or cooled dyed water to a container of room-temperature water (Figure 1). When hot water (red) is introduced to the bottom of the container, the tinted, less-dense water rises. Cold, dyed water (blue) is poured along the side of the container and students observe that the colored water sinks and flows along the bottom.

These traditional models suffer from three fundamental shortcomings. First, movement stops when the introduced fluids mix or density equilibrium is attained. Second, the markers used (typically food coloring) disperse in the fluid of the larger container, which makes the motion difficult to follow. Finally, these models illustrate only a portion of a single convection cell, whereas in nature, convection typically occurs in multiple, complete cells (e.g., global wind patterns, thunderstorms, and mantle circulation associated

Figure 1

Simple model of partial convective circulation using dyed hot and cold water

with plate tectonics). Although the ease of the traditional, food-coloring models is appealing to teachers, students who observe these demonstrations may develop simplistic or incomplete conceptualizations of convection. We have resolved these shortcomings by using an alternative system of fluid and markers in a simple, low-cost apparatus that not only maintains dynamic convective circulation, but also illustrates two adjacent cells (Figure 2).

Figure 2

Apparatus for dynamic modeling of paired convection cells

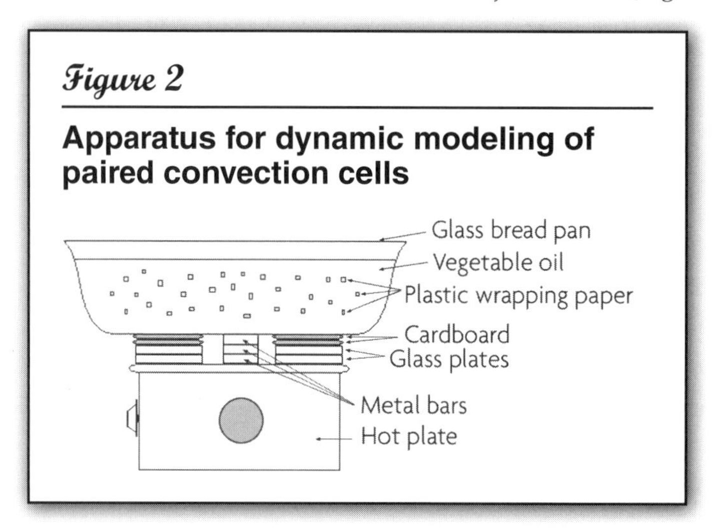

MATERIALS FOR DYNAMIC MODELING OF PAIRED CONVECTION CELLS

- glass loaf pan
- hot plate
- metallized plastic gift wrap, holographic (wrapping paper), cut into small, confetti-like pieces
- vegetable oil
- 4 glass scratch plates (~ 7.5 x 5 cm)
- 4 pieces of corrugated cardboard (~ 7.5 x 5 cm)
- 3 metal bars (aluminum bars from commercially available density kits—stack should be same height as glass/cardboard stack)
- additional pieces of corrugated cardboard

DYNAMIC CONVECTION APPARATUS

The apparatus is comprised of a 13.5 x 23.0 × 7.0 cm glass loaf pan, which is indirectly heated by an electric hot plate. The loaf pan is large enough to allow two convecting cells to develop and has the obvious advantages of transparency and heat resistance.

To develop two opposing convection cells, it is necessary to apply heat in the center of the loaf pan while the ends of the pan remain at lower temperatures. Heat is concentrated at the center of the pan with a stack of metal bars (e.g., three aluminum bars from density kits). Stacks of glass plates and equal-sized pieces of corrugated cardboard are placed under both ends of the pan to insulate the ends from the heat source and maintain stability. The height of these end stacks should be approximately the same height as the center stack of metal bars (in most cases, this will consist of two glass plates and two pieces of cardboard under each end of the loaf pan). The glass plates are placed directly on the hot plate, and the cardboard is placed on top of the glass plates—the cardboard should not come into contact with the metal bars or the surface of the hot plate.

The fluid used in the apparatus to model convection is approximately 700 ml of ordinary vegetable oil (density ~ 0.88g/ml). Confetti-like pieces of metallized plastic gift wrap (approximately 4 × 4 mm) are added to the vegetable oil so that students can actually see convection taking place. "Holographic" varieties of wrapping paper are especially eye-catching. If these markers are stirred into the oil, they remain suspended throughout the fluid column—the suspension is retained at all temperatures so that even when heated, the markers follow the motion of the fluid without floating to the top or sinking. With these markers present, fluid motion and convective circulation are easy to observe.

The loaf pan is placed on the stacks of metal bars, glass plates, and cardboard atop the hot plate, which is set on a low to moderate heat (oil temperature should not exceed 55°C). The use of low heat settings reduces the risk of burning the cardboard and cracking the glass pan, which was not designed to be heated unevenly. Under these conditions we have used the same loaf pan for dozens of trials with no damage. No significant safety concerns exist with this apparatus.

After approximately 15 minutes, two convecting cells are readily apparent as the gift wrap markers rise in the fluid above the metal bars, move toward the ends of the pan, sink, and follow the flow along the bottom of the pan to return to the center (Figure 3, p. 50). Once convection is established, the hot plate may be turned down to a lower setting and the paired convection cells can be maintained for hours.

Provided that a hot plate is already available, the cost for this apparatus is under $10 (the glass loaf pan costs between $3–$5 and vegetable oil is typically less than $2). The remaining materials—metal bars, glass plates, cardboard, and scrap gift wrap—may be gathered without expense from classrooms, homes, or local businesses.

Figure 3

Two convecting cells appear after about 15 minutes

Two-second exposure of apparatus with convection cells in vegetable oil is shown by movement of markers made of metallized plastic gift wrap. Mounting the loaf pan sideways on the hot plate (90° from orientation shown) allows the ends of the pan to remain cooler.

APPLICATIONS

This model can be used to represent air currents in a room or high- and low-pressure systems in the atmosphere. The bits of wrapping paper make it easy for students to see rising and descending "air," and the resulting "winds" at Earth's surface and aloft. Divergence in the upper "atmosphere" can also be modeled when the apparatus is viewed from above.

The rifting of a "continent" can also be modeled by placing pieces of corrugated cardboard (each approximately 5 x 5 cm) on top of the vegetable oil. After convection is well established, the cardboard pieces are floated on the oil in the center of the pan. The outward motion of the oil will cause them to diverge.

When used as a traditional demonstration, the dynamic convection model addresses three parts of the *National Science Education Standards* Unifying Concepts and Processes Standard: evidence, models, and explanation; change, constancy, and measurement; and evolution and equilibrium (NRC 1996, p. 104). Sections of Content Standard B (Physical Science) that deal with density and motion may also be addressed using the model (NRC 1996, p. 154–155).

The most direct application of the model is in Content Standard D (Earth and Space Science), where mantle and atmospheric convection are specifically mentioned (NRC 1996, p. 189). For this standard, the model is ideally suited as a concrete example of the basic process (convection) that underlies more complex Earth processes.

The model may also serve as a starting point for inquiry-based instruction (Content Standard A, Science as Inquiry; NRC 1996, p. 175). Before heat is applied to the model, students can be asked to predict what will happen when the hot plate is turned on. Once convection is initiated, students can formulate explanations and design methods to test their explanations using the model or modified versions of the model.

STRENGTHS AND WEAKNESSES

All models have strengths and weaknesses (Gilbert and Ireton 2003). The dynamic convection model described here has four major strengths. First, the model portrays two actively convecting cells rather than a single or partial cell. Second, convective circulation is maintained, which does not occur with other models. Third, the markers that enable visualization of convection remain discrete and do not disperse. Finally, students can directly observe relationships that are otherwise abstract. For example, convective circulation and upper air winds become easier to visualize as do the connections between mantle circulation and the movement of lithospheric plates.

In addition, the vegetable oil and suspended markers may be reused many times, thus minimizing preparation time after the initial setup. The oil may be stored indefinitely in a sealed container. With time, the markers will float to the surface of the oil. However, they may be stirred into the oil as they were initially and the apparatus functions as well in subsequent uses as it did the first time. The loaf pan is easily cleaned with soap and hot water and can be stored separately, thus minimizing the chances of "oil spills."

The dynamic convection model does have some weaknesses. For example

- vegetable oil is significantly more viscous than the atmosphere, therefore random or chaotic turbulence is suppressed;
- vegetable oil is also significantly less viscous than the asthenosphere, therefore convective circulation is disproportionately rapid relative to the size of the model;
- the markers that enable students to see convection do not have counterparts in nature;
- when used to model plate motion, there is no part of the model that is analogous to oceanic lithosphere; and
- the use of rigid cardboard "continents" does not permit observation of collisional deformation at the margins.

These weaknesses are rather minor in comparison to the model's advantages.

HELPING BUSY TEACHERS

The need for better models of Earth processes—including a dynamic, multicell convection model—was brought to our attention through discussions with veteran Earth science teachers in summer institutes for the preparation of regionally based mentor teachers. The time constraints of teaching at the secondary level prevented these talented teachers from developing and testing such models. Our hope is that this model will provide teachers with a simple, yet effective, way to teach convection in the classroom. At the same time, we hope students will gain a solid understanding of the vital role convection plays in Earth's processes.

References

Gilbert, S. W., and S. W. Ireton. 2003. *Understanding Models in Earth and Space Science.* Arlington, VA: NSTA Press.

National Research Council (NRC). 1996. *National Science Education Standards.* Washington, DC: National Academy Press.

Chapter 8

CONVECTION IN A FISH TANK

By Chris Freeman

As a high school science teacher with limited funds, I am always looking for demonstrations that are both cost-effective and visually stimulating. This demonstration is more captivating than a lava lamp and is a great method of modeling convection using materials most science teachers already have at home or in the classroom.

Understanding convection is fundamental for students to fully grasp the science behind large-scale events on Earth, such as global wind patterns, plate tectonic movement, ocean current patterns, and hydrothermal vent dynamics. Convection can also help students understand how car engines are cooled, how radiators heat a room, how a pot of water boils, and how lava lamp fluid moves up and down. Using a convection demonstration in a classroom emphasizes the importance of convection while captivating students' interest.

MATERIALS AND DIRECTIONS

The following materials are needed for you to lead the convection demonstration:

- 1 38 L (10 gal) fish tank
- 2 dictionaries or 2 stacks of magazines, 5–7 cm high
- 100 pennies
- 3 tea light candles
- 1 bottle of red food coloring
- 1 bottle of blue food coloring
- water
- matches

Figure 1a

Place the two dictionaries 20–25 cm apart on a flat surface and set the fish tank on the dictionaries (Figure 1a). Make sure that the tank is stable, level, and not hanging off the sides of the books. Place the 100 pennies in rows across the center of the tank. Make the

Figure 1b

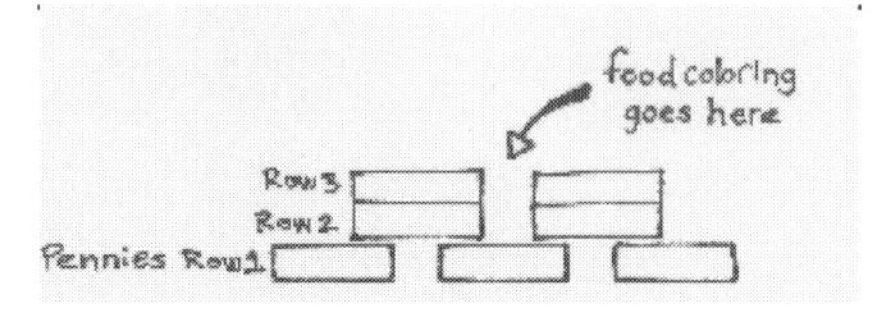

Figure 1c

Figure 1d

Figure 1e

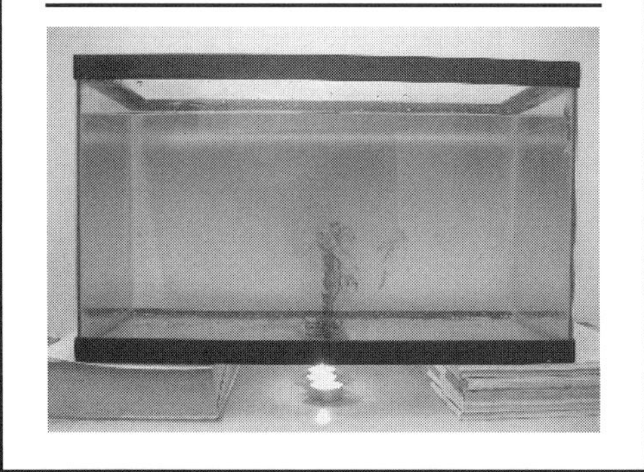

first layer three rows wide, then stack two additional rows of pennies on top of the first layer of coins. The second and third layers of pennies should be stacked to overlap the gaps between the bottom pennies (Figure 1b). This creates a groove between the pennies in the top two rows that will help hold the food coloring in place.

Fill the tank with water to about 5 cm from the rim of the tank. Place the candles in a line below the fish tank directly under the pennies. Take the bottle of red food coloring in hand and, with minimal disruption to the water, move your hand to the bottom of the tank and slowly squeeze the red food coloring into the groove made by the pennies. (If done correctly, the food coloring will stay resting on the pennies, see Figure 1c.) Stop squeezing the bottle and slowly remove your hand and the bottle from the fish tank, trying to disturb the water as little as possible. After the water settles, light the candles under the fish tank. The demonstration is ready to begin (Figure 1d).

Within one minute after being lit, the candles will heat the pennies and start the process of convection. The hot pennies heat the water and the expanded, less-dense water begins to rise away from the heat source. The red food coloring colors the heated water, allowing students to see the water rise. The red color represents the hot, less-dense water rising (Figure 1e).

As the heated water and red food coloring rise away from the pennies, the red-colored liquid is still hot, but it begins to cool as it moves away from the pennies. The red-colored liquid cools down and spreads out as it continues to move up, away from the heat source (Figure 1f).

Place two drops of blue food coloring in rapid succession 2–4 cm from both sides of the rising red food coloring (Figure 1g). The blue food coloring represents the cooled, dense water descending back to the depths of the fish tank.

The dense blue liquid will be pushed away from the center of the tank by the laterally moving, cooling red water at the upper portion of the tank. Then, as the blue food coloring descends, it will be pulled back toward the pennies to replace the water that is rising up away from the heat source (Figure 1h).

Once the blue-colored water comes in contact with the hot pennies, it will rise back up with the red-colored water to begin the process all over again (Figure 1i).

THOUGHTS ABOUT THE DEMONSTRATION

The demonstration of a convection cell is complete once the blue-colored water reaches the pennies. However, to demonstrate how convection moves tectonic plates, a simple addition can illustrate divergence. Float two plastic bottle caps 2–4 cm apart on opposite sides of the uplifting fluid. The bottle caps represent tectonic plates. The current will slowly pull the caps farther apart from each other showing how divergence takes place.

I have developed and modified this demonstration of convection over the past four years. The most noticeable advancement to the demonstration happened when I added pennies to the model. The metal pennies heat up faster than the glass of the fish tank and pennies are an inexpensive, readily available resource. Without the pennies, the convection process takes much longer and is much less dramatic. Once the pennies are set into place, the food coloring rests on and between them, which helps keep the food coloring from spilling out in a messy puddle over the bottom of the fish tank. In essence, the pennies have streamlined this demonstration into a practical classroom tool.

CLASSROOM REACTION

I have used this demonstration in my 9th-grade Earth science classes and my 11th- and 12th-grade oceanography classes. In both settings, students appear captivated by the colors and swirling fluid. They ask numerous questions about how the demonstration works and often want to see what happens when food coloring is added to other areas within the tank. I have even had two students on separate occasions come in the following day and tell me that they had gone home and performed the demonstration for their families.

As part of the final exam in my oceanography class, I have students label and diagram a convection cell.

Figure 1f

Figure 1g

Figure 1h

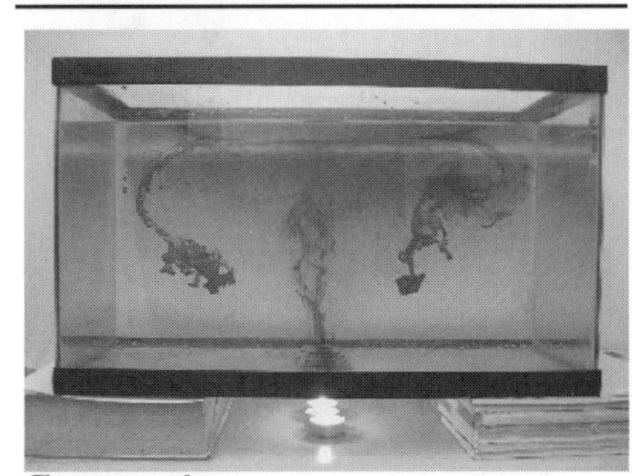

Figure 1i

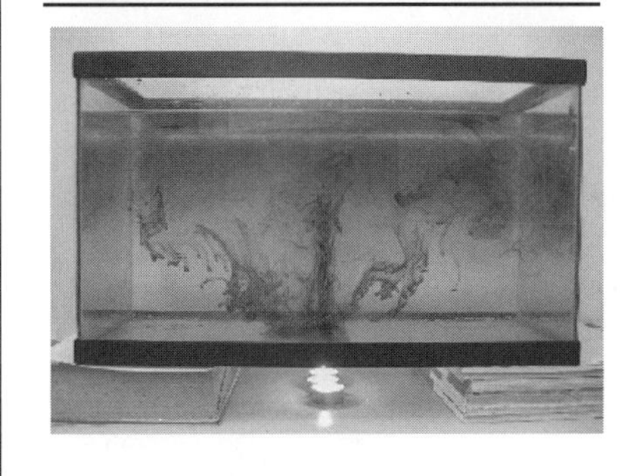

Pre- and postdemonstration questions

Predemonstration

1. Why do hot air and hot water rise?
2. Why do cool air and cool water sink?
3. When air above a flame rises, what happens to the air beside the flame?
4. When air moves up and away from its heat source, does it remain the same temperature?
5. Draw a diagram of the air movement that would happen around the head of a candle. Use arrows to represent air flow and be sure to label the hot and cold air.

Postdemonstration

1. Draw a diagram of the "Convection in a Fish Tank" demonstration. Use arrows to represent water flow and be sure to label the hot and cold water.
2. How does the circulation of water in the fish tank differ from the air flow in your earlier drawing?
3. What factors made the water move sideways in the demonstration?
4. What nonhuman influences make the air beside a candle move sideways?
5. Briefly describe three places in a house where convection might take place.

Many students actually draw a two-dimensional version of this demonstration—the completeness of the diagrams suggests that the demonstration had a lasting effect on these students and enhanced their ability to understand convection. Convection is a recurring concept in Earth science, meteorology, and oceanography. Demonstrations such as this one are an inexpensive, effective way to make a typically invisible concept, such as convection, become visible, tangible, and exciting.

Chapter 9

CELEBRATE WITH SATELLITES

AN INTERNATIONAL POLAR YEAR PARTNERSHIP TO STUDY EARTH'S MATERIALS

By Mikell Lynne Hedley, Kevin Czajkowski, Janet Struble, Terri Benko, Brad Shellito, Scott Sheridan, and Mandy Munroe Stasiuk

Partnerships between scientists, teachers, and students have shown to be one of the most effective forms of science instruction (Wormstead, Becker, and Congalton 2002). Students and Teachers Exploring Local Landscapes to Investigate the Earth From Space (SATELLITES) is one such collaboration that uses the International Polar Year (IPY) as its underlying theme. The fourth IPY (March 2007 to March 2009) was a collaborative, international effort to research the Earth's polar regions (see "A history of the IPY," p. 67).

The SATELLITES program uses geospatial technologies to study surface temperatures of Earth's materials, such as sand, soil, grass, and water. Data are collected using Global Learning and Observations to Benefit the Environment (GLOBE) protocols, which are then used in research projects that are a part of the IPY. Students collect data, conduct field campaigns, design

inquiry-based research projects, and attend and present at a SATELLLITES science conference. One student activity, Heating Things Up, is included in this chapter to help introduce students to the factors that affect Earth's surface temperature.

ABOUT SATELLITES

Since it began in 2000, the SATELLITES partnership has built a community of learners that is beneficial to teachers, students, and scientists around the world (Hedley and Struble 2004). This collaborative program studies Earth using the geospatial technologies of remote sensing, the Global Positioning System (GPS), and geographic information systems (GIS).

Through a five-day Teacher Institute, teachers are trained in GLOBE's surface temperature and cloud protocols, which use Earth's energy budget (Figure 1) as their underlying science concept. Earth's energy budget is a measure of the amount of energy entering and leaving the Earth-atmosphere system in the form of radiative energy, sensible, and latent heat. The amount

Figure 1

The energy budget

This drawing illustrates the dynamics of the heating and cooling of Earth—this is one way to introduce the energy budget to your students.

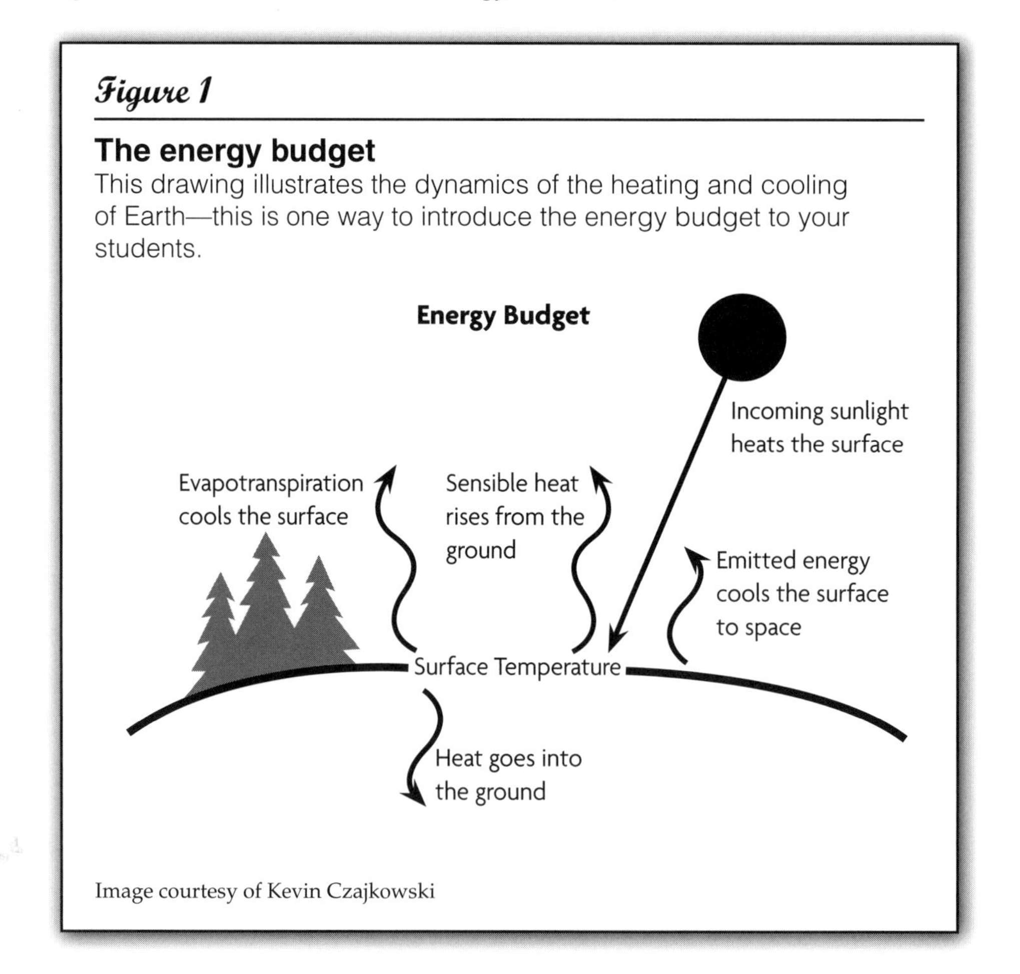

Image courtesy of Kevin Czajkowski

going in must equal the amount going out, or Earth would continually heat up or cool down. At the Institute, teachers learn how to measure Earth's materials with infrared thermometers (IRTs), how to model data-collection methods, and the overall process for students (Figure 2).

Figure 2

A prize-winning project: Arctic ice versus Erie ice

This teacher project used the decrease in Lake Erie ice cover to predict the decrease in Arctic ice cover over a number of years. This particular project was developed by two teachers in a 2007 Teachers' Institute.

Teachers take what they learn back to the classroom and train students to collect surface temperature and atmospheric field data according to these protocols. This involves a field campaign in which students collect data locally during a given time frame, while other students collect data in their locations as a means of comparison. Students then enter their data on the GLOBE website (see "On the web"), which can be accessed by students, teachers, and scientists everywhere (Figure 3, p. 60). GLOBE's focus is to involve these groups in the study of global climate change and other environmental issues (see "More on GLOBE," p. 61).

Figure 3

Participating schools

This map shows the location of some of the schools that have been involved in the collection of surface temperature and atmospheric data as part of GLOBE and the SATELLITES field campaign. Finding Earth's surface temperature at various locations and on different types of surfaces gives students a good picture of how the temperature of Earth is affected by various conditions.

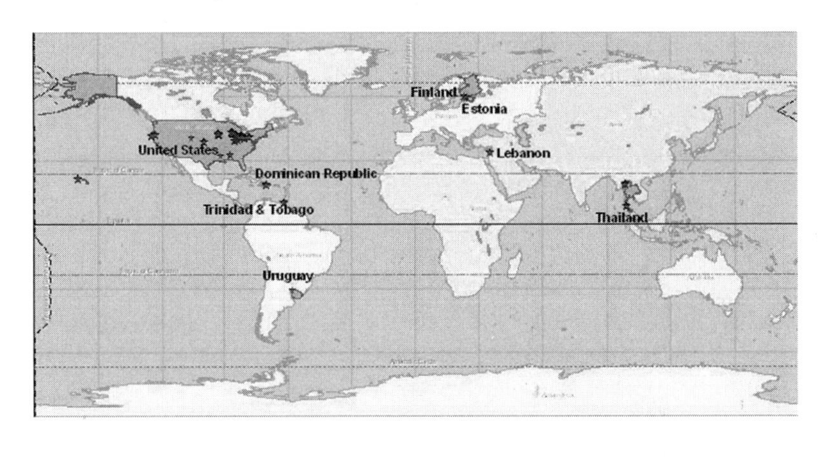

As part of the SATELLITES program, the data students collect is used to develop an inquiry-based research project. Teachers are encouraged to help students brainstorm practical ideas. This helps students arrive at a question they are able to answer using their present skill level, available data or experimental procedures, and time limitations. Since the teachers themselves have done this in the Institute, they can better guide students in coming up with a question and how they will answer it. International Science and Engineering Fair rules are used in the research project.

Student research projects are presented at the annual SATELLITES science conference. This conference brings together the students, teachers, parents, and scientists who have worked together throughout the year. At the conference, students present their projects in person or by video to the judges. Winning students receive medals and ribbons and their schools receive trophies. Some of the winning projects in 2008 had the following titles:

- Take Me Out to the Ball Game (middle school)
- When It's Hot, It's Hot (middle school)
- The Big Stinky Time Bomb (middle school)

- How Clouds Affect Global Warming (high school)
- Global Warming—True or False? (high school)

During the IPY, the conference has included a presentation by a NASA scientist actively involved in IPY research. The IPY theme was chosen to help students understand how the events that happen at the poles have a ripple effect on their own location's weather and surface conditions. This IPY is seeking student involvement in research activities, unlike the previous three.

Students have the opportunity to submit their research projects to GLOBE for inclusion in GLOBE Learning Expeditions. This is a weeklong event that hosts scientists, teachers, students, and other partners from around the world—allowing students to present their projects to the international science community. In 2008, a SATELLITES student project was one of five chosen from the United States to be presented at the South African conference.

More on GLOBE

GLOBE is a worldwide hands-on, primary and secondary school–based science and education program. The program promotes and supports students, teachers, and scientists to collaborate on inquiry-based investigations of the environment and Earth system. The program works in close partnership with NASA and National Science Foundation Earth System Science Projects to research the dynamics of Earth's environment.

Announced in 1994 by then Vice President Al Gore, GLOBE began operations on Earth Day 1995. Today, the international GLOBE network has grown to include representatives from 111 participating countries coordinating GLOBE activities in their local and regional communities. Because of their efforts, there are more than 54,000 GLOBE-trained teachers representing over 24,000 schools around the world. More than 1.5 million students have participated in GLOBE, contributing more than 22 million measurements to the GLOBE database for use in their inquiry-based science projects.

GLOBE brings together students, teachers, and scientists through the GLOBE Schools Network in support of student learning and research. Parents and other community members often work with teachers to help students obtain data on days when schools are not open.

Adapted from the GLOBE website: *www.globe.gov/fsl/html/aboutglobe.cgi?intro&lang=en*

ON THE GROUND

The SATELLITES program uses geospatial technologies—GPS, GIS, and remote sensing—to set up study areas, collect surface temperatures, and portray student data in a visual manner. Students pick both a grassy test area and a paved area such as a parking lot and use their GPS units to pinpoint the exact latitude and longitude of each site—information that is necessary for students to compare their data with other locations. The grassy test area is used to check surface temperatures of natural materials, and the parking lot is used to measure the temperature of human-made materials. Students can observe and compare the difference in surface temperature of these two areas.

Once the study sites have been allocated, students use an IRT—a handheld device that takes the temperature of Earth in much the same way as satellites do—to take surface temperatures. (**Safety note:** Use of IRT controls, adjustments, or performance of procedures other than those specified here may result in hazardous radiation exposure.) Students hold the instrument in their hand with their arm extended and press the lever on the instrument. The IRT gives nearly instantaneous observations using the long-wave radiation emitted by the surface and records the temperature on its display panel. (**Note:** Participants of the SATELLITES Teacher Institute receive a free GPS unit, IRT, and GIS software.)

Several factors can affect surface temperature, such as

- atmospheric conditions;
- percentage of cloud and contrail (or vapor) cover;
- types of clouds at various heights;
- amount of snow cover;
- precipitation; and
- any obscuring factors, such as smoke, volcanic ash, or haze.

These are all noted with the surface temperatures as part of the data. Data is collected for 10 consecutive school days in the weeks between the Monday after Thanksgiving and the beginning of most schools' winter breaks. GIS software can be used to map data-collection areas and compare data.

Students enter their surface temperature and atmospheric data on the GLOBE website, which scientists and students can then use in inquiry-based science projects.

In the past, students' work has been used to ground truth NASA's Moderate Resolution Imaging Spectroradiometer (MODIS) satellite data (Ault et al. 2006). Ground truthing is done to compare the data obtained from remote sensing with the data taken at the actual location. It is an important method used to calibrate a remote-sensing instrument. MODIS is the key instrument on the Terra and Aqua satellites (see "On the web"). Terra passes north to south over the equator in the morning, and Aqua passes south to north over the equator in the afternoon. These two satellites give a good picture of the entire Earth every one to two days. Students' data was used to ground-truth the MODIS data recorded by the Aqua and Terra satellites, primarily to help verify snow and cloud data. In this case, ground truthing was needed because the spectral reflectance—the proportion of incident light that is reflected from a surface at any given wavelength of the visible spectrum—of snow and clouds is very similar, and the satellite data sometimes misreads snow and clouds.

IN THE CLASSROOM

Understanding the factors that affect Earth surface temperatures involves information about how Earth is heated and cooled, how energy flows from the Sun, and what variables affect the amount of energy that enters and leaves Earth. Heating Things Up is an example activity that can be used in the classroom to introduce these concepts to students (Figure 4, pp. 64–65). Janet Struble, a teacher with over 20 years of inquiry-based science experience, developed the activity.

CONCLUSION

The entire SATELLITES program emphasizes the importance of students doing science and not just reading about it. The collaboration with scientists and teachers makes science come alive, and the program is suitable for many ages and grade levels. The use of GLOBE protocols, which have been developed by scientists and educators, enables even third-grade students to learn correct data collection.

Through this project, students can use data they collect to answer their own science inquiries. And because data have been collected by students around the world, students everywhere are able to collaborate with one another. The use of geospatial technology tools opens up a world of new possibilities for the types of data students can use in answering the questions they have about the world, especially those related to the IPY.

The IPY theme provides the opportunity for students to study what happens in their own locality and how it is affected by global climate change. The comparison of data from the present time to that in years past introduces students to the idea that Earth is warming up. It raises their awareness of the problems of climate change and what their part might be in the solution to this global problem.

Figure 4

Heating Things Up activity

Purpose
To investigate the different rates of heating and cooling of certain materials on Earth in order to understand the heating dynamics that take place in Earth's atmosphere.

Overview
Working in small groups, students will explore the energy transfer of different Earth materials when they are heated up and cooled down. Students graph the changes in temperatures that occur over a 15-minute period of heating the Earth material up and a 15-minute period of the Earth material cooling down. Class time: two periods; level: middle and secondary.

Science concepts
Earth and space sciences:

- The atmosphere is composed of different gases and aerosols.
- The Sun is the major source of energy for changes in the atmosphere.
- The Sun is the major source of energy for Earth's surface processes.
- The Sun is the source of energy that heats Earth, including the atmosphere.
- The Sun heats Earth's surface.
- That heat is transferred to the atmosphere from the surface by conduction and convection.

Physical sciences:

- Matter exists in three different states: solid, liquid, and gas.
- Heat transfer occurs by radiation, conduction, or convection.
- Light and radiation interact with matter.
- The Sun is a major source of energy on Earth's surface.
- Energy is transferred in many ways.
- Heat moves from warmer objects to cooler objects.
- Energy is conserved.
- Temperature is a measure of the kinetic energy of molecules and atoms.
- Materials heat up and cool down at different rates.
- Science-inquiry abilities.
- Use infrared thermometer (IRT) or regular thermometers.
- Design and conduct scientific investigations. (**Note:** After conducting this particular activity, students can change variables [e.g., time, types of materials, and types of heating method] and conduct their own investigations.)
- Use appropriate mathematics to analyze data.
- Communicate results and explanations.

Materials and tools

- 5 plastic containers of the same shape and volume
- handheld IRT or 5 thermometers
- sand
- soil
- grass
- water
- gravel or rocks
- a sunny day, 5 desk lamps, or clip-on shop lights with 100 W lightbulbs
- stopwatch or timer

(continued)

Figure 4 *(continued)*

Preparation
Put equal amounts of each Earth material in separate containers.

Teacher preparation
Heat is a difficult concept for students to understand. It is not a substance but a form of energy (the movement of energy). So what does the temperature tell us about the amount of heat in the substance? The thermometer does not measure the amount of heat but the level of average molecular kinetic energy. This distinction is hard for even adults to understand. The big ideas of this activity are that

- heat is the movement of energy;
- energy has many forms; and
- energy can move from one place to another by radiation, conduction, or convection.

One form of energy is the motion of molecules (atoms): kinetic energy. Temperature can be used to compare the molecular kinetic energy of different substances; the movement of this energy is measured as heat.

Procedure

1. Ask students to think about a hot, sunny day at the beach: "You are barefoot. What surfaces would you walk on, and why? Why are some surfaces cool to your touch? Why are some hot to your touch? From where did the hot surfaces get their energy?"
2. Tell students that today they are going to investigate what happens to Earth materials when they are exposed to the Sun.
3. A higher degree of inquiry can be achieved by having students develop an experiment that will answer the question, "Do different Earth materials heat up and cool down at the same rate?" Include the materials listed; other Earth materials could be added. Steps 4–7 offer an example. The amount of help students will need in developing these experiments will depend on their grade level. SATELLITES is open to grades 4–12.
4. Ask each group to record the temperature change of each Earth material. If it is a sunny day, this activity can be done outside. If not, desk lamps or shop lights can be used to simulate the Sun's energy.
5. If using thermometers, place them in the Earth material.
6. Record the beginning temperature of each material with the IRT. Turn on the lamps, and then record the temperature every 3 minutes for a total of 15 minutes. Enter the values on the attached data table (Table 1, p. 66). (**Safety note:** Use care with lamps, as the 100 W bulbs can become hot.)
7. After 15 minutes, turn off the lamps (or shade the containers if outside) and measure the temperature every 3 minutes for the next 15 minutes.
8. Have students examine the data, graph it, and then interpret the graphs.
9. Use the following questions for class discussion or as an individual assignment:
   - Did all the Earth materials heat up at the same rate? Explain the evidence for your answer.
   - Did all the Earth materials cool down at the same rate? Explain the evidence for your answer.
   - Which Earth material heated up the fastest? How do you know this? Explain why this may have happened.
   - Which Earth material cooled down the fastest? How do you know this? Explain why this may have happened.
   - Did each Earth material receive the same amount of energy? How do you know?

Table 1

Data table sheet

	Temperatures (°C)				
Time (min.)	**Bare soil**	**Grass**	**Gravel or rocks**	**Sand**	**Water**
Beginning					
3					
6					
9					
12					
15					
Shaded or lights turned off.					
18					
21					
24					
27					
30					

On the web

GLOBE: *www.globe.gov*

MODIS: *http://modis.gsfc.nasa.gov/about*

The International Geospatial Year: *www.nas.edu/history/igy*

A short history of IPY: *http://classic.ipy.org/development/history.htm*

History: *www.arctic.noaa.gov/aro/ipy-1/History.htm*

About IPY: *www.ipy.gov/Default.aspx?tabid=70*

References

Ault, T. W., K. P. Czajkowski, T. Benko, J. Coss, J. Struble, and A. Spongberg. 2006. Validation of the MODIS snow product and cloud mask using student and NWS cooperative station observations in the Lower Great Lakes Region. *Remote Sensing of the Environment* 105: 341–353.

Hedley, M. L., and J. L. Struble. 2004. A community of learners. *The Science Teacher* 71 (5): 46–47.

Wormstead S. J., M. L. Becker, and R. G. Congalton. 2002. Tools for successful student-teacher-scientist partnerships. *Journal of Science Education and Technology* 11 (3): 277–287.

A history of the IPY

Over the last 130 years, there have been four IPYs. The first was the brainchild of an Australian explorer and naval officer, Karl Weyprecht. Weyprecht was a scientist and a co-commander of the Austro-Hungarian Polar Expedition of 1872–1874. He was convinced that the solutions to fundamental problems of meteorology and geophysics could be found at the North and South poles and that it would take the combined efforts of many nations to solve these problems. In all, 12 countries participated in this first IPY, spanning the years 1882–1883. Fifteen expeditions were conducted: 13 to the Arctic and 2 to the Antarctic. Perhaps even more important than the advances in science and geographical information was the fact that this first IPY set a precedent for international scientists to work together to solve problems. Unfortunately, Weyprecht died in 1881 before the first IPY took place.

The second IPY (1932–1933) with the theme of "Jet Stream," was proposed and promoted by the International Meteorological Organization. The Jet Stream—fast flowing air currents in the tropopause—had recently been discovered, and 40 nations joined in the effort to investigate the global implications for meteorology, magnetism, atmospheric science, and the study of ionospheric phenomena. Forty permanent research stations were established in the Arctic during this IPY, and the second Byrd Antarctic expedition, sponsored by the United States, established a winter-long meteorological station—the first research station on the Antarctic coast.

The third international polar year (1957–1958) was named the "International Geophysical Year" (IGY), and celebrated the 75th anniversary of the first polar year and the 25th anniversary of the second. This IPY was established by physicists Sidney Chapman, James Van Allen, and Lloyd Berkner. That year's research centered on advances in understanding the upper atmosphere. During this year, the Van Allen Radiation Belt that encircles Earth was discovered, the first artificial satellites were launched, continental drift was confirmed, and the total size of the Antarctic ice mass was determined. Sixty-seven countries were involved in the research carried out during the IGY, with 65 stations in Antarctica maintained by 12 nations. One of the most memorable accomplishments that followed from this year was the setting aside of an entire continent for peaceful scientific exploration under the Antarctic Treaty of 1961.

The fourth International Polar Year (2007–2009) was organized by the International Meteorological Organization and the International Council for Science. It was unique in a number of ways. Unlike the previous Polar Years, this "year" lasted two years instead of one. This was done so that the time period would cover a complete summer and winter

(continued)

(continued)

season at both the Arctic and the Antarctic and give equal coverage to both regions. Over 60 countries were involved in 200 research projects in this IPY. The largest difference in this IPY is the breadth of topics researched and the active inclusion of the public, students, and their teachers in the program. This "year" focused on research of the atmosphere, ice, land, oceans, space, and, for the first time, people of the polar region. It came at a time in our history when climate change has led to an unprecedented focus on Earth's polar regions and their effects on the rest of the planet.

More information on the history of these four IPYs can be found on a number of websites (see "On the web").

Chapter 10

TAMING ENERGY

By William Robertson

Energy transformations take place all over the Earth without humans ever getting involved. Being the control freaks we are, though, we spend a lot of time trying to direct those energy transformations to make our lives easier. This chapter is about how we manipulate energy transformations to get electricity for important things like running model racing sets and model railroads.

THINGS TO DO BEFORE YOU READ THE SCIENCE STUFF

You've been sitting around reading long enough. Time to get some exercise. Find yourself a set of stairs and walk up slowly. Did your energy change? Answer: Yes. You gained (more precisely, you and the Earth gained) gravitational potential energy. How much gravitational potential energy did you gain? Answer: mgh, where m is your mass, g is 9.8 meters per second per second (m/s^2), and h is the vertical height of the stairs.

Go up the stairs a second time but this time run. Is it easier or harder to run up the stairs than to walk up the stairs? Don't tax your brain too long on this one—it's harder! Compare the gravitational potential energy gained when you run up the stairs with the gravitational potential energy gained when you walk up the stairs. Let's see, m is the same, g is the same, and h is the same. Therefore, the energy you gain is the same whether you run or walk. Why, then, is it more difficult when you run? Think about that for a while.

Because we're going to talk about electricity use, dig out a copy of your latest electric bill (don't look at the total—it will only upset you). If you have a solar-equipped house or live in an apartment with utilities paid, grab a bill from one of your friends. Find the place on the bill where it shows how many joules of electrical energy you used. Keep searching—it's there somewhere. Well, actually no, it's not. What you will find is the number of kilowatt-hours used (this is often abbreviated *kh* or *kWh*). Why is electrical energy measured in these strange units?

Now head outside and find the electricity meter for your house or apartment. There will be either a digital display (like the odometer on your car) or a series of wheels with numbers on them. Whatever kind of meter you have, notice that the units used to measure your energy use are, once again, kilowatt-hours.

THE SCIENCE STUFF

You discovered in walking and running up the stairs that the *rate* at which you gain energy has some significance. If you do work on something, doing it quickly requires more effort than doing it slowly. To account for that, we define something known as **power**, which is the rate at which you do work or change energy.

$$\text{Power} = \frac{\text{work done}}{\text{Time it takes to do that work}} = \frac{\text{change in energy}}{\text{time it takes for the energy change}}$$

Remember that when you do work on something, that amount of work is numerically equal to the change in energy of the object. That's why it's

possible to substitute "change in energy" for "work done" in the above equation. In symbol form, this is:

$$P = \frac{W}{t} = \frac{\Delta E}{t}$$

where the symbol Δ still means "change in." Whatever the symbols we use, power represents the *rate* at which you do work. Because the work you do on an object equals the change in energy of the object, power is also equal to the change in energy divided by the time.

Let's apply this idea of power to you climbing a flight of stairs. To keep things simple, I'm going to assume you head up the stairs at a constant velocity. If your velocity doesn't change, then you don't change kinetic energy. Your total change in energy is just your change in gravitational potential energy. That's equal to *mgh*, where *m* is your mass, *g* is equal to 9.8 m/s^2, and *h* is the vertical height of the stairs. Whether you climb the stairs in a short time or a long time, your change in energy is the same and is equal to *mgh*.

The power involved, however, is different for different times. I'll use a few numbers to show that. Let's suppose *mgh* in this example equals 1,500 joules, which corresponds to a 75 kg (165 lb.) person climbing stairs with a vertical height of 2 m. If you climb these stairs in 5 seconds, the power involved is:

$$P = \frac{\Delta E}{t} = \frac{mgh}{t} = \frac{1500 \text{ joules}}{5 \text{ seconds}}$$

This equals the number 300 (I'll talk about the units of power in just a bit). Now suppose you climb those same stairs in 10 seconds. Then we have:

$$P = \frac{\Delta E}{t} = \frac{mgh}{t} = \frac{1500 \text{ joules}}{10 \text{ seconds}}$$

which equals the number 150. We now have half the amount of power generated. In other words, the faster you change your energy, the more power you generate.

Now let's look at the same situation from a "work" perspective rather than an energy perspective. Exactly who or what is doing the work on whom or what? Believe it or not, the stairs are exerting a force on you.* As the stairs exert a force on you, they do work on you. To calculate the amount of work the stairs do on you, we use our trusty formula for work, which is $W = Fd$. When you use this formula, you find that the work done in moving you up

* This has to do with something called Newton's third law. When you push on the stairs, the stairs push back on you.

the stairs at a constant velocity is just equal to mgh, where h is the height of the stairs, m is your mass, and g is 9.8 m/s^2. This quantity is the same whether you move slowly or quickly, just as the change in energy is the same whether you move slowly or quickly. The power generated is work divided by the time to do that work, or:

$$P = \frac{W}{t}$$

If t is large (moving slowly), the power generated is small. If t is small (moving quickly), the power generated is large.* The faster you do work, the more power you generate.

The unit of power is the *watt*, named after James Watt; 1,000 watts gives you one kilowatt. Okay, we're getting closer to understanding what a kilowatt-hour is. You measure the rate at which you use electrical energy in kilowatts because that's a unit of power. But you don't pay your electrical bill based on the rate you use the energy; you pay on the total amount of energy used.

Before explaining how kilowatt-hours are units of energy, I'm going to take you on a diversion. Suppose you want to measure how many miles you've traveled in a car, knowing how fast you've traveled and for how long. What you would do is multiply miles per hour by the number of hours driven.

$$\text{Distance traveled} = \frac{\text{miles} \times \text{hours}}{\text{Hour}}$$

Once you get your result, you could say that you have measured the distance not in miles but in (miles/hour)-hours. Of course that would be a totally silly thing to do, but you could do it.

Similarly, to calculate the total energy used when you know the rate at which it is being used (which is power, which is the energy used divided by the time), you multiply the rate (energy divided by time) by the number of hours you used the energy at that rate.

$$\text{Energy used} = \frac{\text{energy used} \times \text{time}}{\text{time}}$$

$$= \text{power} \times \text{time}$$

So a kilowatt-hour is just a unit of energy. Why, exactly, power companies use kilowatt-hours instead of joules, I don't know. I'm betting that, like many other things, it's just a tradition that made sense a long time ago and has just

* If that's not obvious to you, just put in some numbers for W and t, just as I did in the previous paragraph with ΔE and t.

hung around. Or maybe energy companies used to concern themselves with the rate at which people used energy, and then an accountant came along and told them to hang the rate of use and just worry about what was really costing them—the total amount of energy used.

MORE THINGS TO DO BEFORE YOU READ MORE SCIENCE STUFF

What I'm going to have you do in this section is a bit more complicated than what I usually request. It's really instructive, though, so you just might find the extra effort worthwhile. Anyway, you have to start by gathering a few materials. You need the following:

- A compass (the kind that tells you what direction north is). The larger the compass the better. The really tiny ones (a centimeter or two in diameter) won't work.
- A strong magnet (the stronger the better—refrigerator magnets won't cut it this time)
- A 2 m (approximate) section of *insulated* wire. This needs to be thin enough so that you can bend it easily, yet strong enough that it will hold its shape. The folks at the hardware store will be able to help.
- A working flashlight battery
- Some kind of tape
- A tube from a roll of toilet paper, sans toilet paper

Start by stripping a small bit of insulation off each end of the wire. Then wrap the wire a bunch of times around the compass. Leave about 10 cm of wire on one end and at least 1 m of wire on the other end. After making sure you can still see the compass needle underneath all that wire, tape the wire in place (Figure 2).

Figure 2

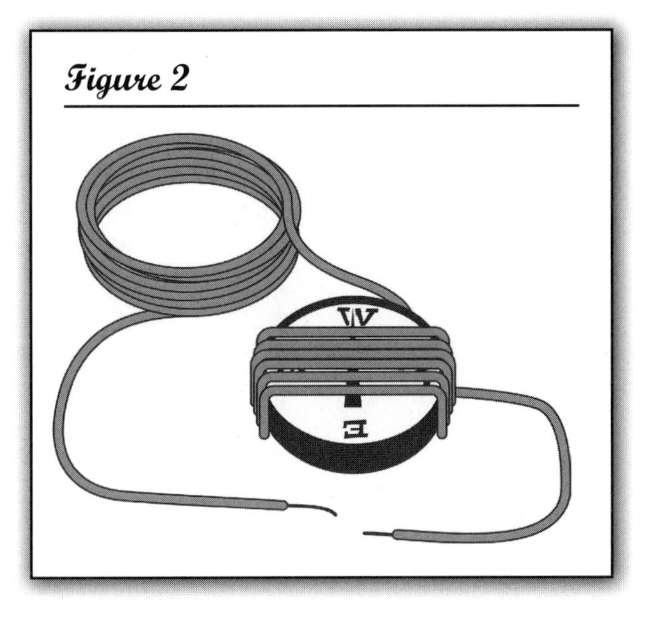

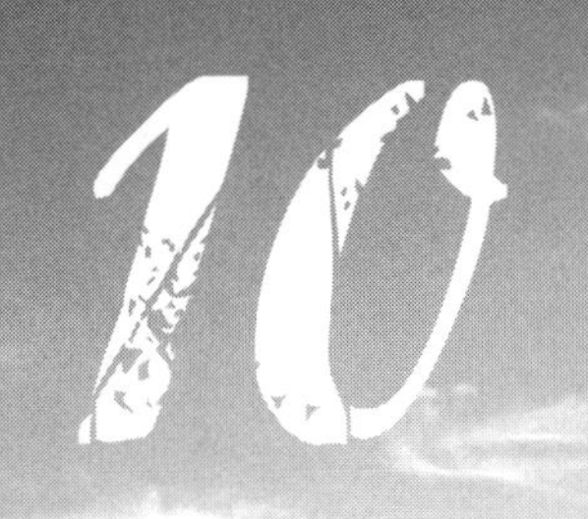

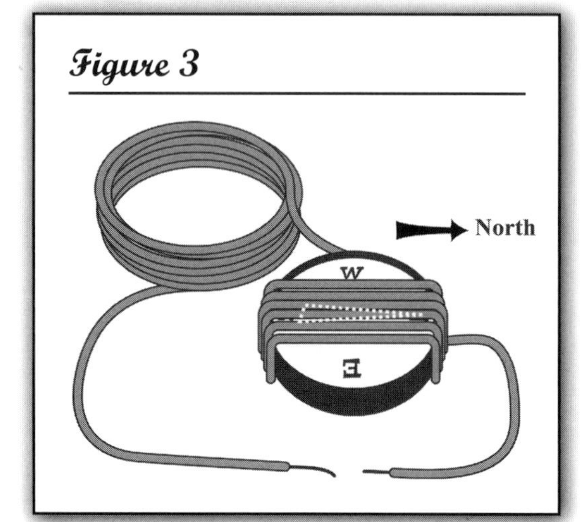

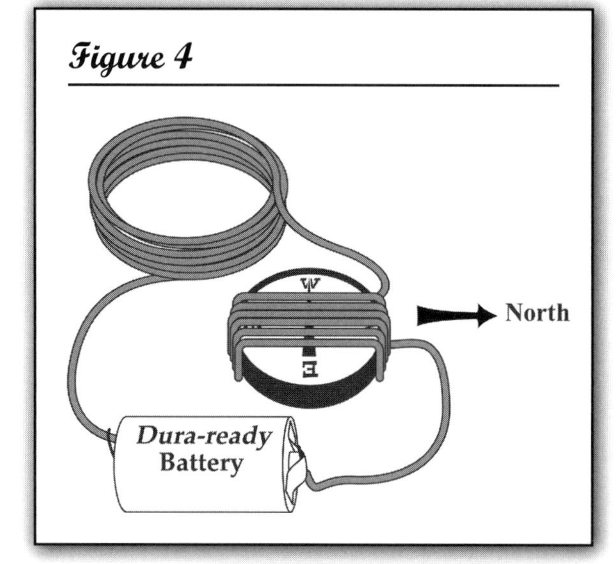

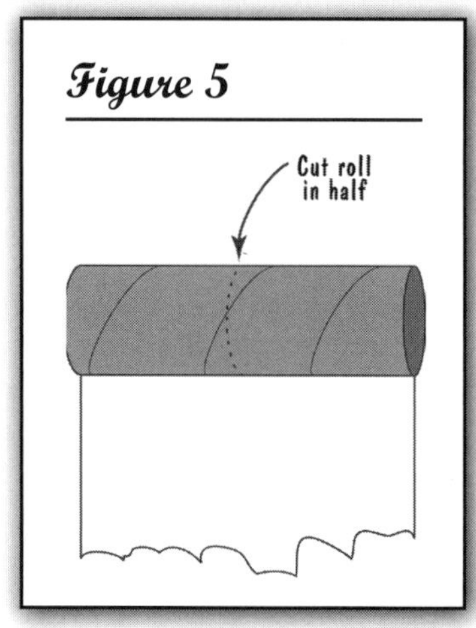

Make sure the compass isn't near your magnet, or any other magnet for that matter (phones and stereo speakers have pretty strong magnets in them). Place the wrapped compass on a flat surface. The compass needle should point north. Rotate the compass until the needle is lined up with the coil of wire (Figure 3).

With the compass in that position, touch the free, stripped ends of the wire to the battery terminals, as shown in Figure 4. Watch what happens to the compass needle when you do this.

If the compass needle doesn't deflect, check to see that the battery isn't dead and that you have the compass lined up as in Figure 3. With a good battery and the correct alignment, the compass needle will deflect as shown in Figure 4. Now batteries cause electrical current to flow in wires, so obviously having an electrical current going through the coil on the compass causes the needle to deflect. Don't worry about why that happens, because it would take half a book to explain it well!

Set the battery aside and cut the toilet paper tube in half. Wrap the long free end of the wire around one of the half tubes as shown in Figure 6. Leave enough wire so you can twist together the stripped ends and be able to place the tube at least 30 cm from the compass.

Move the magnet in and out of the center of the tube. As you do this, watch the compass needle (Figure 7).

The compass needle should move just a tiny bit and only while the magnet is moving. Don't expect the needle to move as much as it did when you touched the wires to the battery.

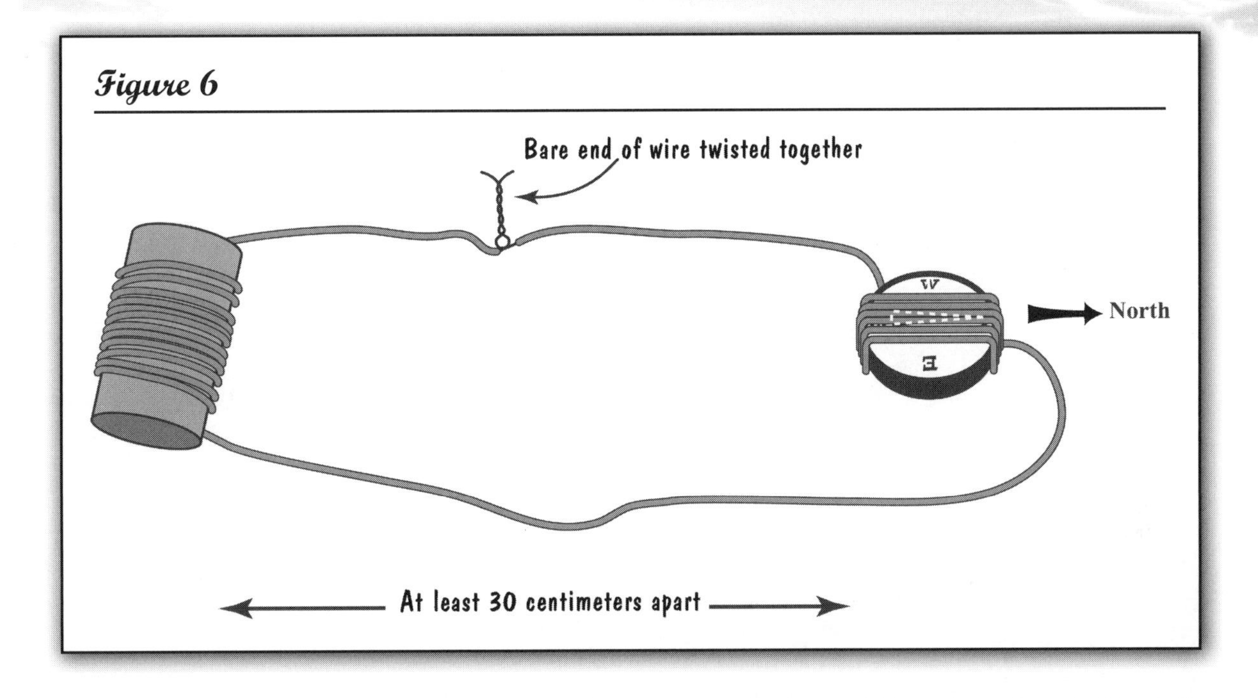
Figure 6
Bare end of wire twisted together
North
At least 30 centimeters apart

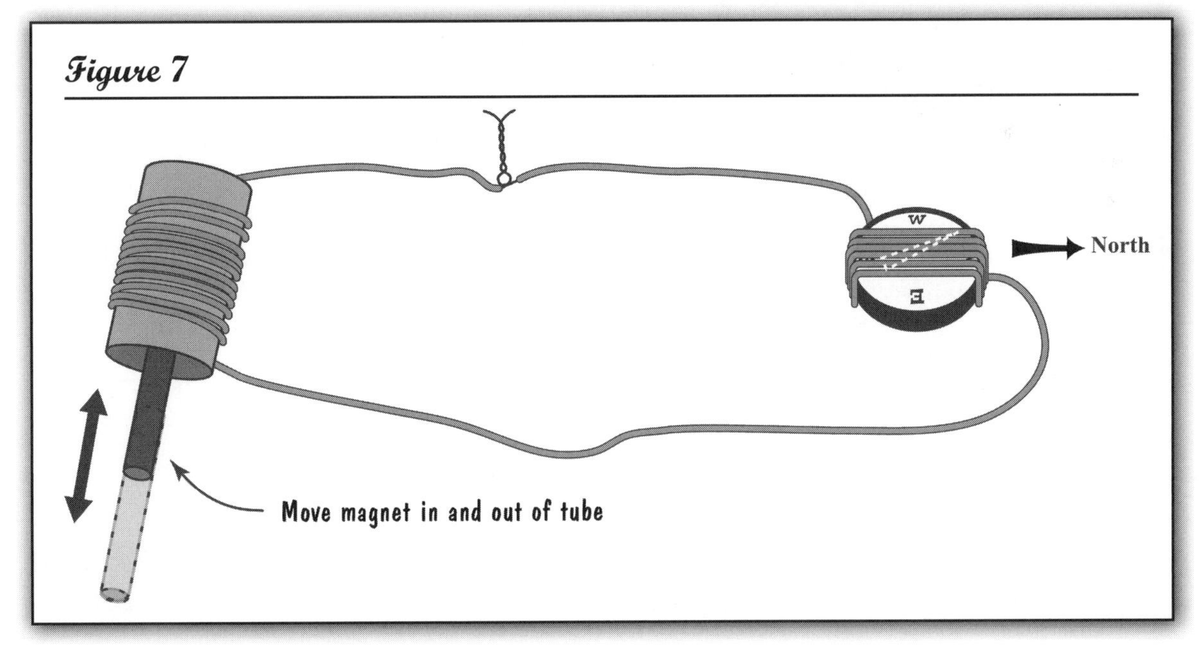
Figure 7
North
Move magnet in and out of tube

MORE SCIENCE STUFF

See if the following reasoning makes sense. The compass needle deflects whenever an electrical current is flowing through the coil of wire that's wrapped around the compass. The needle deflects when you move the magnet in and out of the coil of wire wrapped around the toilet paper tube. Therefore, moving the magnet in and out of the coil of wire causes an electrical current to flow in the wire.

If that reasoning doesn't make sense, well, just pretend that it does. It's a fact of nature that, when you move a magnet in and out of a coil of wire, you will generate an electrical current in the wire. Also, if you move a coil of wire relative to a magnet, you will also generate an electrical current. It turns out that this is the method used to generate just about all the electricity people use. Before going on, I'll mention the other two main sources. One is the battery, which takes chemical potential energy and converts it to electrical energy. The other is the solar cell, which takes radiant energy from the Sun and converts it to electrical energy.

Figure 8

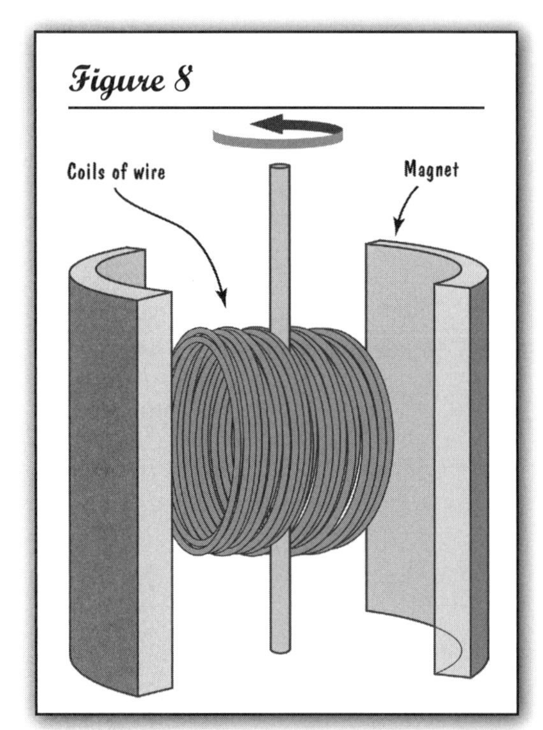

The device used to generate electricity is called a *turbine*. A turbine basically consists of many coils of wire that are free to spin inside stationary magnets (Figure 8).

As the coils spin, they move relative to the magnets, resulting in an electrical current flowing through the wire. That electrical current then can be sent to consumers. More about that later.

So all we need to do is hire a bunch of people at minimum wage to spin turbines, right? Well, no, you couldn't hire enough people to meet the demand for electricity. Better to get an input of energy from some other source. One source is a moving river. Set your turbines up on a dam in a river and let the moving water turn the turbines. What you get is *hydroelectric power* (Figure 9).

Figure 9

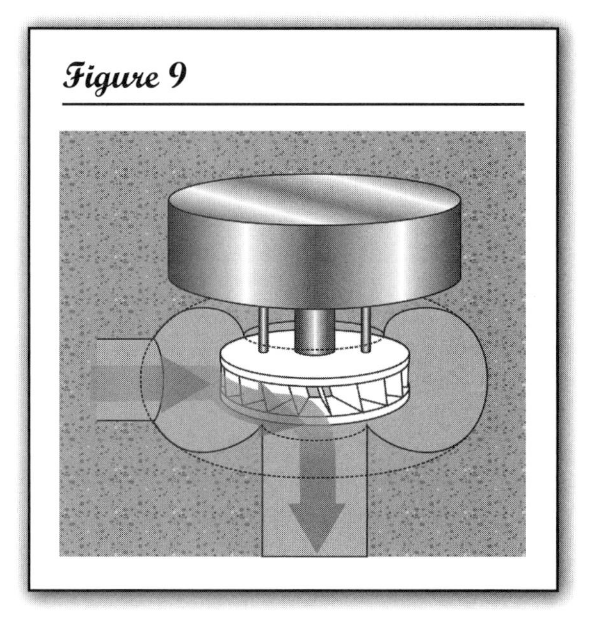

Another source of energy for turning turbines is the wind. Set up a windmill connected to a turbine, and the wind will turn the turbine for you (Figure 10).

One trouble with hydroelectric and wind power is that they can be unreliable. No wind—no electricity. Little rain means lower water levels and less hydroelectric power. To have a more reliable source of electricity, we use the setup shown in Figure 11.

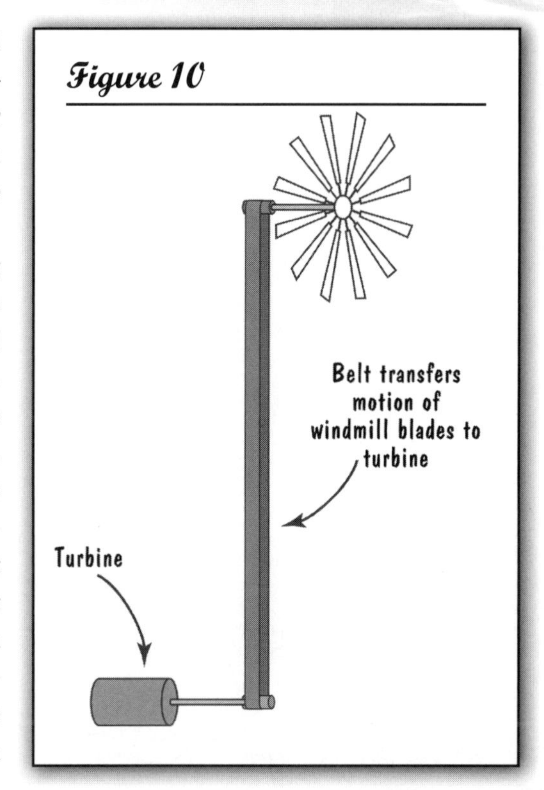

Figure 10

You heat the water reservoir, producing steam. You then force that steam through pipes that lead to—surprise—a turbine! The forced steam turns the turbine, producing electricity. Simple enough. Now all you need is a way to heat the water. There are lots of ways to do that. Burning coal is one way. Burning natural gas is another. Using the heat from nuclear reactions is another. This last method is somewhat controversial, in part because people don't understand what role the nuclear reactions in a nuclear reactor play in generating electricity.* Their sole purpose, though, is to generate enough heat to turn the water to steam, which turns the turbine, which generates the electricity, which goes to the old lady who swallowed the fly.

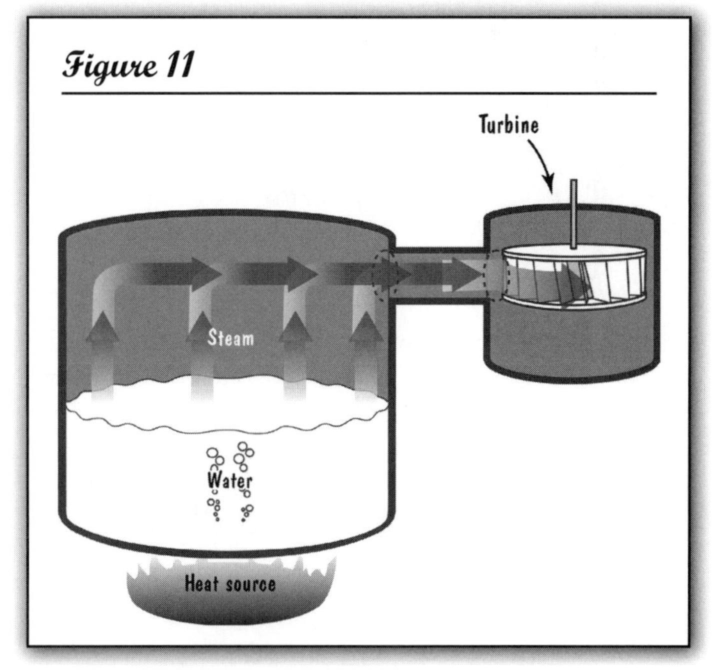

Figure 11

As a summary, let's try to draw a diagram that shows all the energy transformations that result when burning coal to get electricity to the consumer (Figure 12, p. 78). In the diagram, I'm going to put in "losses" to thermal energy in lots of places.

One place you might not expect such losses is in getting electricity from the power plant to the consumer. Any time electrical current flows in wires, however, it generates heat. The scientific explanation for what's going on is that very tiny electrons bang into atoms and produce heat.

---

* Take note of the words "in part." I'm not saying this is the only reason. I just don't want to get into all the other issues, such as waste disposal, here.

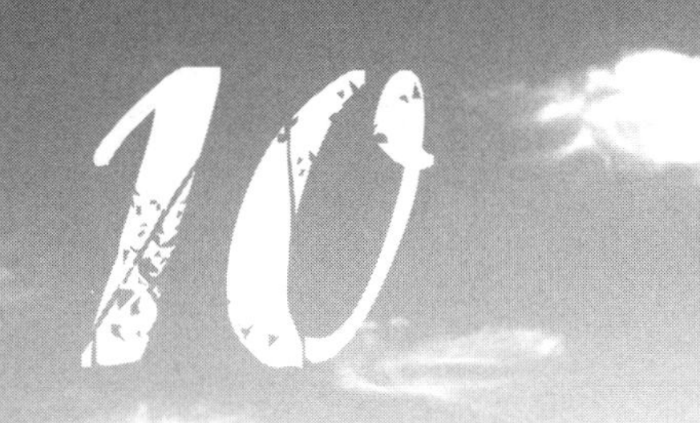

Figure 12

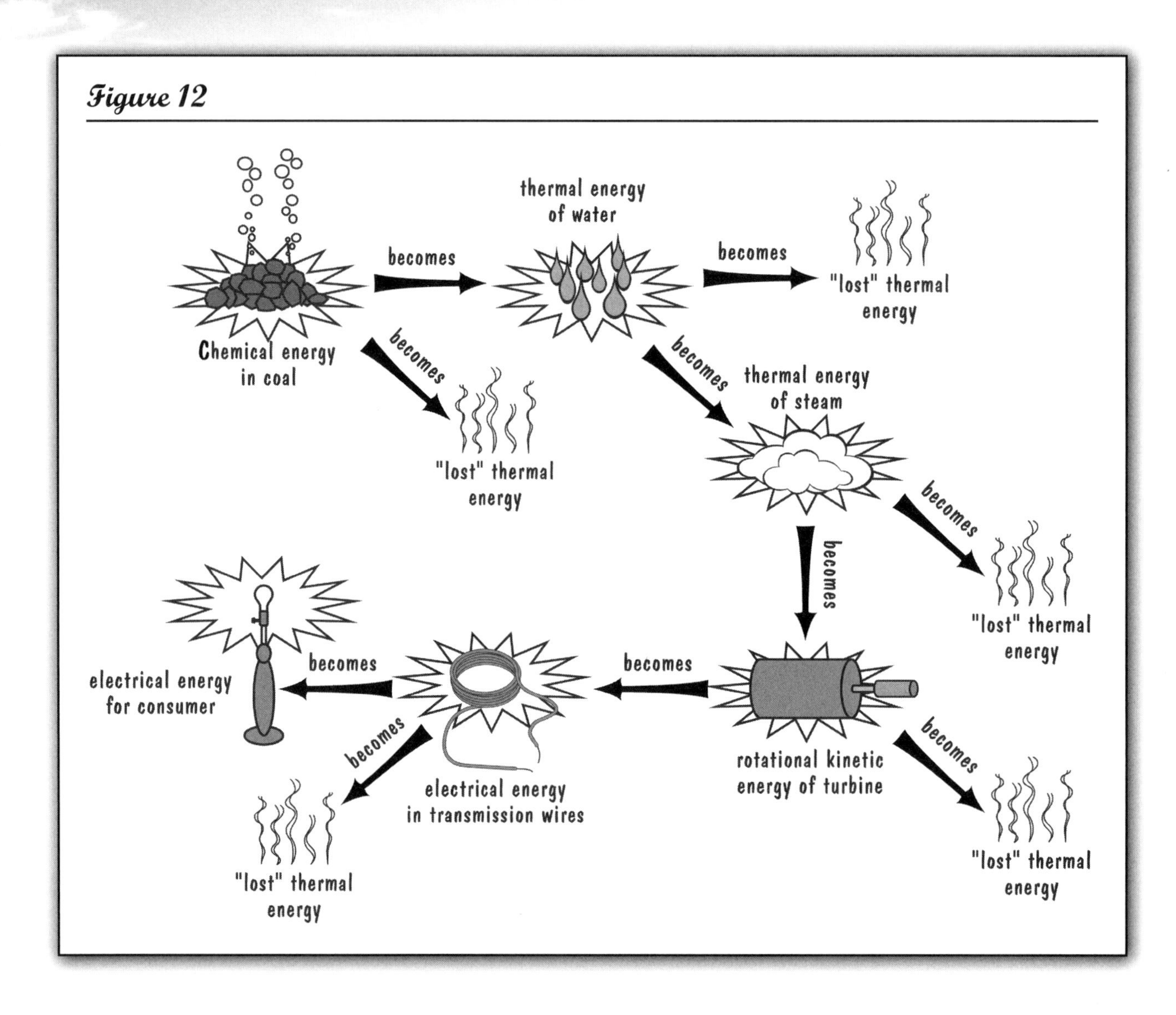

Well, that's about it for energy basics. We certainly haven't covered all the applications, because energy and energy transformations are at the heart of many different areas of physics and many different scientific disciplines.

CHAPTER SUMMARY

- Power is the rate at which you do work or use energy. Power is measured in watts.
- Kilowatt-hours is a unit of energy, not power, and is how the power company measures energy usage by consumers.

- When wires and magnets move relative to each other, an electrical current flows in the wires. This phenomenon is the basis for most of the electricity generated in our country. We use various methods to turn a turbine (coils of wire turning inside a bank of magnets) to generate electricity.

APPLICATIONS

1. What makes electricity cost more in one state than in another, or even in different parts of the same state? Many factors affect that, including the number of nearby power plants, the regulations on those power plants, and just a whole bunch of politics. I won't deal with those, but I will discuss the factor of *efficiency*. If it's possible to get electricity to the consumer with as few energy "losses" as possible, then the electricity should be cheaper. Those energy losses could come at the power plant, where inefficient equipment (machines!) results in more energy losses. Also, the farther you have to transmit the electricity, the more you lose heat as the electricity travels through more wires. That means consumers in remote locations generally pay more for their electricity than urban consumers.

2. Speaking of efficiency, how come gas appliances are so much more efficient (cheaper to operate) than electrical appliances? Consider a clothes dryer. With a gas dryer, you get natural gas piped into your home and burn it to produce the heat to dry your clothes. With an electric dryer, the power company burns natural gas or coal to produce electricity, which you then use to create heat to dry the clothes. It's a whole lot cheaper to use the raw materials (gas or coal) directly rather than have it converted to electricity first. Of course, it's even cheaper to use the ultimate source of energy that created the natural gas and coal. That source is the Sun, but check your neighborhood for ordinances before hanging your clothes outside to dry!

Chapter 11

THE SCIENCE BEHIND NANOSUNSCREENS

LEARNING ABOUT NANOPARTICULATE INGREDIENTS USED TO BLOCK THE SUN'S ULTRAVIOLET RAYS

By Alyssa Wise, Patricia Schank, Tina Stanford, and Geri Horsma

"What are 'nano' ingredients, and how can you tell if your sunscreen has them?" "Why do we need nanosunscreens?" "If a bottle of a well-known sunscreen brand claims that it protects against UVB and UVA rays, should I believe it?" These are just a few of the questions that our Clear Sunscreen nanoscience unit has students asking. Students are excited to learn about science that is applicable to their daily lives and because they find the topic interesting, they are motivated to figure out the complex concepts involved.

In this chapter, we provide a brief overview of the field of nanoscience and why it is an important area for education. We then explain the science behind nanoparticulate sunscreens, describe the different elements of the unit, and reflect on some of the opportunities and challenges of teaching nanoscience at the high school level.

CLEAR SUNSCREEN

Clear Sunscreen is one of four sets of learning materials collaboratively created by educational researchers, science teachers, and nanoscientists as part of the NanoSense project, aimed at helping high school students learn about science concepts that account for nanoscale phenomena. All of the materials are freely available online and are linked to the National Science Education Standards (NRC 1996) (see "On the web"), as well as related chemistry, physics, biology, and environmental science concepts. Although Clear Sunscreen was designed for chemistry classrooms—introductory, Advanced Placement, or International Baccalaureate, depending on whether the advanced materials are used—it has also been used successfully in 9th-grade biology and 11th-grade biotechnology classrooms. Given the interdisciplinary nature of the content, it could also be used in physics or integrated science classrooms.

Throughout the unit, students explore answers to the questions posed in the introduction through presentations, discussions, readings, hands-on activities, and labs. At its conclusion, students synthesize their learning by creating consumer awareness pamphlets that explain the basics of nanoparticulate zinc oxide: how it can appear clear in sunscreens, but still protect against ultraviolet (UV) light; what its benefits are over traditional ingredients; and any potential dangers that might be associated with its use.

ABOUT NANOSCIENCE

Nanoscience is the study of matter on the scale of 1–100 nanometers (nm) in at least one dimension. Given that 10 hydrogen atoms lined up are about 1 nm long, we can loosely describe the field as being concerned with molecules and small clusters of molecules in this size range. One well-known example is Buckminsterfullerene, the soccer ball–shaped carbon (C60) molecule often referred to as a "buckyball."

Nanoscience is an important area of research because matter at this size scale has many unique properties. For example, nanosize carbon tubes are 100 times stronger than bulk steel but are also incredibly flexible, and nanosize substances' melting points decrease as they get smaller. Nanosize zinc oxide—a common ingredient in nanosunscreens—in particular, appears clear (transparent) instead of white. The unique properties of nanosize materials have already enabled new innovations in areas as diverse as textiles (e.g., stain-resistant clothes), the environment (e.g., paint that "cleans" the air), and personal health care (e.g., clear nanoparticulate sunscreen).

Despite their potential benefits, objects at this size scale still present many mysteries. For instance, they are small enough that many of our models for bulk substances do not accurately predict their properties, but large enough that quantum calculations are prohibitively complicated. New models and ways of thinking are being developed to better understand their behavior.

Introducing nanoscience ideas to students presents several exciting opportunities. First, it gives them the opportunity to explore applications that are relevant to their lives, which can serve as a "hook" to get them excited about science. It can also introduce an interdisciplinary perspective. Like most nanoscience topics, the science behind clear, nanoparticulate sunscreen brings together concepts from chemistry, biology, and physics, giving students a chance to see the interconnections between traditional scientific domains and the unity in nature (Roco 2003). Finally, teaching cutting-edge science in which many of the answers (and even the questions) have not yet been formulated provides the opportunity for students to experience science in the making (Latour 1987), helping them develop a better understanding of the nature of scientific knowledge.

ABOUT NANOPARTICULATE SUNSCREENS

Sunscreens are colloidal suspensions of UV-absorbing agents in a lotion. Two kinds of UV-absorbing agents (or "active ingredients") can be used: organic and inorganic. Organic ingredients are carbon based, exist as individual molecules, and absorb specific, narrow bands of UV light. Inorganic ingredients are metal oxides that exist as ionic clusters of various sizes and absorb all UV light whose wavelength is less than a critical value.

Traditional sunscreens often contain "large" inorganic zinc oxide clusters because they effectively absorb the full spectrum of UV light. However, because these clusters also scatter visible light, the cream has an undesirable white color that remains visible on the skin's surface. As a result, people often apply too little sunscreen or choose another, less effective kind (see "Structure and scattering," p. 86, for more on inorganic versus organic sunscreen ingredients).

If, however, nanosize clusters of zinc oxide are used instead of the larger clusters, the sunscreen is transparent because the diameter of each nanoparticle is much smaller than the wavelength of visible light, and thus does not scatter it. The protection remains the same because reducing the particle size does not change its absorptive properties. Given our increased awareness of the dangers of long-wave ultraviolet (UVA) light—which many organic ingredients used in sunscreens do not block—a full spectrum sunscreen that people are willing to apply in sufficient quantities is an important tool for preventing skin cancer. Sunscreens that use nanosize inorganic ingredients can provide this protection. However, there are some concerns about using nanosized ingredients in sunscreen since they more easily cross membrane barriers, though no adverse effects have been found to date.

NANOSUNSCREENS IN THE CLASSROOM

Sun protection basics

Our Clear Sunscreen materials are designed to be modular and adaptable to a variety of classroom levels and orientations. We begin by connecting with students' current science knowledge. In chemistry classes, this usually means asking students to share what they know about using and choosing a sunscreen, and then to examine the kinds of chemicals that are used as active ingredients in the sunscreens they have at home. The Sunscreen Label Ingredients activity asks students to compare the different combinations of ingredients found in sunscreens and look for patterns. We have found that students are excited to recognize some terms on the labels, and a discussion of chemical names usually ensues. This can then segue into a class discussion of consumer chemistry issues, such as why the same chemicals show up repeatedly—the reason is that a limited number are approved by the Food and Drug Administration—and raises important questions about how sunscreen ingredients protect us from UV light (e.g., "Why do most sunscreens have more than one active ingredient?" and "Why is it that sunscreen makers cannot just put in more of the 'best' ingredient?")

In biology classes, we take a slightly different approach and often begin with a discussion of what students know about the dangers of skin cancer and the need for sun protection. In a physics class, the electromagnetic spectrum and the energy of different wavelengths of UV light can serve as anchors.

Regardless of the entry point, to help students think about the different kinds of sun rays that reach Earth and how they interact with our bodies, we use the information, images, and embedded discussion questions in the interactive Sun Protection: Understanding the Danger PowerPoint. These slides are best used to support the development of understanding after students have participated in an activity, such as Sunscreen Label Ingredients, which cultivates a "need to know."

We have consistently found that students (at all levels) are fascinated by the issues related to protecting their bodies and invariably have a lot of questions and personal experiences to share. Interestingly, we have also found that while most students are familiar with the terms UVA and UVB, few actually understand what they refer to: that they are wavelengths of light falling within certain ranges along the continuum of the electromagnetic spectrum (UVB ≈ 280–320 nm, UVA ≈ 320–400 nm). Instead, students often talk about them as singular entities that can be blocked (or not) in a binary fashion, and a surprisingly large number mistakenly believe that they are actually two types of protective ingredients that can be put into sunscreens. Our introduction to the unit helps students reveal and remedy these ideas; supports them in making connections between traditional chemistry, physics, and biology topics; and lays the foundation for understanding the chemistry behind new nanoparticulate sunscreens.

At the end of the introductory section, we pose three essential questions to help students monitor their learning throughout the rest of the unit:

- What are the most important factors to consider in choosing a sunscreen?
- How do you know if a sunscreen has "nano" ingredients?
- How do "nano" sunscreen ingredients differ from most other ingredients currently used in sunscreens?

We ask students to consider their initial ideas about these questions, emphasizing that this is not a test of what they know and encouraging them to make guesses that they can later reevaluate based on what they learn in the unit. If there is time, we have students share their answers—reminding them that there are no "bad" ideas at this stage—and discuss as a group which statements they think are true and the rationale or evidence supporting their assertions.

Appearance and UV blocking

In the second part of the unit, the focus is on chemical and physical properties of sunscreens and their interactions with light. The core activity is the UV Protection Lab which invites students to investigate whether there is a correlation between a sunscreen's appearance—its degree of transparency—and its UV-blocking ability. (**Note:** They later find that there is no relationship, since one depends on interactions with wavelengths of light in the visible range [which determines appearance] and the other depends on interactions in the UV range [which determines blocking ability]. Thus, while there are some materials that are opaque and block UV light, there are also examples of materials that appear clear, but are equally good or better blockers.) The exploration of the counterintuitive nonrelationship between opacity and UV blocking differentiates this lab from many sunscreen labs, which simply have students test the blocking ability of different substances.

In the UV Protection Lab, students select a variety of protective agents (e.g., different sunscreens; t-shirts; inexpensive, clear UV-blocking glass) to test using beads that change color when exposed to UV light. Using a color guide, students compare the color change of a bead covered by a protective agent to that of an uncovered bead. (**Note:** UV-sensitive beads and color-change guides are readily available from several commercial suppliers.) Students also evaluate the appearance of the substances using an opacity guide that we print on a sheet of transparent acetate.

Using the opacity and UV-protection ratings for each substance tested, students create a scatter plot to determine whether there is a relationship between protection and opacity. Assuming students select a wide enough variety of substances to test, the plot should have points in all quadrants, indicating no relationship. If some groups test a restricted range of substances, we use it as an opportunity to talk about sampling issues and then combine the whole class's results to create a more complete data set.

The main safety consideration in this lab is to remind students not to put the sunscreen on their bodies in case of an allergic reaction. Additionally, although no adverse effects attributed to nanoparticles in commercially available sunscreens have been reported, because nanoparticles are still being studied, it is better not to apply them to our bodies in situations where they are not needed.

Structure and scattering

We follow up the lab with the All About Sunscreens interactive PowerPoint. (**Note:** For advanced classes, there are two additional interactive PowerPoints that go into greater depth on absorption and scattering mechanisms, which are freely available on the website). The information, images, discussion questions, and activities embedded in the slides help students explore the different structure and UV-absorption patterns of organic and inorganic sunscreen ingredients.

Their understanding of the different absorption patterns can then be used to explain one of the findings from the Sunscreen Label Ingredients activity: When sunscreens use organic chemicals as UV blockers, they are always formulated with multiple active ingredients; but when they use inorganic chemicals as UV blockers, they can be formulated with just a single one. This occurs because although these inorganic compounds absorb strongly across almost the entire UVB and UVA spectrum, the organic molecules have narrow UV-absorption ranges, thus different ingredients are needed to provide full spectrum coverage.

One concept we have found that students consistently have difficulty with is the difference between molecules and ionic compounds. While our students were familiar with the difference between molecular formulas and formula units, when it came time to use these concepts in practice, they often confused the two. We would encourage other teachers to dedicate extra time and attention to this issue, since it relates to concepts that are fundamental to chemistry.

The second half of the All About Sunscreen slides addresses interactions with visible light, helping students understand that traditional inorganic sunscreen ingredients appear white on the skin because of the way the larger zinc oxide clusters scatter light. Students can explore the scattering mechanism in

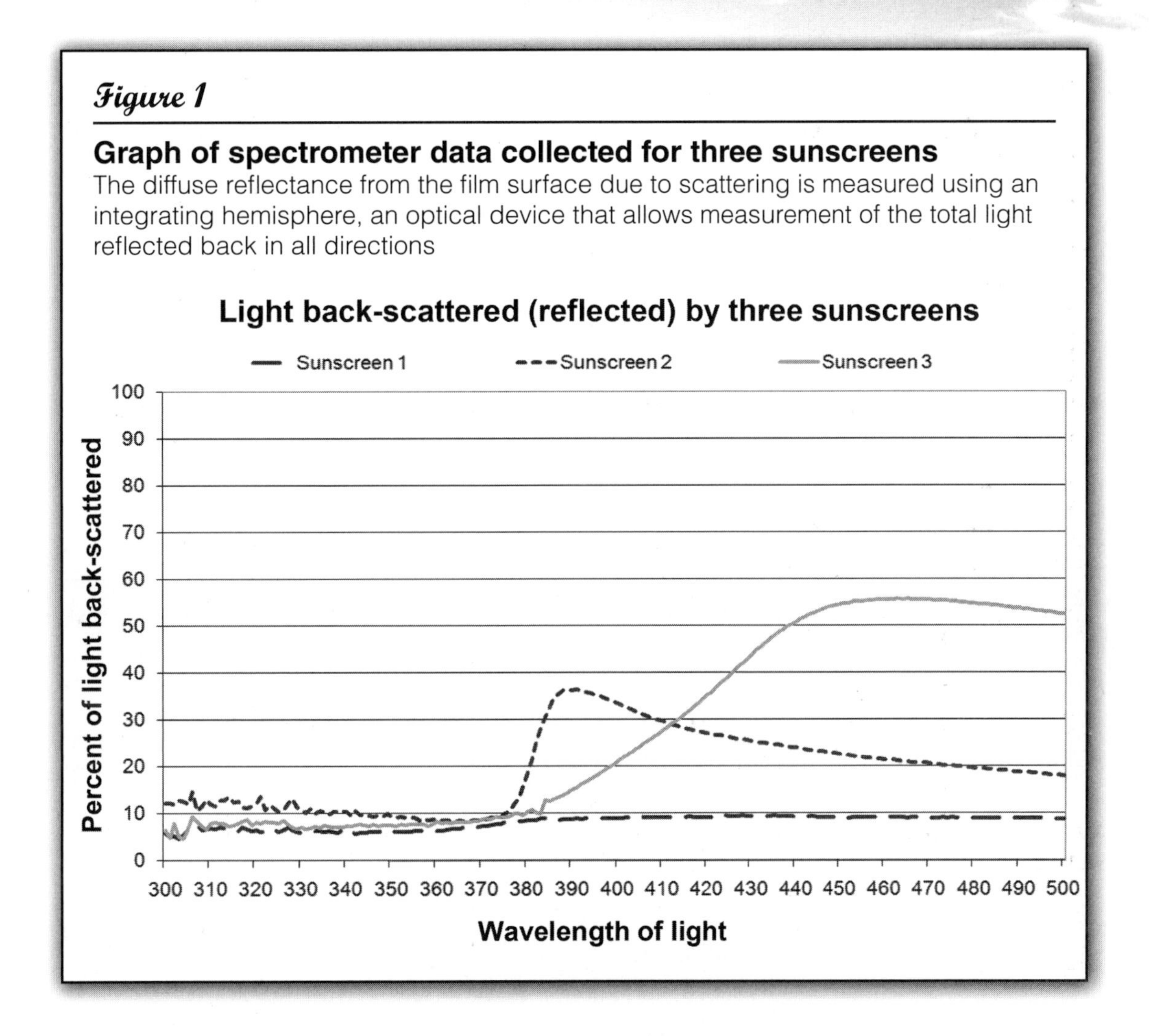

Figure 1

Graph of spectrometer data collected for three sunscreens

The diffuse reflectance from the film surface due to scattering is measured using an integrating hemisphere, an optical device that allows measurement of the total light reflected back in all directions

more depth through the Sunscreens and Sunlight Animations activity. This activity has two variations. Students can use the ChemSense Animator tool, which is freely available online (see "On the web"), to create their own animations of how visible light interacts with different kinds of sunscreen ingredients. Or, if time is limited, they can explore a premade interactive animation of visible light interacting with different kinds of sunscreen ingredients using a guiding worksheet.

The interactive All About Sunscreens PowerPoint also allows us to introduce the size-dependence of light scattering—the central principle of nanoparticulate sunscreens. Maximum scattering of light occurs for clusters whose diameter is half as large as the wavelength of the light (scatterMAX of λ occurs when $d = \frac{1}{2}\lambda$). Thus, zinc oxide clusters whose diameter is less than 100 nm are too small to scatter visible light. These concepts are further explored through the interpretation of spectrometer data we have collected on three kinds of sunscreens containing organic ingredients, nanosized inorganic ingredients, and traditional inorganic ingredients, respectively (Figure 1, p. 87). We ask students to think about four questions for each sunscreen (answers provided in Table 1):

1. Will it appear white or clear on your skin?
2. What size (approximately) are the molecules or clusters?
3. Can we tell how good a UV blocker it is from this graph? Why or why not?
4. Which one of the sunscreens is it? How do you know?

Finally, we raise the question of the potential health risks of nanoparticulate sunscreens, including both scientific aspects (e.g., very small clusters are more likely to cross membranes and get into unintended parts of the body) and socially situated ones (e.g., as new substances, nanoparticulate sunscreen ingredients are not yet fully studied and thus possible harmful effects may still be unknown).

Pulling it all together

At the end of the unit, students solidify their understanding by creating a consumer awareness pamphlet that explains the benefits of nanoparticulate ingredients—the combination of good UVA and UVB protection with clear appearance—and potential dangers associated with their use. The required scientific information we ask for in the pamphlet and a rubric we have created for judging its completeness, depth, and accuracy allow us to use it as a performance assessment tool. We find that this task gives students a meaningful opportunity to show what they have learned, and that situating this task in the context of social awareness helps students to see science as something that is relevant to their lives.

OPPORTUNITIES AND CHALLENGES

The teachers we have worked with to implement the Clear Sunscreen unit have been extremely pleased with the results. In our own estimation, students seem genuinely interested and concerned about sunscreen issues. It is apparent that the issues hit home in that students ask real-life questions (e.g., "How am I going to get something that protects me?") and quickly realize that there is a lot they do not know.

Despite high levels of student enthusiasm, we encountered several challenges in teaching the unit. One set of challenges relates to the specific content

Table 1

Answers to three sunscreen spectrometer graph questions

	Sunscreen 1 (contains organic ingredients)	**Sunscreen 2** (contains nanosize inorganic ingredients)	**Sunscreen 3** (contains traditional inorganic ingredients)
Appearance	No scattering in the visible range. Sunscreen appears clear on the skin.	Very limited scattering in the visible range. Sunscreen appears clear on the skin.	Significant scattering in the visible range. Sunscreen appears white on the skin.
Size	Since no scattering is seen, it is not possible to estimate the size of the molecule from the information in the graph.	The sharp drop in the curve at 380 nm (and the low scattering below 380 nm) is actually due to UV absorption (if light is absorbed, it cannot be scattered). Therefore, we cannot know the exact size of the cluster, only that the curve would have peaked below 380 nm. So the cluster size is smaller than 190 nm.	Because the graph peaks around 450 nm, we would estimate the cluster size to be about 225 nm (well above nanoscale dimensions).
UV blocking	The graph shows very little scattering in the UV range; however, this does not tell us anything because absorption is the main blocking mechanism for UV. Because the UV light is not scattered, we know that it must either be absorbed or transmitted, but we do not know which. We would need an absorption or transmission graph to determine the UV blocking ability of the sunscreens (T + R + A = 1).		
Identity	Virtually no scattering in the visible range indicates organic ingredients. This sunscreen contains the organic ingredients octinoxate and oxybenzone.	Low amounts of scattering in the visible range indicate inorganic ingredients with nanosizedclusters. This sunscreen contains nanosize zinc oxide.	Significant amounts of scattering in the visible range indicate inorganic ingredients with large clusters size. This sunscreen contains traditional titanium dioxide.

involved, and we have consistently found these issues—such as the definitions of UVA and UVB as wavelength ranges along the electromagnetic spectrum and the distinction between molecules and ionic compounds—across all grade levels.

The other set of challenges we have encountered is pedagogical in nature. Nanoscience is on the cutting edge of interdisciplinary scientific research and is expanding the limits of our collective scientific knowledge. Existing models do not always apply, and there are still many unanswered questions. This provides an exciting opportunity to help students experience science in the making, but also requires us to firmly embrace an inquiry approach to science learning. In support, we have created a guide to help other teachers use these challenges as opportunities to model the scientific process (see "On the web").

CONCLUSION

The high levels of student interest and engagement we observe with the Clear Sunscreen unit leads us to believe that teaching the science behind nanosunscreens can be a useful hook to get students excited about learning chemistry. While challenges exist, addressing them provides opportunities to reinforce core chemistry concepts and help students gain a deeper understanding of the nature of science.

Acknowledgments

The NanoSense project was supported by NSF Grant No. ESI-0426319. Any opinions, findings, and conclusions or recommendations expressed in this material are those of the authors and do not necessarily reflect the views of the NSF. All figures have been adapted from the NanoSense unit with permission under a Creative Commons Attribution 3.0 License. All materials for the Clear Sunscreen unit described in this article, as well three other nanoscience units, are freely available on the NanoSense project website (see "On the web"). Anyone is welcome to use or adapt the materials, as long as they acknowledge the NanoSense project.

On the web

Nanoscience project materials: *http://nanosense.org/activities.html*

Alignment with National Science Education Standards: *http://www.nsta.org/highschool/connections/200909AlignmentWithTheStandards.pdf*

ChemSense tool: *www.chemsense.org*

Premade interactive animation: *http://nanosense.org/activities/clearsunscreen/sunscreenanimation.html*

Summary of challenges and opportunities in teaching nanoscience: *http://www.nsta.org/highschool/connections.aspx*

References

Latour, B. 1987. *Science in action.* Cambridge: Harvard University Press.

National Research Council (NRC). 1996. *National science education standards.* Washington, DC: National Academies Press.

Roco, M. C. 2003. Converging science and technology at the nanoscale: Opportunities for education and training. *Focus on Nanotechnology* 21 (10): 1247–1249.

Chapter 12

JUAN'S DILEMMA

A NEW TWIST ON THE OLD LEMON BATTERY

By Timothy Sorey, Vanessa Hunt, Evguenia Balandova, and Bruce Palmquist

When life hands you lemons, make a battery! In this chapter, we describe an activity that we call Juan's Dilemma an extension of the familiar lemon-battery activity (Goodisman 2001). Juan's Dilemma integrates oxidation and reduction chemistry with circuit theory in a fun, real-world exercise. We designed this activity for a ninth-grade physical science class, and our students have found it to be intriguing and challenging.

THE DILEMMA

We begin this activity by introducing Juan and his dilemma. Juan's team has just arrived at the National Radio-Controlled (RC) Car Championships when he realizes that he has left the battery for their battery-powered car at home. It is a two-hour drive to the local RC shop, and the qualifying heat begins in one hour—so his team has to think fast. They must be able to light up the car's light-emitting diode (LED), a small electronic device, to avoid being disqualified. If they can find a way to create a makeshift battery for this device, they might be able to make it to the next round. In the

meantime, Juan's father can drive to the nearest RC shop and pick up a new battery before the second heat begins.

Juan knows that only 3 volts (V) are required to light up the car's LED, but he is not sure how much current he needs. It is a hot day in July, so he and his team head to Sloppy Sorey's Lemonade Stand to cool off and think. Tim, the goofy vendor, is squeezing lemons for fresh lemonade. When the team tells him about their predicament, he jokingly offers them some of his lemons, saying he once saw a lemon-battery "trick" at a children's party.

Although skeptical, Juan's team decides that the lemons are worth a shot. They take the fruit and search through Juan's toolbox for ideas, where they find screwdrivers, wrenches, fishing line, and a digital multimeter (DMM). Juan's sister, Yarisel, contributes items from her jewelry-making kit, including pieces of metal and wire, needles, pliers, and beads. The team has two hours to rescue its entry for the RC championship!

If we suspect that the class will find the scenario too improbable, we bring a few props (e.g., a toolbox, jewelry kit, an RC car, a sign for Sloppy Sorey's Lemonade Stand, and so on) and have students act it out.

LAB ACTIVITIES

After presenting students with this scenario, we assign two lab activities that can be completed in one 2-hour lab session or two 50-minute classes. Teacher explanation is minimal; students measure voltages produced by combinations of three different metals—using a lemon to complete the galvanic cell—and order these voltages hierarchically. They then solve Juan's dilemma by creating a lemon-battery system to power the LED. These activities are designed to help students develop a conceptual understanding of oxidation–reduction reactions and the electrochemical series for metals.

We recommend that students have a basic knowledge of simple electrical circuits; the ability to connect batteries in series and parallel; and the use and operation of a DMM before completing the activity. They should also understand that the LED is a polar device, and be able to connect it to a battery. All materials—such as lemons, LEDs, and small pieces of common metals—are inexpensive and easily obtained. Adequate DMMs are available for less than $20.

In terms of safety, chemical-splash goggles are required to prevent citrus juice and sharp metal edges from contacting the eyes. Students should wash their hands after handling metals and lemon juices, reducing the risk of skin irritation and exposure to heavy metals, such as lead.

Which metals make the best lemon battery?

In the first activity, students are guided to puncture a single lemon with three dissimilar metals—zinc (Zn), lead (Pb), and copper (Cu). A DMM is then used to confirm that small voltages are produced across different combinations of these metals (Figure 1; see "On the web" for step-by-step lab instructions). Students are guided to systematically measure the voltage produced by each combination, using each metal in turn as a common ground or *reference.* The

Figure 1

Lemon-battery diagram

Below is a digital multimeter (DMM) and a lemon battery with zinc as a reference, and two rows of metal pieces, placed 1.5 cm apart.

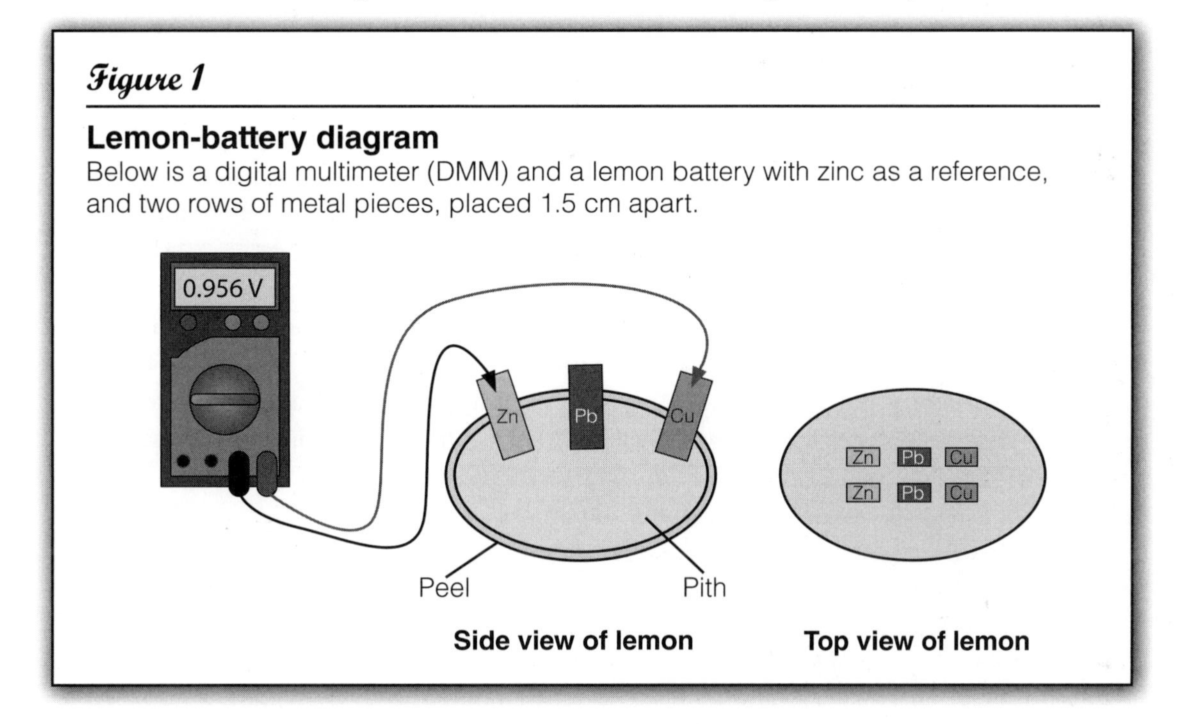

different combinations yield a hierarchy of voltages, which may be positive or negative (Table 1, p. 94).

We then explain that production of voltage implies the capacity for current flow and the transfer from one metal to another, and that reactions in which electrons are transferred are called *oxidation–reduction reactions.* A positive voltage on the DMM implies that the reference metal is being reduced (i.e., gains electrons); a negative voltage implies that the reference metal is being oxidized (i.e., loses electrons). This direct instruction provides valuable scaffolding for the activity and allows students to understand whether the reference metal is gaining or losing electrons for each of the voltages they measure.

Students then produce three data tables that illustrate the relative tendency of each metal to oxidize, or "give away" electrons, and are asked to consider the implications of any pattern that emerges. In the process, students discover that identical metals—copper and copper, for example—do not undergo electron transfer reactions (i.e., oxidation and reduction) or produce a voltage between them.

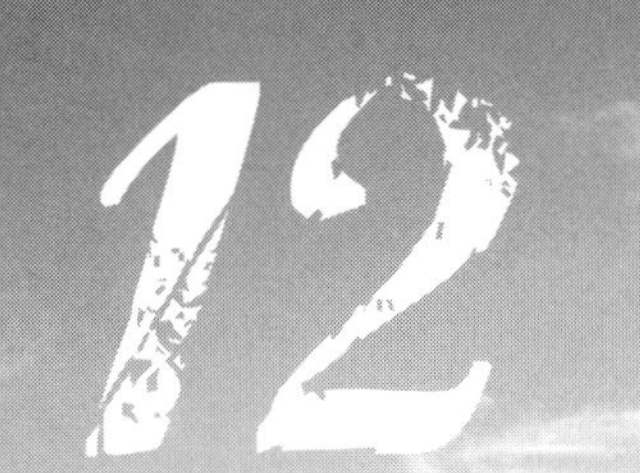

Table 1

Experimentally produced voltages

Reference	Zinc (Zn)	Copper (Cu)	Lead (Pb)
Volts	Pb = 0.512 V	Pb = -0.435 V	Zn = -0.492 V
	Cu = 0.956 V	Zn = -0.925 V	Cu = 0.432 V
	Zn = 0.000 V	Cu = 0.000 V	Pb = 0.000 V

When students record their experimentally produced voltages (Table 1), they observe that zinc always produces a negative voltage with the other metal, and copper always produces a positive voltage. From these observations, students deduce that zinc is losing electrons (i.e., oxidizing) and copper is gaining electrons (i.e., reducing) (Table 2). In ordering these metals hierarchically, students are led to the experimental derivation of a basic electrochemical series (Chang 2007).

Table 2

Hierarchy of voltages

These voltages were produced with respect to the electrochemical series, using lead as the reference metal.

Lead (Pb) as a reference	Voltage (V)	Metal	Gains electrons	Loses electrons	Neither
Most positive voltage	0.432	Cu	X		
	0.000	Pb			X
Least positive voltage	-0.492	Zn		X	

Formally developing the idea of the electrochemical series and oxidation numbers at this grade level is optional, so we sometimes choose not to do this if the majority of the class lacks adequate chemistry knowledge. However, we believe that the conceptual understanding developed by the activity provides a valuable foundation for future science courses.

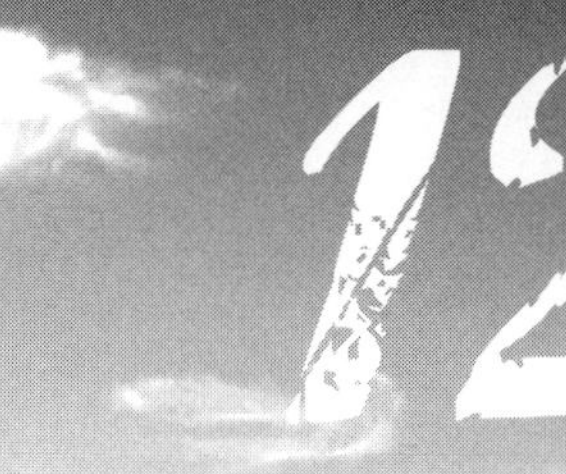

Lighting an LED with a lemon battery

Once students have a working knowledge of the relative effects that dissimilar metals play in electrochemical voltage production, they are ready to design and construct a solution to Juan's dilemma (see "On the web" for step-by-step lab instructions). It can be fun to make this a competition between student teams, but students can also work individually.

One lemon battery is unlikely to yield a voltage greater than 0.9–1.0 V (Table 3), so we reiterate that Juan's dilemma requires lighting an LED with a minimum of 3 V and an adequate amount of current (we tell students that 10–15 milliamps is generally sufficient to produce a "glow," but they can also research this question—online or in the library—as a preactivity homework assignment). We remind students how to use multiple batteries in series and parallel configurations to meet these minimum power requirements (Figure 2, p. 96).

Students choose the metals they wish to use for electrodes—using data from the first activity—and then experiment with different configurations for their lemon batteries. The goal is to produce sufficient voltage and current to light the LED. In our experience, three copper–zinc lemon batteries connected in series—with copper connected to the positive lead of the LED and zinc connected to the negative—works best (Table 3).

We conclude with a conceptual explanation of how an electrochemical battery works, the role of dissimilar metals, and how the lemon battery mimics a "regular" galvanic cell battery in terms of oxidation and reduction processes. The depth and detail of the explanation varies according to the class's science background, but we always cover the essential components of a functioning

Table 3

Lemon battery voltages

The following voltages were produced by lemon battery combinations of zinc and copper (Zn–Cu).

Number of lemons	Parallel		Series	
	Voltage (V)	Current (mA)	Voltage (V)	Current (mA)
1	0.933	0.045	0.933	0.045
2	0.800	0.800	1.736	0.040
3	0.731	0.132	2.58	0.038
4	0.770	0.143	3.27	0.038

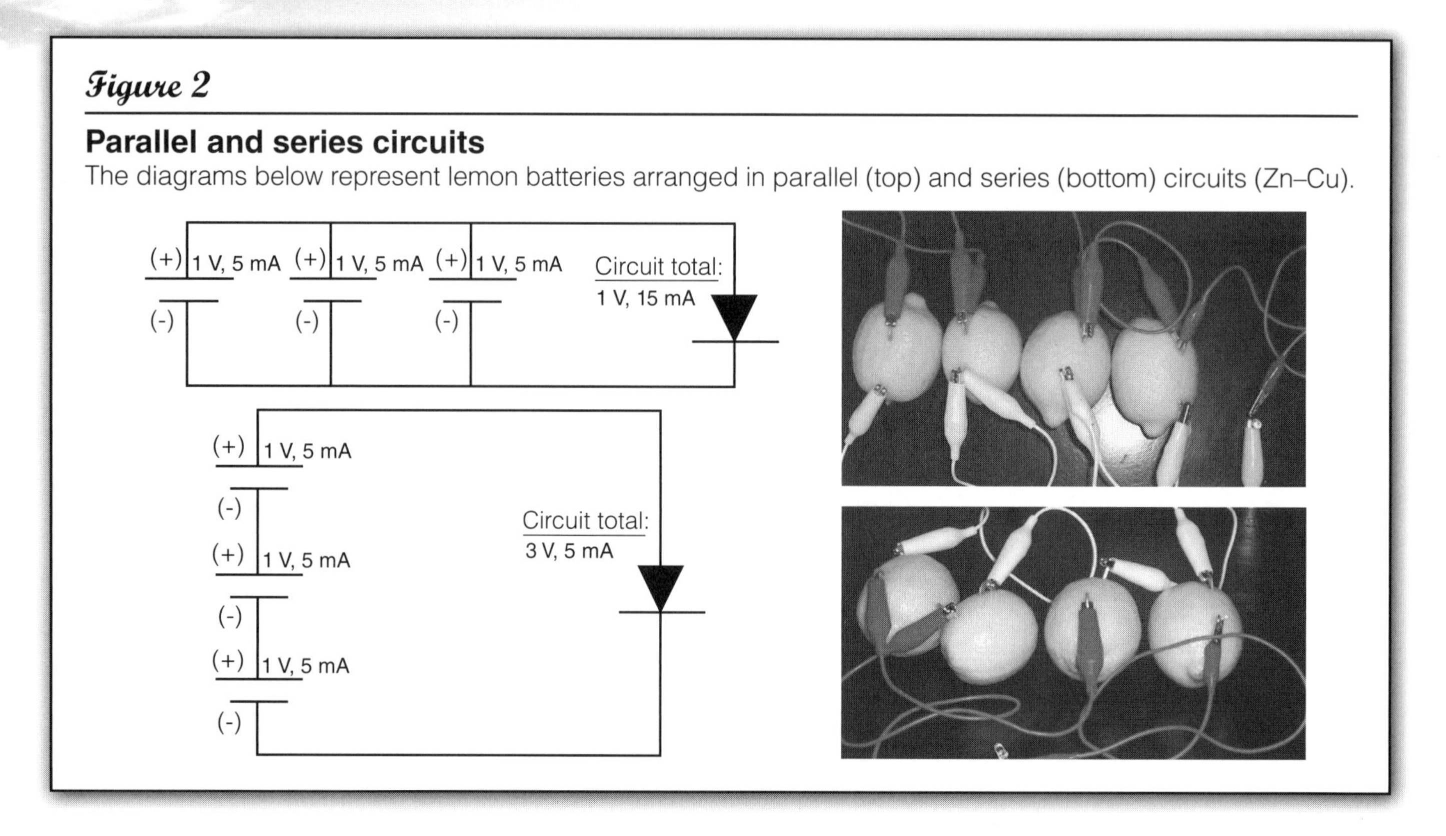

Figure 2

Parallel and series circuits

The diagrams below represent lemon batteries arranged in parallel (top) and series (bottom) circuits (Zn–Cu).

electrochemical cell: two dissimilar metals (an anode and a cathode) as electrodes or *terminals,* and a salt bridge to chemically connect them.

The salt bridge provides a solution for the ions involved in the oxidation–reduction reaction between the dissimilar metals, so that electrons can be released and current can flow. The salt-bridge connection enables the anode (a metal that oxidizes) to give up electrons to the cathode (a metal that reduces), and these electrons flow through the external circuit, creating a current. This current can power an electronic device that is connected to the two metal terminals.

In a lemon battery, the pith and citric acid act as the salt bridge. Once the connection is made between the two metals, the oxidation–reduction reaction continues until

- the anode is completely oxidized;
- the salt bridge is depleted of ions as the citric acid is used up; or
- the cathode is depleted of ions.

Figure 3 provides a list of problems we have encountered with these activities and our proposed solutions.

Figure 3

Observed problems and solutions

1. Initial voltage readings are imprecise. Select a low enough digital multimeter (DMM) voltage range to read millivolts to three significant figures.
2. The lemon battery fails to provide consistent voltages. Rolling the lemon firmly between your palms releases the acid-containing juice within the lemon. Clean metal electrodes of any corrosion with sandpaper and ensure that they are inserted deep enough into the lemon to pierce the pith, and are not touching each other either above or below the surface of the lemon.
3. Do not substitute wires for metal plates. Flat metal plates provide greater surface area for metal ion release.
4. Students make mistakes in connecting multiple lemons in series and parallel. Encourage students to determine and follow the current path of their circuit carefully (Table 3).
5. Students cannot light the light-emitting diode (LED) and become frustrated and convinced that their LED is faulty. Two AA batteries set up in series in the classroom allows for testing of the LED. To help students recognize that the LED is operating though its light is dim, lower the light level in the classroom.
6. If students wish to make current measurements, review use of the DMM as an ammeter.
7. Challenge early finishers to light two LEDs in series with lemon power. This extension activity requires students to configure a new lemon battery circuit that doubles the voltage supplied by the original.
8. Use fresh fruit for each lab. To reduce cost, fruits may be halved or other citrus fruits (e.g., oranges, grapefruit, or limes) may be substituted.

ASSESSMENT

We assess understanding on an "acceptable" or "unacceptable" basis for the first lab activity—students should be able to state whether each metal is gaining or losing electrons (i.e., reducing or oxidizing) and identify the electrochemical series from the voltage hierarchies produced. For the second lab activity, students produce series and parallel configurations of lemon batteries with enough voltage and current to light the LED. For this activity, we assess

students' success as "excellent," "acceptable," or "unacceptable." Rubrics for both activities are available online (see "On the web").

CONCLUSION

Oxidation and reduction reactions are an important component of students' chemistry understanding, yet high school students frequently fail to grasp this abstract process (Schmidt, Marohm, and Harrison 2007). The hands-on experiment presented in this chapter actively guides students through an interesting problem that delivers a conceptual understanding of the electron transfer process, while integrating chemistry with the study of electric circuits. To increase the inquiry level of the activity, students can develop their own methods to determine two metals that produce the most effective batteries, or design data collection sheets.

On the web

Grading rubric: *www.nsta.org/highschool/connections/201010GradingRubric.pdf*

Which metals make the best lemon battery?: *www.nsta.org/highschool/connections/201010LabActivityOne.pdf*

Lighting an LED with a lemon battery: *www.nsta.org/highschool/connections/201010LabActivityTwo.pdf*

References

Chang, R. 2007. *Chemistry.* 9th ed. New York: McGraw-Hill.

Goodisman, J. 2001. Observations on lemon cells. *Journal of Chemical Education* 78 (4): 516–518.

Schmidt, H. J., A. Marohm, and A. G. Harrison. 2007. Factors that prevent learning in electrochemistry. *Journal of Research in Science Education* 44 (2): 258–283.

Chapter 13

A VIRTUAL CIRCUITS LAB

BUILDING STUDENTS' UNDERSTANDING OF SERIES, PARALLEL, AND COMPLEX CIRCUITS

By Matthew E. Vick

Circuits are at the heart of cell phones, gaming systems, and even that old lamp in your room. Teaching about circuits can help students understand the underlying principles of many of today's electronic devices and require them to use quantitative thinking skills.

The University of Colorado's Physics Education Technology (PhET) website (see "On the web") offers free, high-quality simulations of many physics experiments that can be used in the classroom. The Circuit Construction Kit, for example, allows students to safely and constructively play with circuit components while learning the mathematics behind many circuit problems. This chapter describes my experience using the Circuit Construction Kit with my 11th- through 12th-grade physics students.

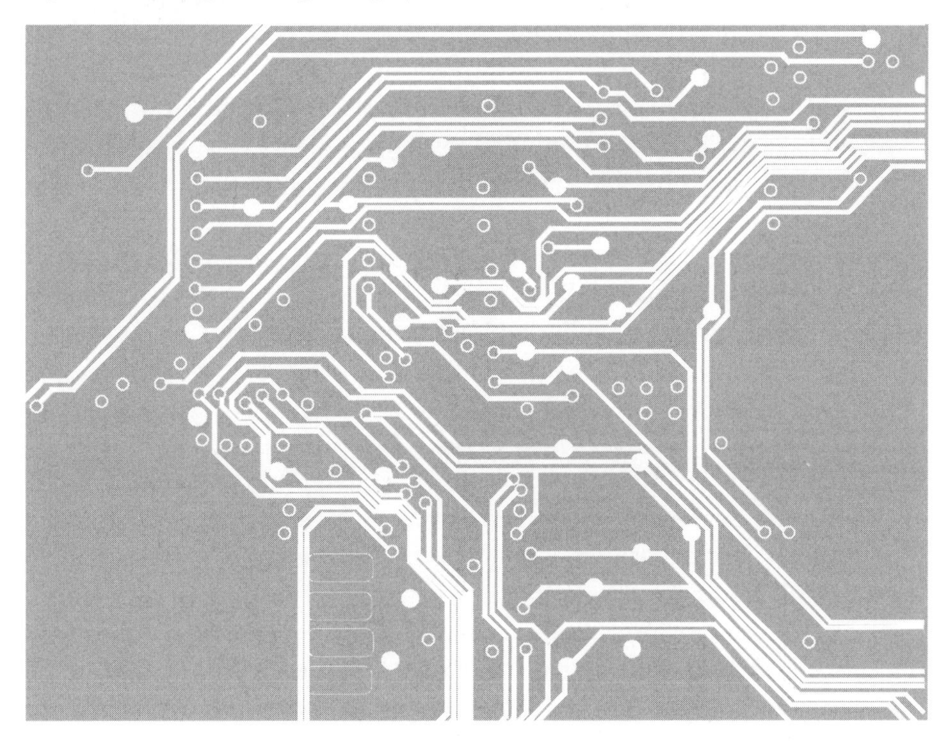

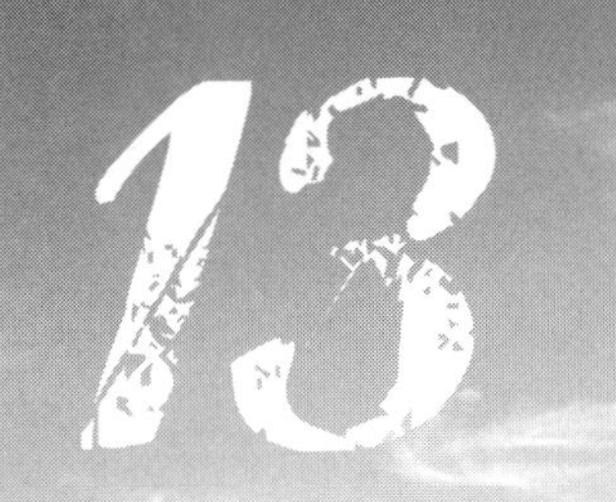

SIMULATIONS VS. HANDS-ON LABS

Building circuits often entails the use of breadboards (a construction base for an electronic circuit). Although breadboards present a tactile opportunity for physics students, their use is not always well understood. This is often because the physical layout of resistors does not look much like the diagrams students draw in class. Also, students who fail to measure current in series often blow the multimeter's fuses. This adds stress for both students and the teacher, who must continually replace these devices.

Finkelstein and colleagues (2005) studied the effects of using the PhET Circuit Construction Kit in place of a traditional circuit-building lab in a college-level introductory physics course. They found that students who used these simulations scored higher on a conceptual set of circuit questions and built physical circuits faster than students who participated in a traditional hands-on lab. Teaching assistants reported that the simulations allowed students to focus on content questions rather than the mechanical questions that result from blown fuses or loose connections. I have found the same to be true in my physics classes: The PhET simulations allow me to focus on the inquiry elements of the lab, rather than the mechanics of circuit construction.

Finkelstein and colleagues (2006) also identify several characteristics that the simulations illustrate in correlation with the National Science Education Standards (NRC 1996). Content Standard A, for example, emphasizes that students develop the "abilities necessary to do scientific inquiry" (NRC 1996, p. 173); using the PhET simulations, students draw conclusions based on their own data. Content Standard B encourages students to develop an understanding of the "structure of atoms" and "conservation of energy" (NRC 1996, p. 176); the PhET simulations help students visualize the invisible world of electrons and address the misconception that electrons are "used up" in a circuit.

The important thing to remember, though, is that the simulations themselves do not make for a constructivist, inquiry-based lesson—the teacher must use these simulations as a tool for exploration and discussion. Lessons should allow for creativity and problem solving, instead of simple observation.

A CIRCUITS LEARNING CYCLE

The PhET Circuit Construction Kit allows students to create circuits that closely resemble schematic diagrams—symbolic representations of the resistors, batteries, and other items in a circuit. Creating circuits that resemble schematic diagrams provides scaffolding for students to connect the symbolic and physical worlds. I use the Atkin and Karplus (1962) learning cycle in my physics class, which can be adapted for other physical science courses. The cycle consists of three stages: concept exploration, concept introduction, and

concept application; the BSCS 5E Instructional Model—engage, explore, explain, elaborate, and evaluate—is a similar model.

Before using the PhET Circuit Construction Kit with my class, I have students physically connect a single resistor to an adjustable voltage power supply. They then plot the voltage versus the current to introduce the concepts of Ohm's law and resistance. This physical experience is the part of the lesson designed to "hook" students. It also introduces Ohm's law in an inquiry-based manner.

The lesson then begins with three days of virtual exploratory activities that help students discover Kirchhoff's laws: one day for series circuits, one for parallel circuits, and one for complex circuits. Screenshots from the PhET website help guide students through the construction process, but many choose to "play" and build circuits on their own. These students add more components (e.g., switches) or items (e.g., pencil or coin) from a virtual grab bag. I have found this to be encouraging, as less confident physics students in previous classes did not show interest in "playing around" with the circuits.

After completing the virtual explorations, students physically build the circuits on breadboards. Several students have remarked that this experience got them excited about electronics. When class time is limited, I make sure that students build at least one physical circuit so that those who enjoy working with electronic devices have the opportunity to do so. (**Safety note:** Students should be reminded to keep the voltage reading on their power supply at 4 V or less and asked not to touch uninsulated wires; safety glasses are required for this activity.)

Important terms and concepts

Ammeter: An instrument used to measure the electric current in a circuit.
Breadboard: A construction base for an electric circuit.
Current: A flow of electric charges through a conductor; measured with an ammeter.
Kirchhoff's laws: Rules for finding current and voltage in a series or parallel circuit.
Multimeter: An electronic device used to measure voltage, current, and resistance that combines several measurement functions into one unit.
Noncontact ammeter: A bull's-eye that is placed over any area of a circuit to read its current.
Ohm's law: The current through a conductor between two points is proportional to the voltage across the two points and inversely proportional to the resistance between the two points, or *Current = voltage/resistance or* $I = V/R$.
Resistance: The property of a material that resists the flow of electric charges through it.
Voltage: A measure of the difference in electric potential between two points in a space, material, or electric circuit.
Voltmeter: A device used to measure the voltage in an electric circuit.

SERIES CIRCUIT EXPLORATION

The lesson's first exploration requires the most class time, since students are still learning how to use the PhET website and Circuit Construction Kit. In this

activity, students construct a virtual series circuit with three resistors. I provide them with short written instructions to guide them through the simulation, since previous classes complained that they did not have enough support to use the PhET site. Instructions for changing the default values of the resistors are particularly important—this helps students develop accurate generalizations. If all resistance values are left at the default of 10 Ω, the voltage across each resistor will be equal, which can lead to the misconception that voltages across series resistors are always equal. The written instructions for each exploratory activity and an application lab can be found online (see "On the web").

The interface in the Circuit Construction Kit includes a voltmeter with leads, so students can connect opposite sides of the resistors with a power supply for measuring voltage. Internal resistance and resistance in the wires are modeled to make the simulation more realistic.

To measure current, students use either the ammeter or the "noncontact ammeter," a bull's-eye that is placed over an area of the circuit to read its current. I use the noncontact ammeter to avoid the difficulty of "breaking" circuits when inserting the ammeter. Other instructors may choose to use the ammeter so that their students can model the skill of inserting this device.

Students explore the lesson's concepts by drawing conclusions from their data, instead of through open-ended circuit building. The exploration instructions I provide ask them to construct a data table of voltages across and currents through each of the resistors. Students share their data tables with the class and discuss the concepts that voltage drops in one circuit add up to the total voltage drop in an entire circuit and that current is the same throughout a series circuit.

When using the Circuit Construction Kit, teachers should be aware that the simulation shows the charges moving faster with higher current. To avoid misconceptions, ask students to explain the ways in which the animated charge carriers are not like actual electrons. For example, current is the measure of the movement of charge, not of individual electrons—electrons continue to drift in a random pattern that moves more in the direction of the current than against it.

PARALLEL CIRCUITS EXPLORATION

The second virtual exploration, on parallel circuits, moves faster than the first since it "parallels" the format. It is important to include screenshots in the instructions for this activity because the parallel circuits need "extra wire" to allow for the measurement of current in each branch of the circuit. It is also important that the resistance values are different so that the current through each branch will also be different. This helps avoid the misconception that current is always equal in each branch of a parallel circuit.

Once again, students form their own data tables for voltage and current. I ask them to use the noncontact ammeter to record the total voltage across the battery and the total current right next to the battery. When measuring resistance, I have students use Ohm's law with the total voltage and current, rather than focusing on the complex relationship between total resistance and the individual values. My intention is for students to realize that the total resistance is less than any of the individual resistance values.

Figure 1

Breadboard with a series circuit and multimeter

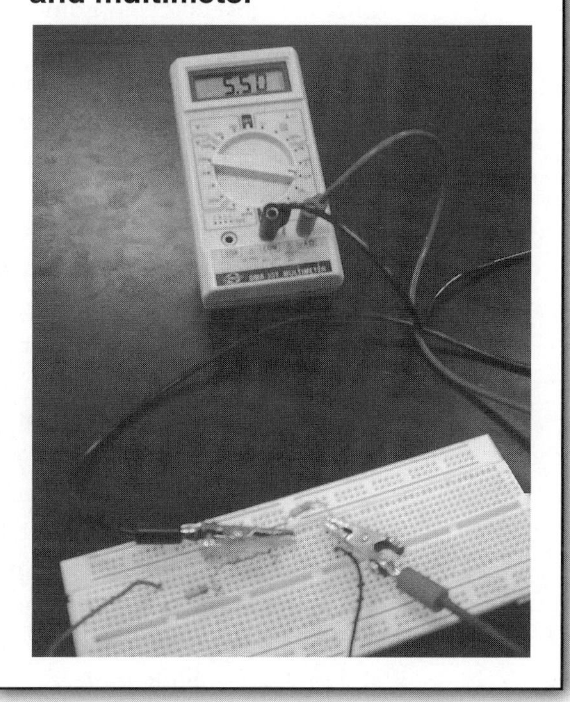

As students share their data with the class, they quickly realize that voltage drops are equal in each branch of a parallel circuit and that the current through each branch adds up to the total current. I have students discuss why they initially think that the total resistance is less than the individual resistors. Students often complain, "But that doesn't make sense!" The visualization of electron flow that the simulations provide aid in this discussion; it allows me to ask questions about how the simulation shows the charge flow in each branch.

COMPLEX CIRCUITS EXPLORATION

During the third exploration, students construct a virtual complex circuit. At this point, they are often proficient with the PhET program. In this activity, students construct two complex circuits and then measure voltage and current values. They often look for broad generalizations, so the lab should include questions that focus on the series and parallel components of the circuit separately.

This lesson requires the most scaffolding on the teacher's part. Students want to be "given" the complex circuit rules, rather than work through the realization that series circuit rules are used in some parts of the circuit and parallel circuit rules are used in others.

CONCEPT DEVELOPMENT

After at least three days of building virtual (and possibly physical) circuits in series, parallel, and complex structures, students are prepared to work on

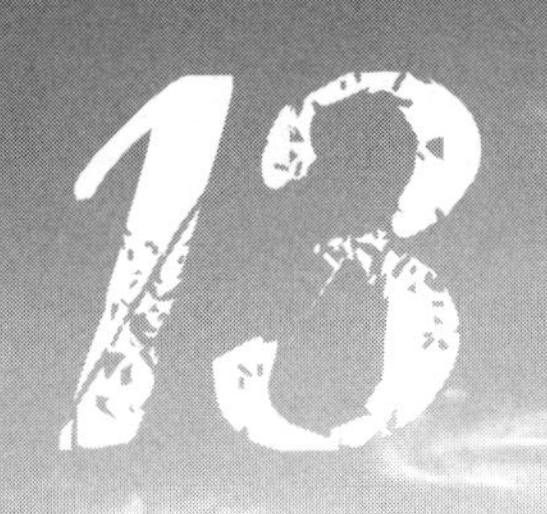

more abstract problems. At this point, they often want to memorize specific circuit arrangements. I give students a labeled list of "series" and "parallel" circuit equations each time we do this activity. Undoubtedly, one student will ask why the complex circuit equations are not provided. Giving students ample practice, both in groups and as individuals, is vital to helping them discover how to apply the series and parallel voltage and current relationships in a non–simple circuit case. I find it necessary to lead a daily discussion to remind students that complex circuits do not have one set of equations. As a class, students fall back on the expectation for a single set of equations for complex circuits almost every day.

APPLICATION LAB: BUILD A COMPLEX CIRCUIT

Finally, students are assigned a specific total resistance for which they are to construct their own complex circuit given certain resistance values. I have created a spreadsheet that generates possible values from the combinations given. As an extension, students can physically create the circuit on a breadboard and then check the total resistance using a multimeter.

The PhET simulations focus students on connecting the concepts that resistors in series add to the total resistance and that resistors in parallel generally decrease the total resistance. The benefit of the physical construction is that it often serves as a "hook" for students who want to study electronics further. A formal lab report can be required, but I like to emphasize the "play" and problem solving inherent in the activity, so students receive credit when they complete the task successfully and show me their final product.

CONCLUSION

The PhET website has many simulations that allow students to create, observe, measure, and analyze—instead of simply watching an animation over which they have no control. This learning cycle guides students through experiences that help them develop an understanding of how current is divided through parallel circuits and how voltage is divided in series circuits.

The virtual lab allows students to construct circuits that look like the diagrams found in textbooks and tests; the physical lab allows students to manipulate real objects and deconstruct electronic circuits—so they become more than just "black boxes." The use of both a virtual lab and a physical lab provides scaffolding for students, allowing them to connect the physical and symbolic worlds.

The Atkin and Karplus learning cycle approach (1962) works well with the virtual labs provided by PhET. The Circuit Construction Kit makes the exploration less repetitive for students since they can easily construct new circuits. The application phase allows students to create a circuit using their own creative thinking.

In my classes, students gave positive feedback about the use of the simulations and showed much less frustration than previous students who had only worked with breadboards. In this series of activities, virtual labs can be used to support inquiry-based instruction in concert with a physical lab—allowing teachers and students to reap the benefits of both methods.

On the web

Complex circuit challenge: *www.nsta.org/highschool/connections/201007VickEXTENSION5.pdf*

Complex circuit exploratory: *www.nsta.org/highschool/connections/201007VickEXTENSION3.pdf*

Ohm's law and resistance exploratory: *www.nsta.org/highschool/connections/201007VickEXTENSION4.pdf*

Parallel circuit exploratory: *www.nsta.org/highschool/connections/201007VickEXTENSION2.pdf*

Physics Education Technology website: *http://phet.colorado.edu/index.php*

Series circuit exploratory: *www.nsta.org/highschool/connections/201007VickEXTENSION1.pdf*

References

Atkin, J. M., and R. Karplus. 1962. Discovery or invention? *The Science Teacher* 29 (5): 45.

Finkelstein, N. D., W. K. Adams, C. J. Keller, P. B. Kohl, K. K. Perkins, N. S. Podolefsky, S. Reid, and R. LeMaster. 2005. When learning about the real world is better done virtually: A study of substituting computer simulations for laboratory equipment. *Physical Review Special Topics—Physics Education Research* 1 (1): 010103(1)–010103(8).

Finkelstein, N. D., W. K. Adams, C. J. Keller, K. K. Perkins, C. Wieman, and the Physics Education Technology Project Team. 2006. High-tech tools for teaching physics: The physics education technology project. *Journal of Online Learning and Teaching* 2 (3): 110–121.

National Research Council (NRC). 1996. *National science education standards.* Washington, DC: National Academies Press.

Chapter 14

MEASURING WAVELENGTH WITH A RULER

By Paul Hewitt

In the late 1960s, Arthur Schawlow, coinventor of the laser, visited my school's physics department to share some intriguing laser demonstrations. He popped colored balloons inside transparent ones, and showed interesting diffraction patterns with laser light. The most memorable demonstration began with the following question: "Can the wavelength of laser light be measured using only a common ruler?" He proceeded to illustrate that it could be, based on the familiar formula $m\lambda = d \sin \theta$. The ridges of a plastic ruler—the raised millimeter marks—were used as the diffraction grating to disperse the laser light. The 1 mm distance between ridges (d) seemed enormously large for a diffraction grating. The solution to this problem was to use *grazing incidence* (where the laser light is at a small angle with the ruler) and to position the ruler several meters away from a viewing screen (Figure 1).

Schawlow's explanation can be simplified and used to easily measure the wavelength of laser light in the classroom (Schawlow 1965). The formula $m\lambda = d \sin \theta$ applies to a transmission grating when the light is at a right angle to the grating; however, the formula works when the laser is pointed toward a ruler at a small angle (grazing incidence), and measurements are

Figure 1

Laser positioned at an angle of 1:10 (5.7°) and reflected off the smooth side of the ruler onto a viewing screen (no diffraction occurs)

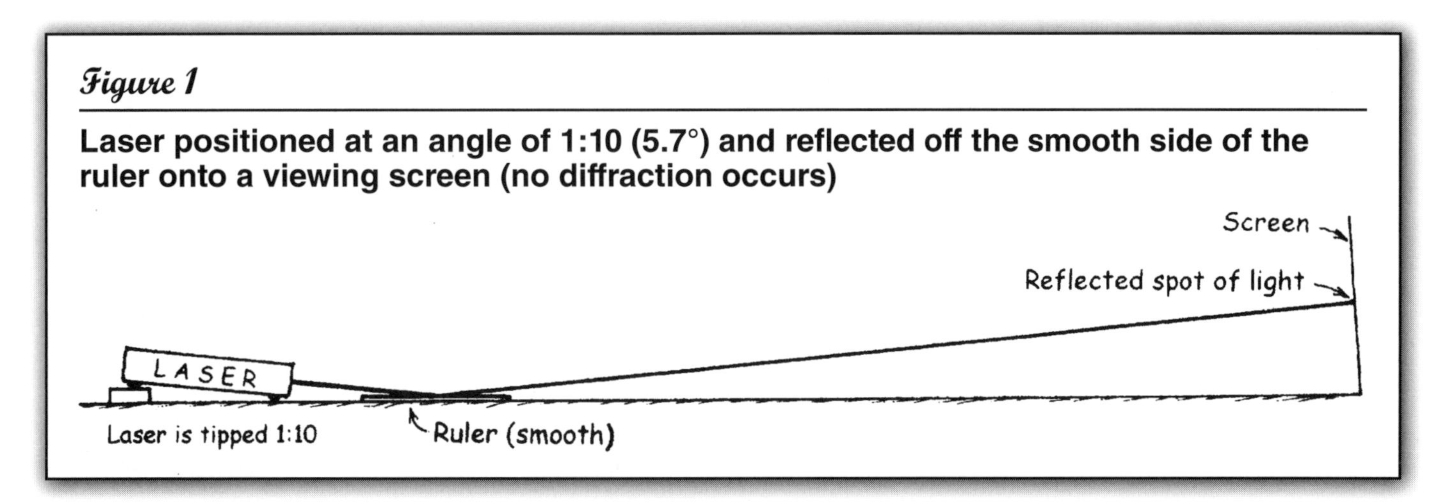

made close to the angle of specular reflection (the defined beam that reflects off a smooth surface). Then d, as seen by the laser beam, is less than the 1 mm spaces between the ruler ridges. With the screen a few meters away, diffraction fringes centimeters apart can be observed.

To begin this demonstration in class, teachers should elevate the laser at an angle of 1:10 (5.7°) and shine the laser light on the smooth side of the ruler (Figure 1). In this manner, no diffraction occurs and specular reflection shows the dot of laser light on the distant screen. The vertical distance of the dot above the ruler will be 1/10 of the horizontal distance to the reflection point on the ruler. That distance should be several meters, which is not shown to scale in Figure 1. The smooth surface demonstration should be done first so that students better appreciate what occurs when the light is next diffracted from the ridges.

When the ruler is flipped to allow the light to encounter the millimeter ridges, diffraction occurs (Figure 2). Diffraction is evident by the array of fringes above and below the line of reflection. The ruler effectively acts as a suitable diffraction grating. Figure 2 shows the angle (θ) between the reflected beam, the zeroth fringe, and the first diffraction fringe ($m = 1$).

Figure 2

Laser positioned at an angle of 1:10 (5.7°) and reflected off the ridged side of the ruler onto a viewing screen (shows diffraction fringes)

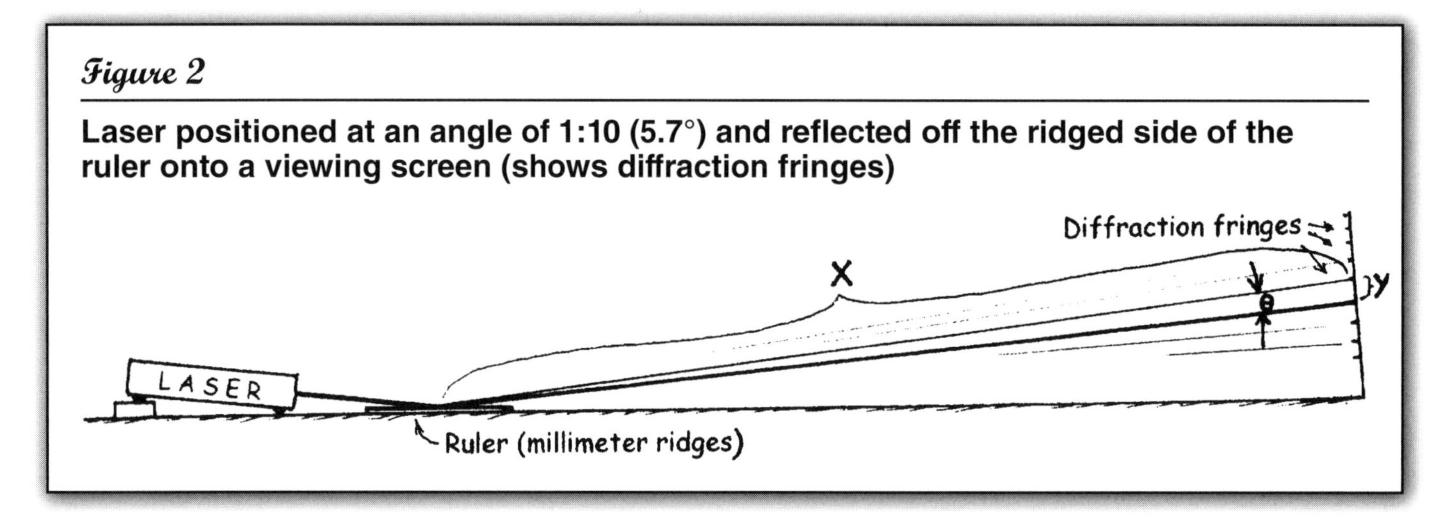

The wavelength of light can be found by measuring $\sin\theta$; $\sin\theta = y/X$, where y is the distance between the zeroth and first fringe. (The fringe spaces are nonlinear and are closer together at higher fringes. Therefore, the distance between the first two fringes should be measured on either side of the zeroth fringe, $m = +1$ and $m = -1$, and divided by 2 to find the value of y.) The distance the reflected beam travels from the point of diffraction on the ruler to the viewing screen is equal to X. Because of the 1/10 tilt of the laser beam, $d = 1/10$ of 1 mm = 0.0001 m. (Strictly, d is the *average* spacing of the millimeter ridges as

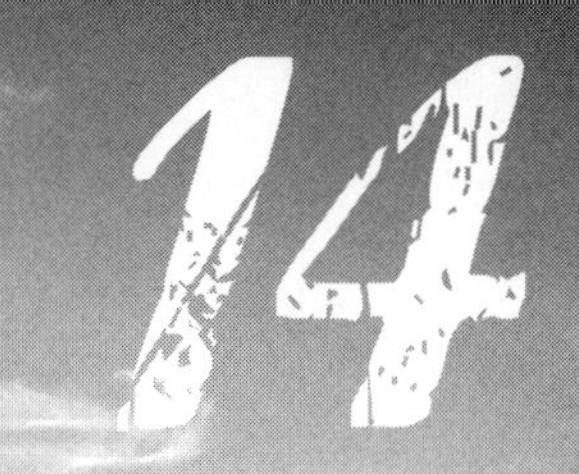

"seen" by the incoming and outgoing beams. Working just between the two fringes at $m = +1$ and $m = -1$ means that this average is very close to the spacing seen by the incoming beam, 1/10 of 1 mm.) If $d = 0.0001$ m, then $\lambda = (1/m)(d \sin \theta) = (1/1)(0.0001 \text{ m})(y/X)$. So, in nanometers, $\lambda = (100{,}000 \text{ nm})(y/X)$.

These two measurements can be made with a tape measure or a meter-stick, both common "rulers." However, if the fringes are marked on the screen, all measurements can be made with the same traditional plastic ruler used as the diffraction grating. With some care, the measurement of the wavelength of laser light will closely agree with its known value (633 nm for the light of a helium-neon laser).

Acknowledgments

For insightful suggestions, I thank Ken Ford, David Kagan, and City College of San Francisco physics instructors Tsing Bardin and Norman Whitlatch.

Reference

Schawlow, A. L. 1965. Measuring the wavelength of light with a ruler. *American Journal of Physics* 33: 922.

Chapter 15

SHEDDING LIGHT ON THE INVERSE-SQUARE LAW

STUDENTS DEMONSTRATE THE QUANTITATIVE RELATIONSHIP BETWEEN LIGHT INTENSITY AND DISTANCE

By Richard E. Uthe

Many students in introductory science courses at both the secondary and tertiary levels learn science as a miscellaneous collection of facts, concepts, and equations that must be memorized to pass examinations. One way to show students that they actually can "do" science is to have them use an observable event to generate a relationship that can be used as a predictive tool. If that relationship can be quantified using "curve-fitting" and simple algebra, an equation describing the relationship can be created. In this way, students realize that algebraic equations in science do not usually begin the process of discovery; rather, these equations are one of the natural outcomes of the process.

To demonstrate this in the classroom, teachers can have students investigate the inverse-square law with the following activity. Through a simple experiment using readily available and inexpensive apparatus, students learn about the law by demonstrating the quantitative relationship between light intensity and distance.

EXPLAINING THE RELATIONSHIP

The inverse-square law describes many quantitative relationships in science, such as how gravitational attraction between two bodies decreases as the distance between their centers of gravity increases, or how

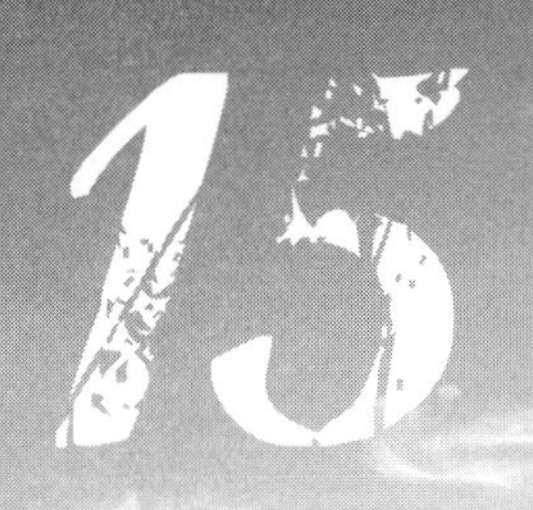

attraction increases between two particles of opposite charge (e.g., a proton and an electron) as the distance between them decreases. For example, gravitational attraction between two celestial bodies can be described by the relationship $F = (Gm_1m_2)/d^2$, where F is a measurement of the amount of attraction, G is the gravitational constant, m_1 and m_2 are the masses of the two attracting bodies, and d is the distance between their centers of gravity. If the distance between the two bodies increases from one "distance unit" to two "distance units" to three "distance units," the attraction between them changes from $1/1^2 = 1$ to $1/2^2 = 1/4$ to $1/3^2 = 1/9$, and so forth. The inverse-square law applies not only to gravity and electricity, but also to optics. One example of this is how the headlights of an approaching automobile appear brighter as its distance decreases. The classroom experiment described in this chapter focuses on optics.

Several authors have published brief descriptions of experiments that concern the quantitative investigation of the inverse-square law applied to light sources. George and Doebler (1994) use a laser beam through a two-dimensional diffraction grating to create a square diffraction pattern on a screen and measure area illuminated at different screen positions. Blair and Eaton (1981) use a small transformer to supply variable current to a lightbulb, the illumination from which is measured at various distances using a semiconductor and a digital multimeter. And Narayanan and Narayanan (1999) use a laser beam sent through a small, circular hole (an Airy's disk), in which the size of the illuminated circle on a screen increases as the screen is moved away from the disk.

Three authors use a grease-spot photometer (Bunsen photometer), which compares intensities of illumination from two light sources. Hastings (1965) uses a grease-spot photometer mounted on a meterstick and compares a lighted candle to various other light sources to find their light intensities. Freier and Anderson (1981) describe qualitatively the placement of a Bunsen photometer between two lightbulbs to show equal illumination. Miller (1969) uses a similar procedure but uses candles instead. Miller (1969) and Freier and Anderson (1981) describe the use of paraffin blocks (a Joly diffusion photometer), but only for investigating equal illumination of the blocks. Only one article describes a stepwise laboratory investigation whereby students collect data to verify the inverse-square relationship (Kruglak 1975), but the procedure involves using a high-intensity lightbulb, diffusing screen, and silicon solar cell on an optical bench. The bulb is moved toward the solar cell and photocurrent readings are recorded in arbitrary units. Many physical science books (e.g., Payne, Falls, and Whidden 1992) discuss the inverse-square law for optics. The books describe the law qualitatively and use geometry to derive an illumination equation for spherical radiation from a point source of light, such as a lightbulb, but do not include a stepwise procedure for experimentally determining the relationship. As a result, students must accept the textbook relationship without ever discovering it for themselves.

DEMONSTRATING THE LAW

For our experiment, the quantitative relationship between light intensity and distance can be demonstrated with two lightbulb sockets mounted on two small pieces of wood, a selection of bulbs of varying labeled wattage, a meterstick, and two rectangular chunks of paraffin. To make the paraffin apparatus, students place a piece of aluminum foil between two paraffin blocks, and then secure the package with two rubber bands. We use paraffin slabs that are 13 × 6.5 × 2 cm. The slabs are available at many grocery stores and usually come prepackaged for under $4 per pack of four paraffin blocks. The aluminum foil should be cut twice as big as the paraffin blocks and then folded in half to allow the shinier side of the foil to face both paraffin blocks.

When the paraffin apparatus (sandwich) is placed along its 13 cm side on a table between two lightbulbs, perpendicular to the line between the bulbs, the paraffin blocks will be illuminated separately by each bulb. The basic procedure requires students to find the place between the two lightbulbs where the two paraffin blocks of the sandwich are illuminated equally. Students then compare the position of the paraffin apparatus to the difference in luminosity between the two bulbs.

To minimize errors in measurements, students should use all unfrosted (or all frosted) lightbulbs from the same manufacturer. Students should record lumens for each wattage used as written on the cardboard sleeves of the bulbs because luminosity is the actual measure of the light energy emitted by the bulbs. (Although the term *wattage* is an obsolete and non-SI way to refer to power, it is used here as a convenient label for each of the lightbulbs used.) Students work in groups of three in a darkened room. Opaque baffles, such as large pieces of thick cardboard, are placed between the groups to minimize extraneous light impinging on the paraffin blocks. Light reflected off the ceiling and walls is a source of error, however.

Students use bulbs of 15, 25, 40, 60, 75, 100, 150, 170, 200, and 250 W. Students use the 15 W bulb as a standard and compare the other bulbs to it. The first trial run compares two 15 W bulbs, the second trial run compares a 25 W bulb to the standard 15 W bulb, the third compares a 40 W bulb to the standard 15 W bulb, and so forth. Students collect 10 sets of data to provide 10 points for graphing the relationship.

The bulbs are placed about a meter apart; their positions must remain fixed for the entire experiment. One student in the group places the paraffin

block sandwich between the two lightbulbs and moves it until the two sides of the sandwich are equally illuminated. The distance in meters from the filament of the 15 W bulb to the aluminum of the sandwich is measured. Each person in each group of three students repeats this procedure twice for each bulb combination, resulting in six measured distances for each bulb combination. The distances are then averaged. The difference in luminosity (lm) between the two lightbulbs is graphed against the average measured distance between the bulbs. Luminosity values for the bulbs are taken directly from the cardboard sleeves of the bulbs.

Radiated power, in watts, cannot be used because luminous intensity of a light source is defined as the visual comparison of the source with some arbitrarily chosen standard source, and the human eye is not equally sensitive to equal amounts of radiation of different colors. In addition, some of the power of a lightbulb, in watts, is converted to heat energy as well as light energy, and this heat-to-light ratio is not constant for bulbs of different labeled wattage. Distances are always measured from the 15 W bulb when bulbs of other wattages are used. Because paraffin blocks may vary in quality, rotating the sandwich around 180° between the two readings might help increase accuracy.

The teacher must instruct students on how to correctly handle the electrical apparatus and the hot lightbulbs. The bulb-socket apparatus itself is constructed so that no bare wires are exposed and its base is large enough to prevent the apparatus from tipping over easily. Students must use insulated gloves when handling lightbulbs, especially those of higher wattage, which can quickly become extremely hot. Lightbulb sockets made of porcelain, with the porcelain having a wide base and with concealed screws for attaching electric cords, are readily available at most home improvement stores. Pieces of plywood can be cut in squares about 15 cm on each side for mounting the porcelain sockets. The apparatus can then be clamped onto a table to ensure that it does not move during the experiment. Any lightbulb sockets on already existing small lamps may be used instead, as long as exchanging lightbulbs does not move the lamp or the bulb socket position. Care must be taken to ensure that the paraffin sandwich always lies in a direct line between the two lightbulbs, rather than be positioned noticeably higher or lower than the line between the bulbs.

ANALYZING THE DATA

Numerical results of a typical experiment are listed in Table 1. If the difference in luminosity is plotted on the vertical axis of a graph and distance is plotted on the horizontal axis, a typical $y = k\,(1/x^2)$ curve results (Figure 1, p. 116). However, if difference in luminosity (vertical axis) is plotted against the inverse-square of the distance (horizontal axis), the result is a linear graph

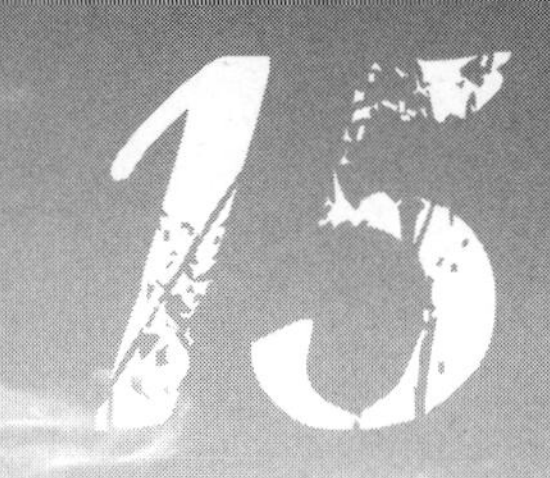

Table 1

Data table of lightbulb intensity and distance to a standard 15 W lightbulb

Lightbulb labeled wattage (W)	Lightbulb luminosity (lm)	Difference in luminosity from 15 W bulb[a] (lm)	Distance from 15 W bulb to paraffin[b] (m)	Inverse of distance (m^{-1})	Inverse square of distance (m^{-2})
15 (standard)	110	0	0.527	1.90	3.60
25	210	100	0.490	2.04	4.16
40	490	380	0.330	3.03	9.18
60	855	745	0.275	3.64	13.2
75	1170	1060	0.248	4.03	16.3
100	1660	1550	0.199	5.03	25.3
150	2790	2680	0.161	6.21	38.6
170	3100	2990	0.154	6.49	42.2
200	3910	3800	0.138	7.25	52.5
250	4500	4390	0.129	7.75	60.1

a. compared to 110 lm for the 15 W standard bulb = test bulb luminosity-110 (lm)
b. average of six trial runs

(Figure 2, p. 117), verifying the inverse-square nature of the relationship. If only the inverse of the distance, rather than the inverse-square, is plotted on the horizontal axis, the result is not linear. Using a standard computer spreadsheet and graphing program makes these relationships very clear, but good results also can be obtained with calculators and hand-drawn graphs. Note that the linear graph in Figure 2 does not intersect the origin (0,0) because the experiment compares equal illumination of the apparatus from two bulbs, not just one. When the two 15 W bulbs are compared, their difference in luminosity is zero, but the distance from the paraffin block to the standard 15 W bulb is not.

DISCUSSION

The light from a bulb is radiated almost uniformly in all directions. The filament of the bulb is not quite a point source but can be considered as such at some distance away. This may result in some minor errors for the shorter distances measured.

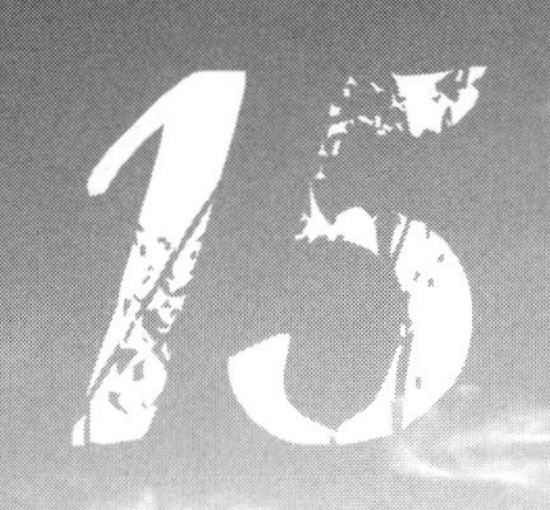

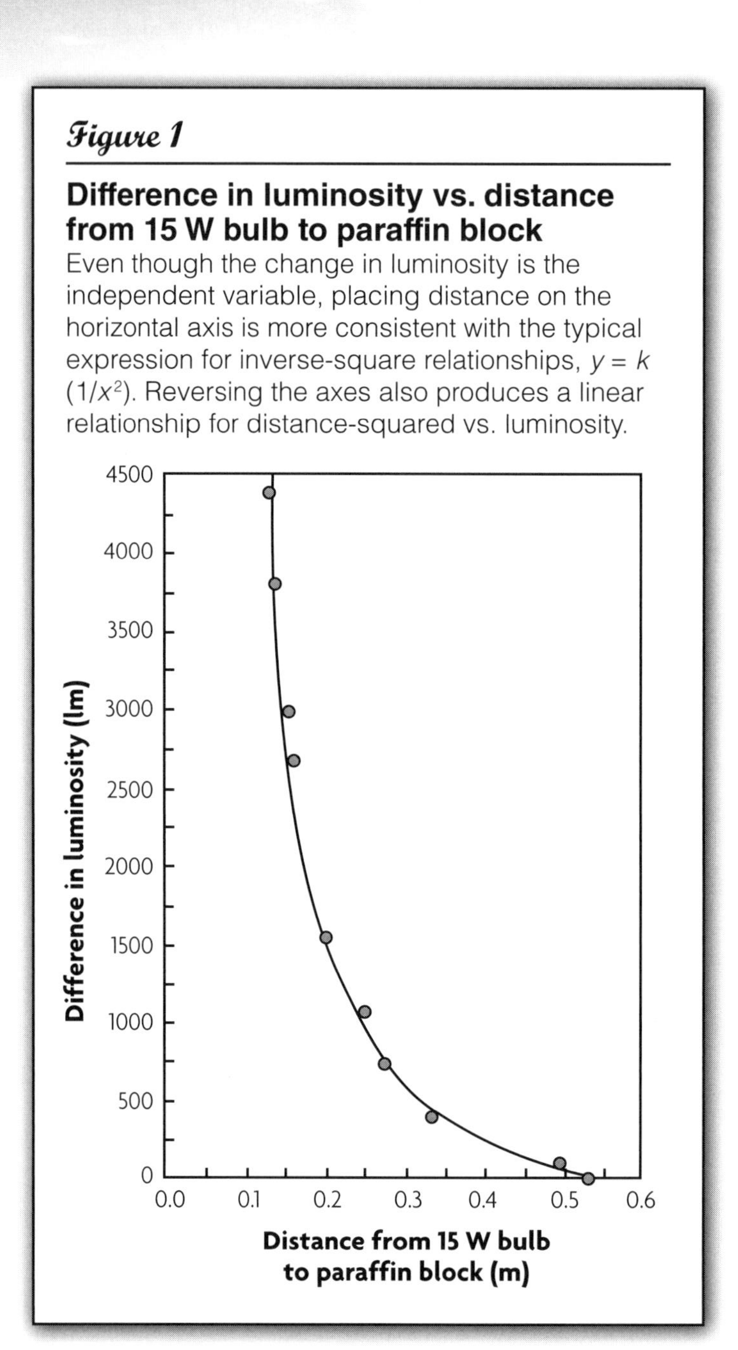

Figure 1

Difference in luminosity vs. distance from 15 W bulb to paraffin block

Even though the change in luminosity is the independent variable, placing distance on the horizontal axis is more consistent with the typical expression for inverse-square relationships, $y = k(1/x^2)$. Reversing the axes also produces a linear relationship for distance-squared vs. luminosity.

The inverse-square nature of relationships such as gravitational and electric attractions and luminosity can be understood through simple geometry. In the case of light, the amount of illumination, the luminosity, is a measure of light flux (energy received over time) per unit area of surface. Since light is radiated from the bulb in all directions, it can be thought of as falling on the surface of a sphere of radius r, with the light distributed evenly

(diluted) over the entire area of the sphere. Because the surface area of a sphere is $4\pi r2$, the illumination (E) falling on one "unit area" of the sphere is the luminosity (light flux) of the bulb (Φ) in lumens divided by the entire area illuminated, or $E = \Phi/4\pi r^2$ (Payne, Falls, and Whidden 1992). If r is measured in centimeters, the illumination units will be "lumens per square centimeter" of the surface area of the sphere. So the observed illumination (E) at any point away from the bulb is proportional to $1/r^2$. This equation pertains to light from a single source. Even though this experiment uses two light sources, the inverse-square relationship between luminosity and distance can still be discovered and quantified.

Figure 2

Difference in luminosity versus inverse-square of distance from 15 W bulb to paraffin block

Difference in luminosity (lm)

4500
4000
3500
3000
2500
2000
1500
1000
500
0

0 10 20 30 40 50 60 70

Inverse-sqare of distance from 15 W bulb to paraffin block (1/m²)

In addition, using the same experimental setup, students can now predict the "equal illumination" distance for any unknown bulb (relative to the 15 W standard) given its luminosity or predict an unknown bulb's luminosity given (or experimentally determining) the "equal illumination" distance. Teachers might wish to withhold one of the bulbs while students collect and analyze data, and then use it to verify a prediction students make after they have ascertained the quantitative relationship.

The inverse-square relationship plays an important role in understanding many physical science phenomena, from the amount of light that reaches Earth from distant stars to the forces that hold electrons within the atom. Too often, students must simply accept this relationship as presented by the teacher or textbook. This investigation, using a simple apparatus, allows students to discover this mathematical relationship themselves.

References

Blair, J. M., and B. G. Eaton. 1981. *Laboratory Experiments for Elementary Physics.* Minneapolis: Burgess Publishing.

Freier, G. D., and F. J. Anderson. 1981. *A Demonstration Handbook for Physics.* Stony Brook, NY: American Association of Physics Teachers.

George, S., and R. Doebler. 1994. Rainbow glasses and the inverse-square law. *The Physics Teacher* 32: 110–111.

Hastings, R. B. 1965. *Laboratory Physics.* Saint Paul, MN: Bruce Publishing.

Kruglak, H. 1975. Laboratory exercises on the inverse-square law. *American Journal of Physics* 43(5): 449–451.

Miller, J. S. 1969. *Demonstrations in Physics.* Sydney: Ure Smith Pty. Limited.

Narayanan, V. A., and R. Narayanan. 1999. Inverse-square law of light with Airy's disk. *The Physics Teacher* 37: 8–9.

Payne, C. A., W. R. Falls, and C. J. Whidden. 1992. *Physical Science—Principles and Applications.* Dubuque, IA: William C. Brown Publishers.

Chapter 16

A 50-CENT ANALYTICAL SPECTROSCOPE

By John Frassinelli

During a recent unit on astronomy, my students asked, "How do we know all that we know about the stars, since they are so far away and no one from Earth has ever visited them?" A fair question, to be sure. While astronomers believe they have decent estimates of the mass, temperature, size, and composition of "nearby" stars, it can boggle the student mind to think of how this type of distant science is even possible. The activity described here will not fully answer the question, but it should give students an insight into this process. In keeping with the Benchmark, "Understanding the Nature of Scientific Inquiry," students can use a 50-cent analytical spectroscope to gather, analyze, and interpret scientific data (Kendall and Marzano 1996).

Students can construct a simple, inexpensive instrument that demonstrates some key differences among light sources in the classroom—an analytical spectroscope. With the spectroscope, students can observe homemade "stars," graph results, and draw appropriate conclusions. The materials are basic: an empty paper towel tube, black construction paper, rubber bands, a hobby knife, masking tape, and sheets of diffraction grating. The only material not likely found around the house is a sheet of diffraction grating—one sheet can make dozens of spectroscopes, and it is available from many science supply stores for about $15.

SETTING THE STAGE

Before students begin building the spectroscope, a primer of stars is probably in order (Fraknoi 1995). Students seem naturally fascinated with the sky, and a little preparation on the teacher's part can go a long way. For example, a slide show on stars, a good introductory video, or a discussion of the Hertzsprung-Russell diagram can serve to show students that classifying stars is an interesting activity that dates from the early days of keen-eyed ancient shepherds to modern astronomers' use of such technological marvels as the Hubble Space Telescope.

BUILDING THE SPECTROSCOPE

To build an analytical spectroscope, each student should receive an empty paper towel tube, which will serve as the body of the instrument. Each student should also have access to scissors, two rubber bands, one half-sheet of black construction paper, copy paper, masking tape, a hobby knife, and a piece of 25 × 15 mm diffraction grating. As for the hobby knife, I have found it useful to hold a clinic before the activity with strict safety guidelines discussed and posted. (**Safety note:** Students must wear safety goggles. It is a good idea to inventory all sharp items distributed for student use, to be sure they all are properly returned when the activity is completed.)

Figure 1

Endpiece mounting for diffraction grating (eyepiece end of spectroscope; not actual size)

Cut out all straight lines from the circle, but don't cut out the circle itself. Tape the diffraction grating over the rectangle, like a window.

The endpiece is made of black construction paper, and measures about 8 cm on a side.

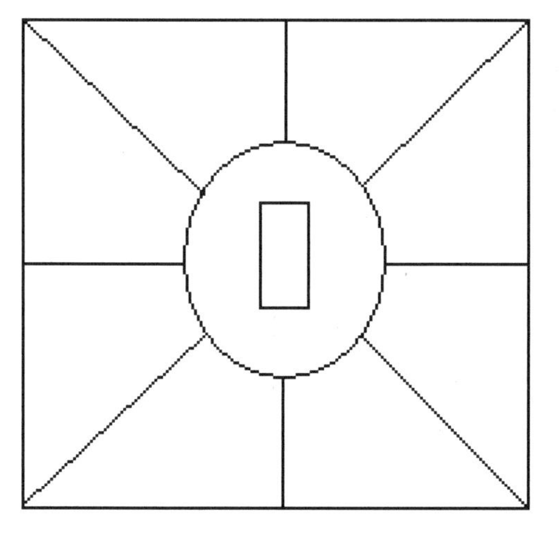

To build an analytical spectroscope, students should follow these instructions:

1. Using black construction paper, cut the endpiece mounting for the diffraction grating. A black square measuring about 8 cm on each side is suggested (Figure 1). This will become the eyepiece for the spectroscope.
2. Place the end of the paper towel roll on top of the center of the black construction paper square in a perpendicular fashion and trace its circle on the paper. Draw 8 lines (rays) on the black paper, extending

from the circle out to the edge of the paper. Cut a small rectangular window in the middle of the traced circle; it should be about half the size of a postage stamp. Carefully tape a slightly larger piece of the diffraction grating over the small window around its edges. You should be able to see through the window you have made.

3. Cut each of the rays as they extend from the circumference of the circle all the way out to the edge of the paper. Do not cut the circumference itself.

4. As you face the taped diffraction grating, carefully bend each of the cut flaps in toward you, with each flap crease along the circumference of the circle. As they bend, they will overlap each other, creating a sort of baffle, or light-shield. Take the whole endpiece assembly and place it on one end of the paper towel tube, like a cap.

5. Secure it gently with a rubber band or two along the outside of the black paper. The endpiece should rotate areound the tube and should be snug, not loose. The main light entering the tube now should be from the open, opposite end.

6. The opposite end of the spectroscope will contain the nanometer scale and also a slit for light to enter. This end is made from white copy paper, and embedded within it is the black nanometer scale, shown actual size in Figure 3. This endpiece will eventually be securely taped in place and will not move about the tube like the eyepiece. This rectangle, which can be photocopied, measures 9.5 × 6.5 cm.

7. Take the open end of the whole tube and gently squeeze it into a noticeable oval, which remains oval-shaped after you stop applying pressure. Place the 9.5 × 6.5 cm white rectangle from the previous step, flat on the table.

8. Within the oval outline, carefully cut a slit for light to enter using the hobby knife. A piece of corrugated scrap cardboard placed underneath the white rectangle may be helpful in obtaining a clean, crisp cut. The width of the slit should be around 1–2 mm, and its height is about the same as that of the black nanometer scale (make a photocopy of the true-to-size Figure 3).

9. Lay the Figure 3 rectangle on the table, face up, and position the open, oval-shaped end of the tube on top of it. The tube and rectangle are now perpendicular. Note that each numeral on the

scale represents that number times 100 nanometers, so the 7 for example, represents 700 nm.

10. Take the tube away and carefully cut the rays starting from the periphery of the white rectangle, in toward the oval, but do not cut the oval. Like the eyepiece, each of the small flaps created by the rays will be folded and creased in toward you. After all creases are made and the slit is cut, the nanometer-end scale is ready to be installed.
11. Reposition the open end of the tube onto the endpiece and flaps you just created, being careful not to disturb the delicate slit and nanometer scale. Gently but firmly tape the endpiece into position; unlike the other end of the tube, it will not move after being installed. The whole tube should now be light safe except for the light entering through the slit you have made on one end and the window on the other.

Figure 2

Partially constructed spectroscope (this end points toward light source and does not rotate; not actual size)

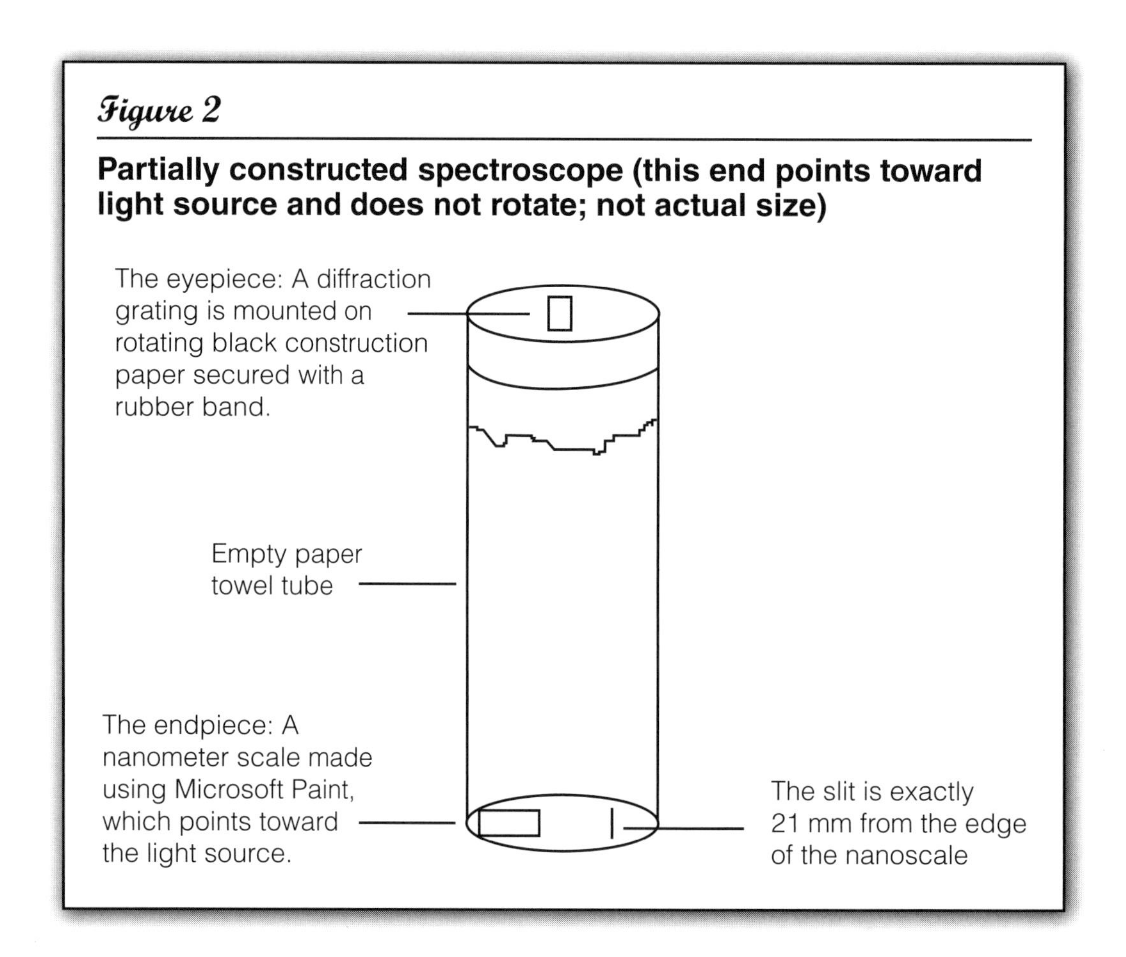

Figure 3

Nanometer-scale end of a spectroscope (this end is pointed toward the light source; actual size)

Slit is exactly 21 mm from the black edge of the nanometer scale, is parallel to it, and should be roughly the same height as the scale itself. Light will enter here and appear as color bands along the scale.

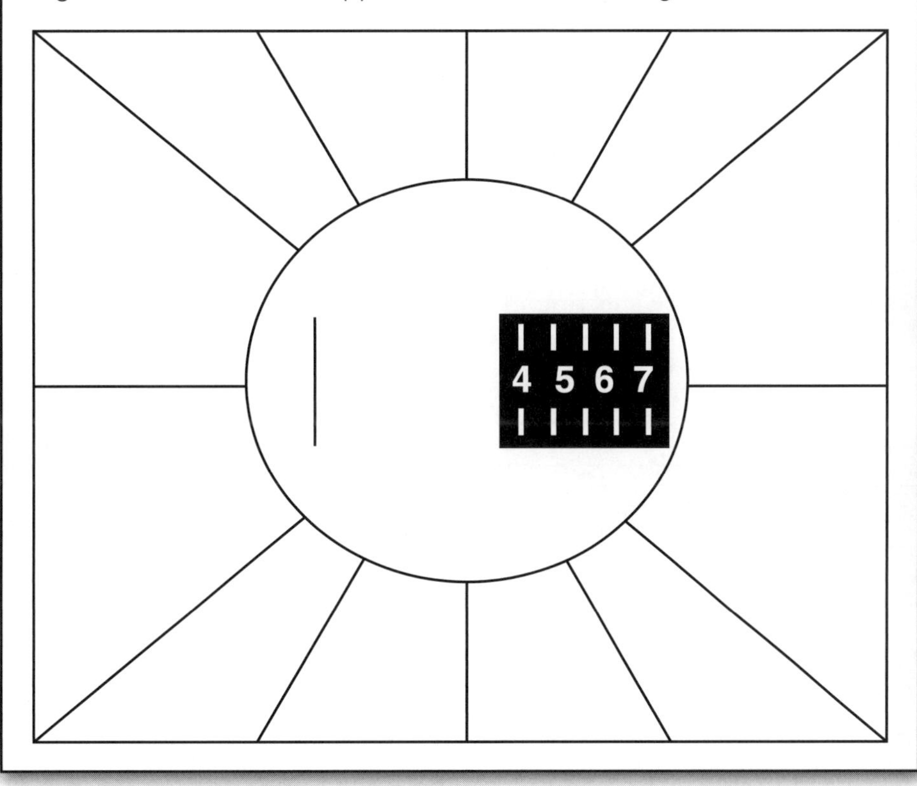

To test their spectroscopes, students need a fluorescent light, such as those typically used as classroom overhead lights. Students should look through the eyepiece and, pointing the nanometer scale end toward the fluorescent light, see colored bands of light. Students turn the eyepiece until these vertical rainbow bands line up horizontally along the nanometer scale. There should be a noticeable discreet bright-green line almost exactly halfway between the 5 and the 6. Since each numeral on the scale actually represents that number multiplied by 100 nanometers, the bright-green line is actually showing the "signature" of a trace of mercury used in the manufacture of the fluorescent lights. The mercury is actually showing up at a wavelength of 546 nm (Vogt 2001). Students should take note of the rest of the colors—the widths of each

color band and where each falls along the scale. The blue hues have shorter wavelengths; the reds are longer.

WHAT'S HAPPENING?

Visible light is only a small part of the radiation from the Sun and stars, but it is the only part that the human eye can detect. Visible light is a very narrow band of radiation ranging from roughly 400 to 700 nm in wavelength. The diffraction grating in the student-made spectroscope breaks up the white light entering through the slit; the nanometer scale shows the approximate wavelengths of the colored light in billionths of a meter.

A CLASSROOM TOOL

The spectroscope can be used to examine the spectral signature of the fluorescent lights found in most classrooms. As students peer into their spectroscopes, they will notice, first and foremost, the rainbowlike patterns that appear. The colors will seem to be superimposed on the nanometer scale, a reference to compare the widths, intensities, and possible bright lines or dropouts that may be present for a given light source. Students can draw graphs of what they see with colored pencils.

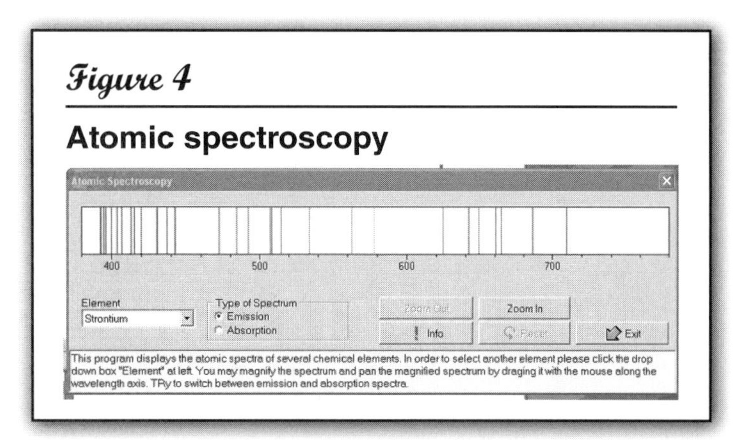

Figure 4

Atomic spectroscopy

Other lights can be examined and graphed, for example, incandescent lightbulbs, krypton flashlights, and candles. Each source of light will have a unique spectrum falling along the nanometer scale. Some aspects will be the same for each light source, but some will not. Observant students will notice the subtle differences that appear in their spectroscopes for each light source displayed. (**Safety note:** Students should be cautioned to never look directly at the Sun.)

I usually set up four "stars" in a central location in the classroom and have students gather in a semicircular pattern with their spectroscopes, graph paper, and colored pencils. A small fluorescent lamp, an incandescent bulb, a krypton flashlight, and a candle are easy sources of different types of light that will yield different views inside the spectroscopes. If teachers have access to the physics lab, they can also use a transformer and light up discreet tubes of glowing gas, such as nitrogen, oxygen, and hydrogen. Each will look different inside the darkness of the spectroscope.

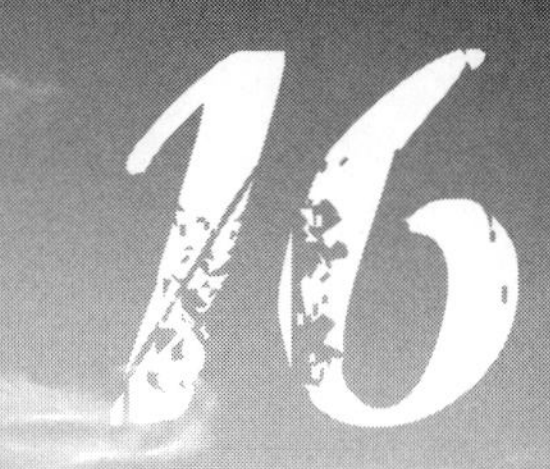

The spectroscope described here is fairly easy to build and costs about 50 cents. Students may be interested in knowing that much more expensive and accurate spectroscopes are found in observatories. These spectroscopes are attached to sophisticated telescopes, and serve as aids in studying starlight coming to us from many light years across empty space. Since the invention of the telescope in the 17th century, astronomers have used visible starlight to learn much about the temperature, movement, distribution, and chemical makeup of the stars.

The stars, planets, meteors, and moons are fun to observe and inspiring to study. This simple activity may cause students to pause, look up at the stars, and perhaps wonder (Sagan 1995). And such wonder, said Albert Einstein, is the first true source of all knowledge.

References

Fraknoi, A., ed. 1995. *The universe at your fingertips: An astronomy activity and resource notebook.* San Francisco: The Astronomical Society of the Pacific.

Kendall, J., and R. Marzano. 1996. *Content knowledge: A compendium of standards and benchmarks for K–12 education.* Aurora, CO: Mid-continent Regional Educational Laboratory.

Sagan, C. 1995. *The demon-haunted world: Science as a candle in the darkness.* New York: Random House.

Vogt, G. L. 2001. *Space-based astronomy.* Washington, DC: U.S. Government Printing Office.

Chapter 17

FUELING THE CAR OF TOMORROW

AN ALTERNATIVE FUELS CURRICULUM FOR HIGH SCHOOL SCIENCE CLASSES

By Mark Schumack, Stokes Baker, Mark Benvenuto, James Graves, Arthur Haman, and Daniel Maggio

It is no secret that many high school students are fascinated with automobiles. Cars are a big part of their lives—evoking excitement, curiosity, and perhaps even a certain amount of anxiety. The activities in Fueling the Car of Tomorrow—a free high school science curriculum, available online (see "On the web")—capitalize on this heightened awareness and provide relevant learning opportunities designed to reinforce basic physics, chemistry, biology, and mathematics principles. The curriculum consists of 17 activities that can serve as a resource for science classes or as the basis of an entire course; each activity is also explicitly tied to state and national science standards (NRC 1996).

THE PROCESS

We—a team of five faculty members in biology, chemistry, and mechanical engineering and the director of precollege programs in the College of Engineering and Science at the University of Detroit Mercy—developed the Fueling the Car of Tomorrow curriculum. We began by compiling a list of topics we thought the curriculum should cover; these included

- biological mechanisms governing the production of biofuels;
- the chemistry of combustion, pollution, and biofuel manufacture;
- hydrogen generation; and
- various technological aspects of energy conversion systems, internal combustion engines, and hybrid and electric vehicles.

Figure 1

Format for curriculum activities

Section	Description
List of outcomes and linkages to standards (teacher version only)	Specific outcomes for the activity are listed and tied explicitly to state and national science and math standards.
Narrative	A short one- to three-page narrative introduces the principles demonstrated in the activity. Ample connections with students' likely experiences are made.
Questions for students	Several questions are presented to the class as a whole, and serve as a starting point for a discussion on the topic.
Activity description and procedure	The activity is summarized. Material lists and detailed procedures lead students through the activity in a step-by-step manner.
Postactivity analyses	Students are presented with exercises that prompt reflection about what they have done. These include performing calculations, making graphs, writing reports, preparing presentations, developing further questions, and so on.
Postactivity assessment	An assessment tool is provided—usually in the form of a short quiz—that determines the extent to which students have met the outcomes. Each quiz question is linked explicitly with the outcomes identified at the beginning of the activity.

Each of us identified key learning outcomes for several topics and then—often with the help of our undergraduate students—embarked on a multiyear effort to develop the activities. To ensure uniform presentation, we used the format indicated in Figure 1 for each of the activities.

We then piloted the curriculum through the Detroit Area Precollege Engineering Program (DAPCEP), which offers classes for K–12 students at the University of Detroit Mercy on Saturdays during the school year and on weekdays in the summer. Over the last few years, several of the Fueling the Car of Tomorrow activities have been used in these classes; based on these offerings, curriculum adjustments were made as needed.

Evaluations of one of the DAPCEP classes, Powering the Car of Tomorrow, indicate that students were pleased with their experience (Figure 2). In addition to these offerings, two local public high school teachers tested most of the activities and provided valuable formative feedback.

Figure 2

Student ratings for the precollege class offerings, on a scale of 1 (strongly disagree) to 5 (strongly agree)

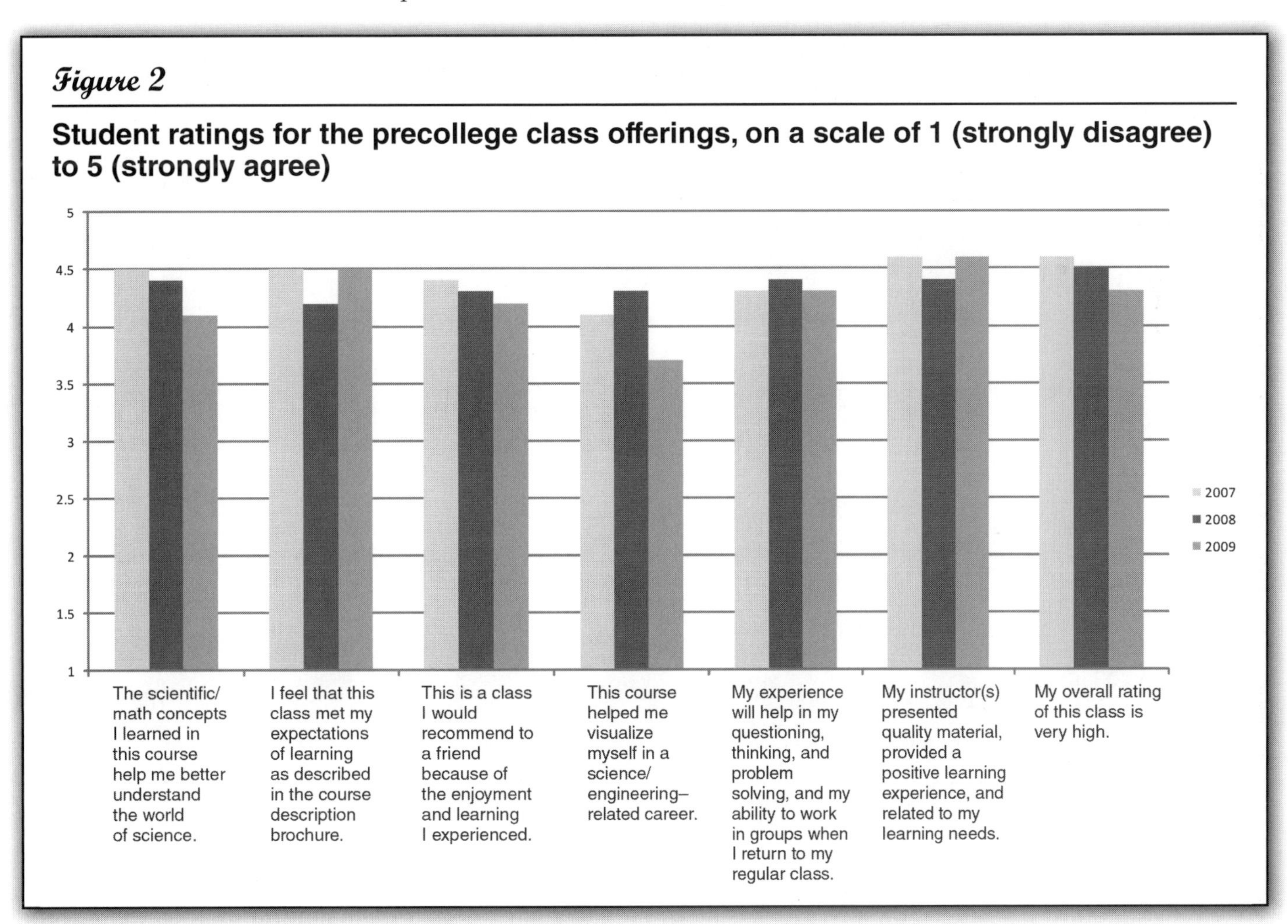

THE CURRICULUM

The 17 activities in the Fueling the Car of Tomorrow curriculum were designed for students with rudimentary algebra and general science knowledge. Most of the activities are laboratory exercises, but some are analytical explorations using spreadsheets, computer simulations of "what if" scenarios, group brainstorming sessions, and pencil-and-paper calculations. The activities can be offered individually as supplements to biology, chemistry, or engineering classes and completed in one or two class periods, or the entire series can form the basis of a semester-long or even yearlong preengineering course.

Figure 3 lists the types of activities, their associated disciplines, and brief descriptions. The order given is a suggested—but not essential—sequence for delivery. Each of the activities also has a set of learning outcomes with clear linkages to the National Science Education Standards and state science and mathematics standards (NRC 1996; Figure 4). The state linkages apply to Michigan, but most states have similar guidelines.

The first three activities in the curriculum, Introduction to Alternative Fuels, Nonrenewable Energy Resource Depletion, and Fundamentals of Internal Combustion Engine Operation, are introductory exercises that provide students with the background needed for the remaining activities. The following sections describe these activities in more detail.

Figure 3

The activities

Activity name	Discipline	Activity type	Description
1. Introduction to Alternative Fuels	Engineering	Group brainstorming	Introduces students to fuel and propulsion system possibilities and the concepts of energy source, energy storage, and energy conversion
2. Nonrenewable Energy Resource Depletion	Engineering	Data manipulation and analysis using a spreadsheet program	Establishes an awareness of the urgency and inevitability of fossil fuel depletion and the need for alternative fuels
3. Fundamentals of Internal Combustion Engine Operation	Engineering	Pencil-and-paper calculations and graphing	Provides a basic understanding of the mechanics behind internal combustion engine operation and fuel economy

(continued)

Figure 3 *(continued)*

Activity name	Discipline	Activity type	Description
4. Chemistry of Combustion	Chemistry	Lab experiment	Provides an understanding of combustion fundamentals
5. Pollution From Burning Fuels	Chemistry	Lab experiment	Provides an understanding of the source of unwanted combustion products
6. Making and Testing Biodiesel	Chemistry, engineering	Lab experiment	Introduces students to biofuels (e.g., ethanol and biodiesel) and teaches them how to make and test biodiesel
7. Production of Ethanol and Methanol	Chemistry, biology	Lab experiment	Teaches students how ethanol can be produced from sugar and yeast
8. Making Fermentable Sugars From Plant Material	Biology	Lab experiment	A series of procedures that allows students to explore various aspects of cellulose- and starch-based fermentation
9. Where Do Plants Store Energy?	Biology	Lab experiment	Two procedures that illustrate the energy-storage potential of plant matter
10. Ethanol-Producing Microorganisms in Nature	Biology	Lab experiment	Teaches students about microorganisms that can produce ethanol
11. Comparison of Food Compounds for Ethanol Production	Biology	Lab experiment	Teaches students about the suitability of sucrose, starch, and cellulose as feedstock for ethanol
12. Oxygen and the Production of Combustible Gas	Biology	Lab experiment	Teaches students about the anaerobic formation of methane gas
13. Electric Vehicles (two activities)	Engineering	Lab experiments	Two activities that demonstrate how motor speed can be controlled by voltage and how the energy density of batteries can be determined
14. Hydrogen Production	Chemistry	Lab experiment	Demonstrates how hydrogen is generated
15. Engine Performance for Alternative Fuels	Engineering	Group research	Links engine performance to various fuel characteristics
16. Fundamentals of Hybrid Vehicles	Engineering	Computer simulation	Enables students to compare conventional and hybrid vehicle fuel economy
17. Which Fuel Is Best?	Engineering	Group research and presentation	Provides students with an opportunity to assimilate knowledge and present findings

Figure 4

State and national standards addressed

State science standards (Michigan Department of Education 1996)
• Design and conduct scientific investigations • Gather and synthesize information • Communicate findings of investigations • Explain social and economic advantages and risks of new technology • Explain how multicellular organisms grow • Explain the process of food storage and food use • Compare ways that living organisms have adapted to survive and reproduce in their environments • Measure and describe things around us • Identify properties of elements • Explain chemical changes • Explain why mass is conserved in physical and chemical changes • Describe energy transformations • Analyze patterns of force and motion in the operation of complex machines • Explain energy conversions in moving objects and machines • Explain how current is controlled in simple series and parallel circuits • Evaluate alternative long-range plans for resource use • Explain impact of human activities on the atmosphere
State mathematics standards (Michigan Department of Education 1996)
• Collect and explore data • Organize data using tables, graphs, and spreadsheets • Make predictions based on data, including interpolations and extrapolations • Compute with real numbers and algebraic expressions

(continued)

Figure 4 *(continued)*

National Science Education Standards (NRC 1996)
• Be able to do scientific inquiry • Understand structure and properties of matter, chemical reactions, motions and forces, and conservation of energy • Understand matter, energy, and organization in living systems • Be able to perform technological design and understand science and technology • Understand natural resources and science and technology in local, national, and global challenges

Introduction to alternative fuels

In the first activity, students are prompted to think about how a vehicle is propelled and brainstorm possible fuel and energy conversion systems. Up until a few years ago, progress in automotive propulsion technology focused mainly on the internal combustion engine. Recently, however, with rising fuel prices and increased concern about future gasoline supplies, engineers have explored new paradigms for transportation, such as hybrid vehicles, electric cars, hydrogen-powered cars, cars fueled by ethanol and biodiesel, and even solar-powered cars.

This activity invites students to join the new climate of inventiveness with their own creative input. Along the way, they are exposed to the conceptual distinction between energy source, energy storage, and energy conversion. For instance, in the brainstorming exercise, typical student responses include "electricity," "batteries," and "electric motors." After the brainstorming session, the instructor helps students categorize their responses by explaining that all three must be present in an electric vehicle: The electricity is generated using an energy source such as coal; batteries store the electricity for onboard use; and the motor converts the stored electrical energy into motion. Figure 5 shows a diagram for the activity, which helps students draw connections between fuels, onboard storage systems, and propulsion devices.

Nonrenewable energy resource depletion

The second activity sets the stage for why we need alternative fuels. Students use a spreadsheet program to plot the U.S. oil production data available to geologist M. King Hubbert in the 1950s when he made his now-famous

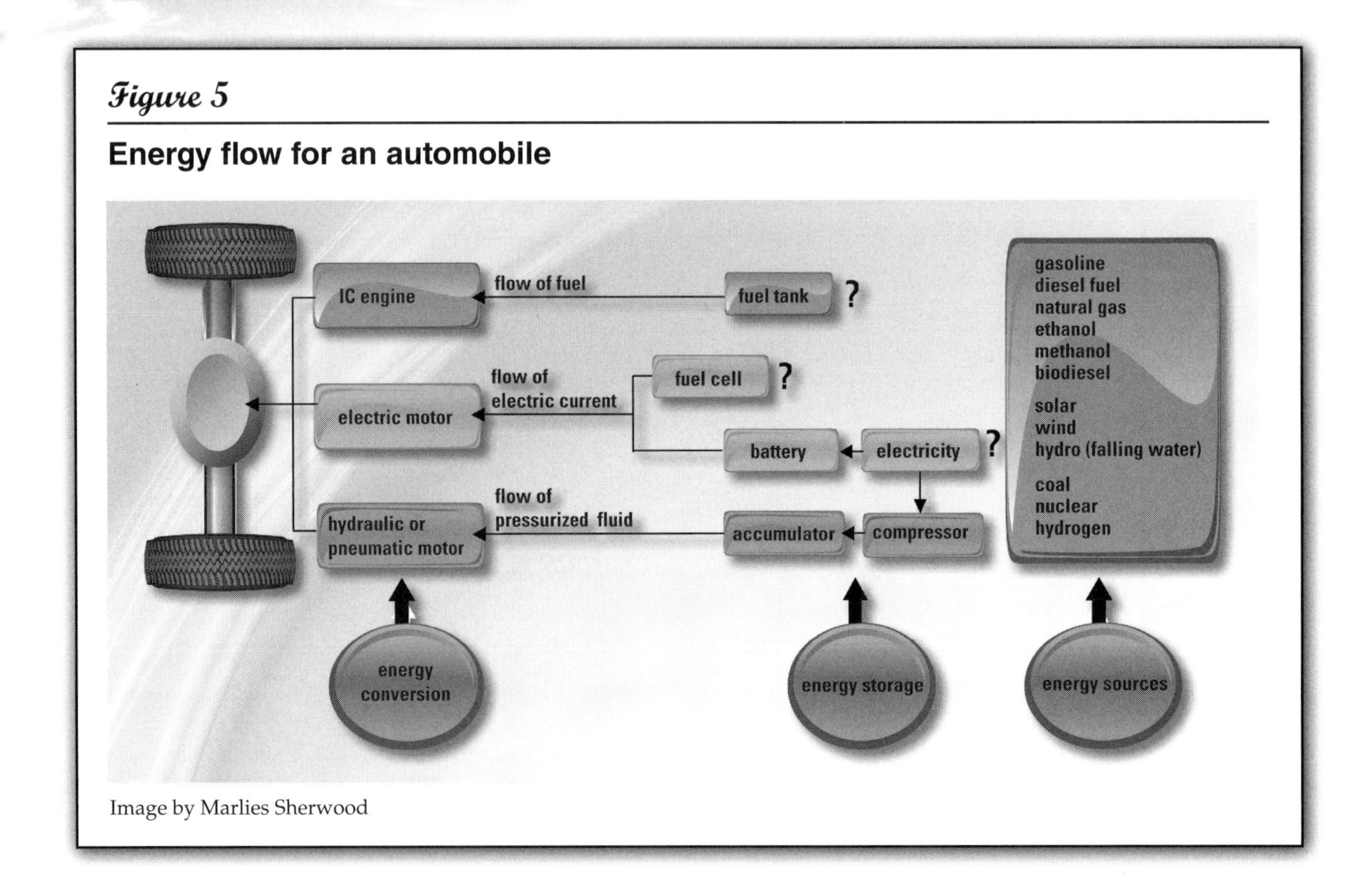

Figure 5

Energy flow for an automobile

Image by Marlies Sherwood

prediction that production would peak in the early 1970s—no one believed him at the time (Deffeyes 2001). They then compare Hubbert's model predictions with actual data through 2006 and discover that Hubbert's predictions were dead-on (Figure 6).

Applying the model to world oil production, students then predict when global production will peak (around 2012) and estimate at what point production will decrease to levels seen in 1900 (2050), when oil production was just beginning its exponential rise. This activity not only allows students to graphically visualize an impending world shortage, but also provides the opportunity to think about model limitations and why predictions may not come to pass (e.g., we may discover other energy sources making oil irrelevant, political situations could disrupt markets, and so on).

Fundamentals of internal combustion engine operation

Before studying propulsion alternatives, students should be familiar with the principles of conventional systems. The third activity provides students with an understanding of the fundamentals of internal combustion engine

Figure 6

U.S. oil production

Triangles show actual data; solid line shows M. King Hubbert's model predictions

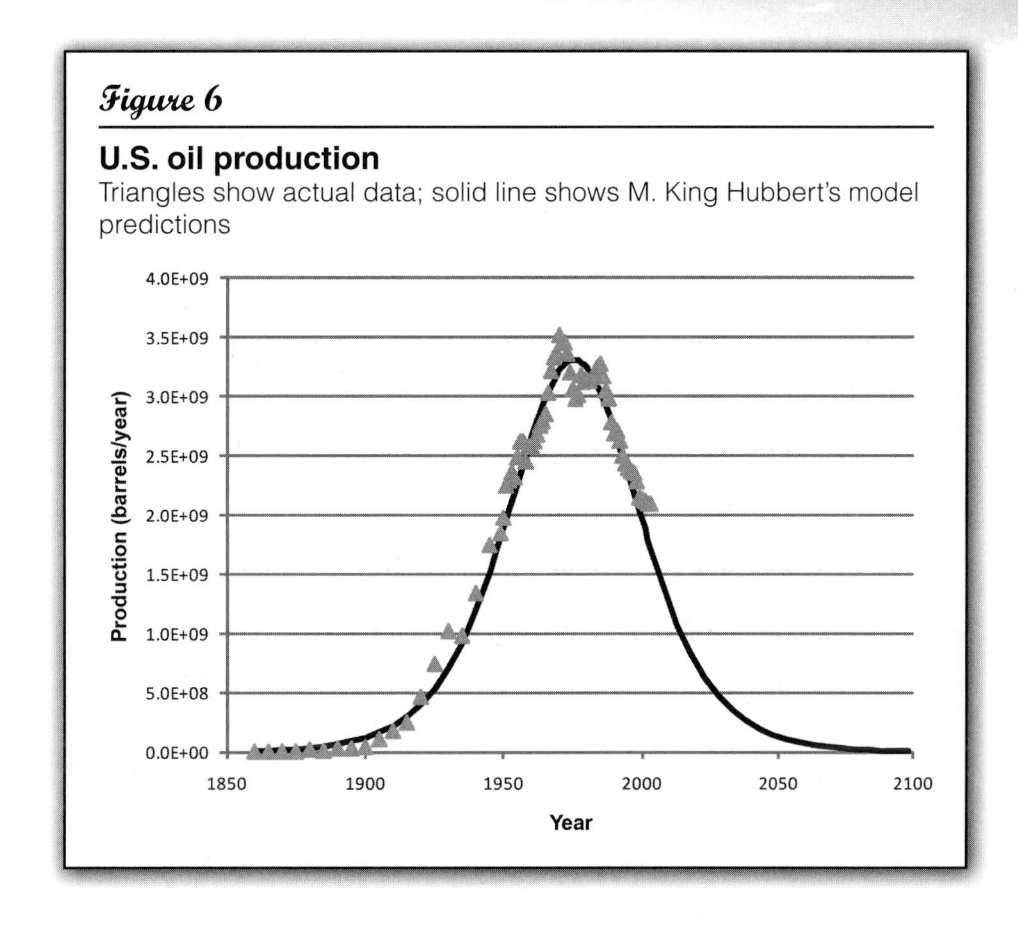

operation. Using actual data collected from an engine dynamometer—a device for measuring force, torque, or power—students perform pencil-and-paper calculations to gauge the relationship between fuel economy and vehicle speed. They calculate—among other things—the fuel expense and travel time for a trip from Detroit, Michigan, to Washington, DC, at 55 mph (89 km/h) and at 75 mph (121 km/h).

Where do plants store energy?

After these introductory activities, students are exposed to a series of chemistry and biology activities covering topics such as combustion; pollution; and the production of ethanol, methanol, biodiesel, and hydrogen. For example, the ninth activity, Where Do Plants Store Energy? leads students through a series of experiments to determine the amount of energy contained in plant seeds (as in the production of ethanol from corn) compared to that contained in stems and leaves (as in the production of ethanol from cellulosic material).

Students construct and use a simple calorimeter to measure the energy released from burning various types of plant material.

Which fuel is best?

The final activity assimilates material from previous activities. Students consider various fuels for a road trip from Detroit to Seattle, Washington. (Instructors can use the materials for a trip between any two cities.) Students are provided with specific fuel characteristics, such as yield (e.g., 2.7 gal. [10 L] of ethanol per bushel of corn), energy content, carbon dioxide production, cost, and conversion efficiency. They then calculate the total fuel consumed, trip cost, carbon dioxide produced, and amount of raw material used for each fuel option (see "On the web" for information provided to students). Based on information from previous activities and internet research, students come up with recommendations for the "best" fuel choice. They then deliver an oral presentation to the class.

CONCLUSION

News outlets regularly broadcast stories about the latest automotive technologies. Although students are familiar with the factors driving these innovations, such as global warming and increasing energy costs, they—and the general population—suffer from an abysmally low rate of "energy literacy" (DeWaters and Powers 2008). In fact, *Time* magazine has suggested that the American driving public shares some of the blame for the Gulf of Mexico oil spill (Time Staff 2010). Perhaps if the public were better educated about how lifestyle choices influence the environment and energy resource depletion, future generations would be more likely to drive sustainably.

Fueling the Car of Tomorrow provides a way to address this situation, and at the same time motivates students to consider careers in science and technology—a development that must occur if the United States is to maintain its lead in global innovation.

Acknowledgments

We would like to gratefully acknowledge funding from the Ford Partnership for Advanced Studies and the Michigan–Ohio University Transportation Center for this project. High school teachers Robert Santavicca and Francis Muylaert provided indispensable advice after piloting the activities with their students. Charles Smartt, Katie Heitchue, and Marlies Sherwood also assisted with crucial aspects of project development.

On the web

Fueling the Car of Tomorrow curriculum (teacher and student versions) and supplementary materials: *http://eng-sci.udmercy.edu/pre-college/alt_fuel_curriculum*

References

Deffeyes, K. S. 2001. *Hubbert's peak: The impending world oil shortage.* Princeton, NJ: Princeton University Press.

DeWaters, J., and S. Powers. 2008. Energy literacy among middle and high school youth. 38th ASEE/IEEE Frontiers in Education Conference Proceedings, Saratoga Springs, NY.

Michigan Department of Education. 1996. Michigan curriculum framework. *www.michigan.gov/documents/Michigan CurriculumFramework_8172_7.pdf*

National Research Council (NRC). 1996. *National science education standards.* Washington, DC: National Academies Press.

Time Staff. 2010. The dirty dozen: Who to blame for the oil spill. Time.com. *www.time.com/time/specials/packages/article/0,28804,1995523_1995491,00.html*

Chapter 18

THE INTERDISCIPLINARY STUDY OF BIOFUELS

UNDERSTANDING QUESTIONS AND FINDING SOLUTIONS THROUGH BIOLOGY, CHEMISTRY, AND PHYSICS

By Philip D. Weyman

From media news coverage to fluctuating gas prices, the hot topic of energy is hard to ignore. However, little connection often exists between energy use in our daily lives and the presentation of energy-related concepts in the science classroom. The concepts of energy production and consumption bring together knowledge from several science disciplines to both enhance student understanding and seek solutions to important global problems.

Students learn the second law of thermodynamics, photosynthesis, and Ohm's law in the classroom, but they may not see the direct application of these concepts to their daily lives—from the electricity that powers their computers to the ethanol-blended gasoline that fuels their cars. Students may

have even more trouble relating to the world's rapidly emerging energy crisis. As global demand for energy increases, supplies of liquid transportation fuels used to power our cars, trucks, and airplanes decrease, leading to a potential crisis in their cost and availability (Hudson 2005). In addition, increasing evidence points toward planetary climate changes resulting from carbon dioxide emissions associated with burning fossil fuels. Substituting biofuel for fossil fuel is one potential solution to these energy problems.

This chapter provides an overview of activities and discussions teachers can use to address the questions raised about biofuels in biology, chemistry, and physics classes. Complete descriptions of these biofuel activities—including lesson plans and student worksheets—can be found on the Engaging Inquiring Minds Through the Chemistry of Energy website (see "On the web").

FOSSIL FUELS VERSUS BIOFUELS

Fossil fuels are the products of plants, animals, and microorganisms that lived millions of years ago—the remains of which have been trapped underground away from the atmosphere. When fossil fuel is burned, carbon that has been trapped for millions of years is suddenly oxidized to carbon dioxide (CO_2) adding to the CO_2 already present in the atmosphere. This net increase in CO_2 is referred to as a "carbon-positive" process.

Biofuels are the products of organisms that have lived recently (within the past few years). Biofuels store the energy from the Sun without increasing atmospheric CO_2 by taking advantage of photosynthesis. As a plant grows, it uses photosynthesis to take in CO_2 from the atmosphere and convert and store the carbon from CO_2 in various chemical components (e.g., starch, oil, and cellulose). Biofuel is created when plants are processed through chemical or microbiological methods. The sugars from starches or cellulose can be used to make fermented fuels such as ethanol or butanol; oils can be turned into biodiesel; and waste material can be converted to methane.

When a biofuel reacts with oxygen through burning, the released CO_2 returns to the atmosphere, but no increase in atmospheric CO_2 is observed—this is because the CO_2 that is returned was recently in the atmosphere, before the plant removed it. This process is referred to as "carbon neutral." As some plants grow, more CO_2 is removed from the atmosphere and stored than is harvested (e.g., perennial plants with deep root systems). When the resulting biofuel is burned, less CO_2 is released than was originally removed from the atmosphere by the plant, and such a fuel is referred to as "carbon negative."

Biofuels include chemicals such as ethanol, butanol, biodiesel (esters of fatty acids derived from fats and oils), methane, and hydrogen. In order to designate a chemical as a biofuel—and determine whether it is carbon positive, neutral, or negative—we must know how it was produced. For example, methane is carbon positive because it is a large component of natural gas, which is a fossil fuel. However, when waste material is processed by microorganisms to produce methane, the resulting fuel could be considered a carbon neutral or negative biofuel. To be truly carbon neutral or negative, the energy used in the processing must also come from biofuels. If fossil fuels are used in the production of the biofuel (e.g., if a fossil fuel diesel tractor is used to cultivate corn), then the resulting biofuel (i.e., corn-derived ethanol) may end up being more carbon positive than carbon neutral.

ADDRESSING FAQS

As society moves toward biofuel alternatives, students often hear conflicting reports concerning their value. Is ethanol produced from corn mitigating global warming? Does it contribute to food scarcity, cost, and world hunger by diverting food sources to energy use? Why can energy-rich cellulose not be used to produce ethanol *now?* Is hydrogen a biofuel?

Ultimately, these questions can be answered by understanding the biology, chemistry, and physics behind these fuels. This chapter takes a look at how these scientific disciplines can be used to address common questions raised about biofuels. Biofuel and global warming activities can be structured to allow the material to reinforce both content objectives and methods of scientific inquiry. Exposure to biofuels can and should be included even in the current educational climate, where a focus on standards is increasingly important.

BIOLOGY OF BIOFUELS

Discussions of biofuels often start and end with how the fuel is used; however, production of biofuels is an ideal topic for the biology classroom. The raw materials used to create biofuel may come from different parts of a plant and can be interesting supplements for discussions of plant anatomy and biology. For example, corn seeds (kernels) contain a starchy endosperm that can be processed and fermented to produce ethanol. Soybean seeds can be processed to remove their oil, which can be further processed to create biodiesel. The seeds on a plant are just a small fraction of a plant's total biomass, which includes leaves, stems, and roots and is largely composed of cellulose (a tough polymer made of glucose sugars). The development of technologies that allow efficient digestion and fermentation of cellulose will greatly increase the efficiency of biofuel production from plants (Buckley and Wall 2006).

Future biofuel production may not involve multicellular plants and may rely instead on microorganisms. Microalgae may be involved in producing fuels such as hydrogen or biodiesel using sunlight, CO_2, and water as the sole primary inputs (Buckley and Wall 2006).

Hydrogen gas

Hydrogen gas (H_2) is one of the simplest fuels chemically, but its designation as a biofuel is often quite complicated. Since many of the activities described in this chapter use it, a brief overview of H_2 may be useful. H_2 can be produced in many ways, and, as a fuel, it is an energy storage molecule. When produced through natural gas reformation, H_2 is derived from fossil fuel resulting in a net CO_2 increase. H_2 can also be produced through electrolysis; in this case, the fuel is only as "green" as the source of electricity. If the electricity comes from a coal-fired power plant, then the H_2 is not a biofuel but merely a means to store fossil fuel energy. If the electricity comes from a hydroelectric dam, the fuel is no longer linked to CO_2 emissions, but it is not a biofuel as it is not derived from a recently living organism. (**Note:** It could be considered an alternative fuel.)

H_2 can be considered a biofuel when derived from microorganisms that make H_2 as a byproduct of their metabolism. These microbes have enzymes called "hydrogenases" that can produce H_2 when given a source of energy and nutrients. Some microorganisms can use sugars from plant-derived starch or cellulose to produce H_2, while certain microalgae can use the energy from photosynthesis. The latter case is perhaps ideal as the power of the Sun can be used directly to make a fuel (H_2) in a continuous process, as long as the microalgae are living.

Investigating biofuel conditions

Because biofuels require growing the plants or microorganisms, students must consider the conditions that favor growth. Biofuels can be introduced as a practical application through discussions of the different modes of nutrient acquisition (e.g., by heterotrophs and autotrophs). The difference between the two is more than just an academic distinction when biofuels are concerned. For example, to produce butanol, a fermented alcohol fuel, in one scenario a microorganism uses refined sugars from plants to grow and produce the fuel (Figure 1a-d). In a second scenario a photosynthetic microorganism makes the fuel using the light from the Sun, water, and CO_2 as input energy and raw materials (Figure 1e-f).

Students can decide which is more desirable from an agricultural and energy efficiency point of view. If it costs money and energy—possibly derived from fossil fuels—to refine the sugars, which then go to feed the microorganisms, perhaps it would be more economical and "green" to grow the microorganisms that can use sunlight for energy instead. The drawback is fermentation technologies that grow microorganisms using sugars as nutrients have been well established, whereas efficient production systems that use photosynthetic microorganisms still need to be developed. A variety of short

Figure 1

Two different hypothetical biofuel production schemes

In the first scenario, crops are grown (A) and processed to yield refined sugars (B) that can be converted to a biofuel in a bioreactor with the addition of the proper microorganism (C and D). In the second scenario (E and F), a photosynthetic microorganism such as a microalga can grow and produce the biofuel directly using the light of the Sun, water, and CO_2 as inputs.

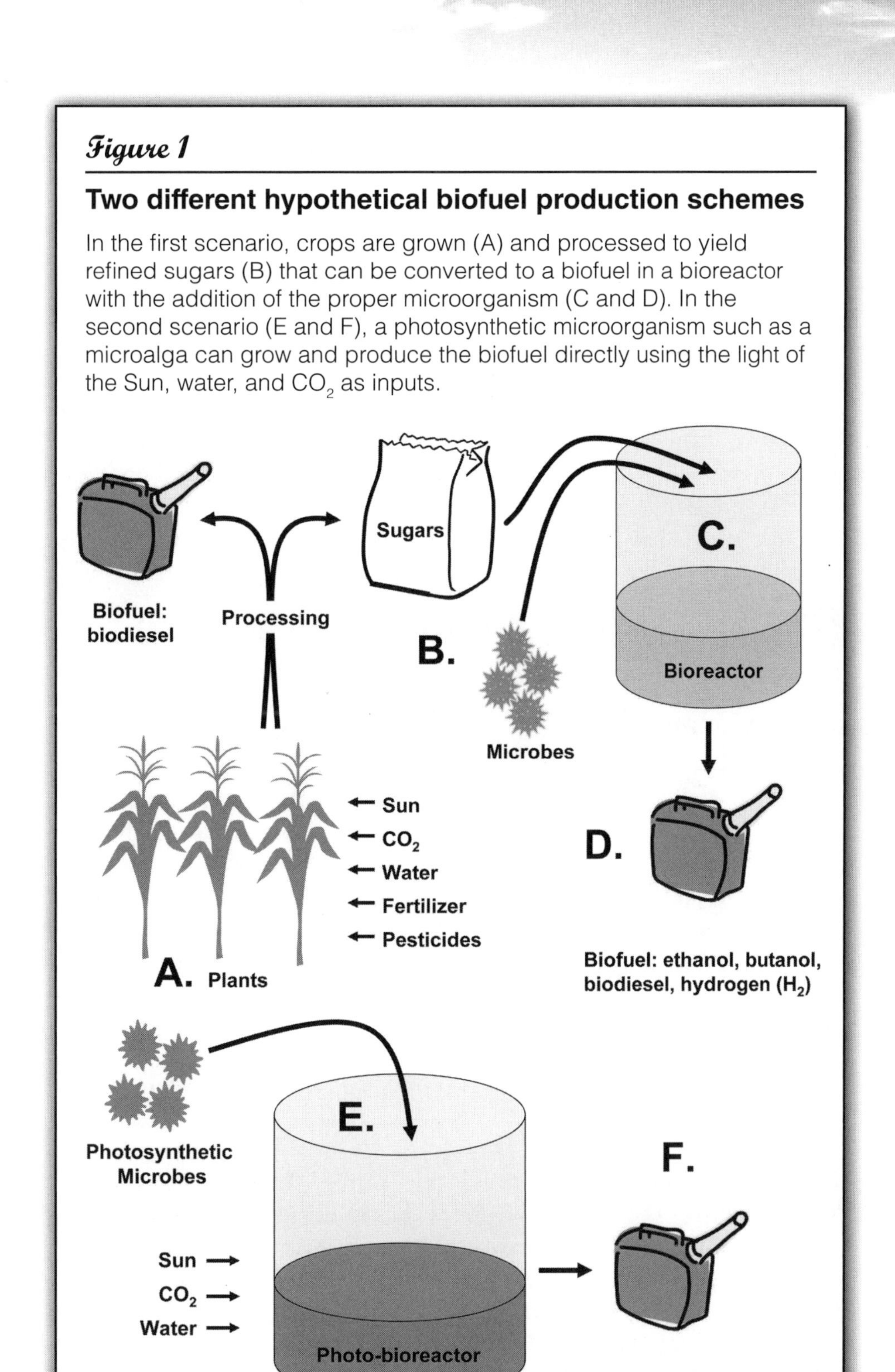

Figure 2

The Kill-a-watt energy meter
The energy meter is connected to an extension cord to allow students to use and read the device more easily at a lab bench.

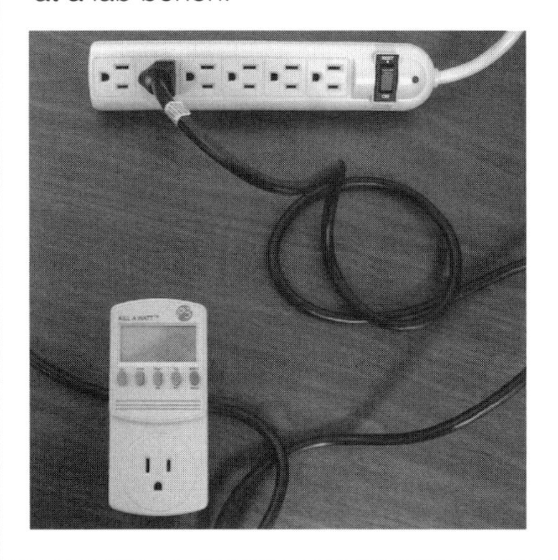

YouTube videos may provide students with an exciting glimpse of how microorganisms are beginning to be used for energy (see "On the web"). This is an area of current research that can be interesting for students to investigate.

Most biology classes have units on the carbon cycle and sometimes on human impact on the biosphere as well. To incorporate biofuels into biology classes, we have been using an energy meter lab (Figure 2; see also Energy meter activity, p. 146) to investigate how electricity use relates to these environmental concerns, and an activity on the greenhouse effect to explore properties of greenhouse gases (Figure 3).

CHEMISTRY OF BIOFUELS

When considering the production of transportation fuels, the related chemistry ranges from simple elements (e.g., H_2) and alcohols (e.g., ethanol) to complex mixtures of hydrocarbons (e.g., petroleum diesel). Each fuel has characteristics that make it more suitable for one application than another. For instance, H_2 stores the most energy per kilogram of any fuel—this makes it ideal for the space shuttle, where fuel mass must be minimized, but problematic for ground-based vehicles, where volume must be minimized through gas compression. Ethanol may be appropriate for local production and use. However, it is not a good candidate for energy-efficient long-distance transport by pipelines—ethanol has corrosive tendencies and can mix easily with water (an unavoidable contaminant during pipeline transport). For ethanol to be used as a transportation fuel, the water must be removed, which is an energy-expensive process.

Fuel cell car activities

To help students begin to explore the chemistry of biofuels, we have developed several fuel cell car kit–based activities that focus on hydrogen chemistry. The discussion of H_2 fuel cells fits well with electrochemistry units, where the movement of electrons in electrochemical cells is reviewed (Pintauro 2004). The important chemistry concept of electrochemistry is reinforced when related to a cleaner transportation technology (e.g., H_2 fuel cells) that could replace combustion engines.

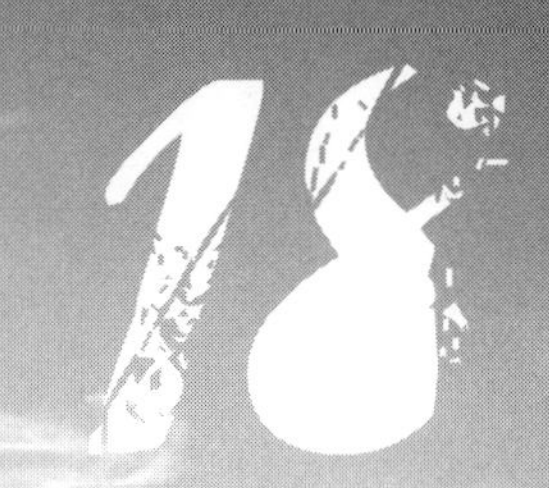

Figure 3

Greenhouse effect activity

To investigate the greenhouse effect and how it might lead to global warming, we modified a previously published activity in which students observe the effect of CO_2 on terrarium temperature (Lueddecke, Pinter, and McManus 2001). In the original activity, students construct a mini-Earth using an aquarium. Baking soda and vinegar are used to make the CO_2, while the Sun is simulated by shining heat lamps down on the aquarium. Students measure the effect of gases on the temperature inside the tank by comparing a tank containing elevated CO_2 levels with a control tank containing classroom air. The lab stresses basic experimental concepts such as formulating a hypothesis, designing an experiment, determining the independent and dependent variables, graphing data, and drawing conclusions from data.

As a cheaper, more transportable substitute—and in order to provide a setup for every two students in each class—we modified the activity to use deep, plastic storage containers in place of aquariums (see "On the web"). Also, the key to success with this lab is to allow a slow and steady release of CO_2 with minimal disturbance to the gases within the tank. This can be difficult given the many classroom air currents and student "fiddling" that can occur. Heat transfer must be allowed to occur between gases inside and outside the tank, but gases should not be disturbed by turbulent air flow. Rather than using baking soda and vinegar—which requires students to reach into the tank and stir the reagents—we keep a constant supply of CO_2 inside the tank with a small CO_2 cartridge (for bicycle tire inflation) hooked to the end of a long hose. Although more expensive than baking soda and vinegar, the ability to add CO_2 continuously maintains a constant CO_2 layer in the tank despite small amounts of air turbulence in the classroom.

The lab is appropriate for biology and environmental science classes in units covering greenhouse gases, renewable fuels, or basic experimental design skills. For chemistry classes, it can be used to study gas laws, extend organic chemistry, or discuss types of reactions (using the reaction of baking soda and vinegar as an example).

Example pre- and posttest assessment or discussion questions:

1. What is the greenhouse effect?
2. List one gas found in the atmosphere that can act as a greenhouse gas.
3. What is one way that this greenhouse gas might be increased by human activity?
4. Is the dependent variable the variable that you measure or the variable that you change? Describe the dependent variable you used in today's experiment.
5. What is one thing you can do to decrease the amount of greenhouse gas from question 2?

The activities we developed outline how hydrogen and oxygen gases can be made from water using the solar cell provided in fuel cell car kits (see "On the web"). The H_2 made in the process is stored in small tanks (provided with kits) and can then be used to power the fuel cell car. This demonstrates how fuel cells can be used to either make or use H_2—depending on the input energy—and how the fundamental chemistry of using or producing H_2 is the same, but the direction of the reaction can change. The activity also demonstrates that H_2 is an energy storage molecule. The manner by which the H_2 is produced is critical in determining whether H_2 is a carbon neutral or negative biofuel, or if it merely stores the energy from fossil fuels (and therefore is carbon positive).

Many fuel cell car kits are available commercially on the web and in hobby stores. The kits can be expensive, ranging in cost from $40–$195, but they are durable, last several years, and serve hundreds of students with minimal upkeep. The kits can also be shared among biology, chemistry, and physics classes. We tested several fuel cell car kits and reviewed the advantages and disadvantages of each type. The kit we ultimately chose is simple enough for a basic introduction on how to make and power a car, but it also allows for a more detailed investigation of how the car works and the efficiencies and barriers encountered. In addition to the H_2 activities, labs with biodiesel are well suited for chemistry classrooms and have been described elsewhere (see "On the web"; Bucholtz 2007).

Energy meter activity

Another activity we have used—that ties chemistry together with physics, biology, and environmental science—requires a class set of small energy meters (called "Kill-a-watts"), which can be purchased from several online vendors for $20–$40 each (Figure 2). As with the fuel cell car kits, these devices can be used for many years in virtually every science discipline. Any appliance can simply be plugged into the meter: The device provides real-time measurements of the appliance's power, voltage, and current use and records cumulative power in kilowatt-hours.

In the online version of this activity, a worksheet is available that explains basic electrical concepts and guides students through calculations to help them grasp the environmental effects of our electricity use (see "On the web"). For example, the energy requirements of a light with a conventional 100 W bulb may not seem significant. But when students consider the hours of use throughout an entire year, the kilowatt-hours begin to accumulate. Students also can investigate the effect of replacing a 100 W bulb with one of lower wattage, or of replacing conventional bulbs with compact fluorescent ones. Students then use the average amount of CO_2 emitted by power plants using fossil fuels (available in the online version of this activity; see "On the web") to convert kilowatt-hours to tons of CO_2 emitted.

For this activity, students measure the watts of power used by a given device (e.g., a desk lamp with a 100 W bulb). From this measurement, students calculate kilograms of CO_2 emitted per kilowatt-hour (kWh) using a conversion factor (0.58 kg CO_2/kWh). The conversion factor was calculated by considering the total amount of CO_2 emitted by U.S. power plants in 2006 divided by the U.S. electrical power output that same year (EIA 2006a, 2006b) (Figure 4a). To further tie in chemistry concepts, students can try to understand the mass of CO_2 from a volumetric perspective (Figure 4b) using the gas laws.

Figure 4

Example chemistry calculations

A. Measured W × 1 kW/1000 W × hr used/day = kWh/day

kWh/day × 365 days/yr = kWh/yr

kWh/yr × 0.58 kg CO_2 /kWh = kg CO_2 /yr from that appliance

B. How much volume does 1 kg of CO_2 occupy at room temperature and standard pressure?

CO_2 has a molecular weight of 44 g/mol

1 kg CO_2 = 1,000 g; 1,000 g × (1 mol/44 g) = 22.7 mol CO_2
$V = nRT/P$, $V = (22.7)(0.0821)(300)/1 = 559$ L CO_2 at 27°C (300K), 1 atm

This is a little more than half a cubic meter, approximately equal to the volume of two bathtubs or the trunk of a large car.

This lab can be taught in

- biology classes by focusing on the effects our machine-driven lifestyles have on the global environment;
- chemistry classes by emphasizing the gas laws and the factor-label method (dimensional analysis) used to convert electricity use to carbon emissions; and
- physics classes by putting an empirical emphasis on electricity concepts such as *current, potential, power,* and *kilowatt-hours.*

PHYSICS OF BIOFUELS

When considering the physics of energy, a fuel cell car may help focus student attention on how models of energy transfer relate to the energy sources that power our lives. Energy meters complement a unit on electricity—the meters emphasize how various devices draw different amounts of current with a constant voltage. When students can relate a given current to the requirements needed to recharge an iPod, they may be better able to grasp the abstract concepts of electricity and also better understand the energy costs of routine daily activities.

The fuel cell car can be used to practice calculations of fuel efficiency and average speed (see "On the web"). The activities on the Engaging Inquiring Minds Through the Chemistry of Energy website integrate biofuels (and the technologies that use biofuels) in the classroom while providing opportunities for students to measure physical variables and perform calculations.

The fuel cell car we used can be converted into a crane with a spindle and string (included with the kit). This crane can be used to lift an object—the mass of the object, distance the object is lifted, and time required can be used to calculate the power of the crane's H_2 fuel cell motor $(P=mgh/t)$. While this calculation could be made with almost any motor, using an H_2 fuel cell reinforces the concept of how the fuel cell works and inspires students to consider alternatives to fossil fuel energy sources.

Efficiency—the energy output of a device compared to the energy that is invested—is a concept that can be explored in several ways using fuel cell car kits. For instance, the efficiency of the solar panel that comes with a kit can be calculated, as can the efficiency of the fuel cell (see "On the web").

CONCLUSION

Many subjects—beyond biology, chemistry, and physics—can be applied when considering an interdisciplinary study of biofuels. Mathematics is used throughout the sciences and is reinforced by the activities described in this article—from calculating the power of an H_2 fuel cell crane to graphing the effects of CO_2 on temperature. The fervent discussions about energy policy witnessed in the 2008 presidential election demonstrate the social, political, and economic significance of resource usage. All aspects of the social sciences and humanities can be engaged in an understanding of biofuel and how it may help solve our energy problems.

Given the importance and growing issues surrounding energy in our lives, it is vital that science teachers raise these issues with students. One avenue to achieve this may involve replacing or enhancing traditional science labs with labs that focus on energy alternatives such as biofuels. When students learn about biofuels and future energy technologies from multiple

science perspectives, they begin to understand the concept as a real-life issue they will confront throughout their lives as citizens and perhaps as scientists.

On the web

Volume: *www.thinkmetric.org.uk/volume.html*

Engaging inquiring minds through the chemistry of energy:

- Biodiesel activities: *www.umsl.edu/~biofuels/biodiesel.html*
- Energy meter lab: *www.umsl.edu/~biofuels/Energymeter.html*
- Greenhouse gas labs: *www.umsl.edu/~biofuels/Greenhousegas.html*
- Hydrogen fuel cell activities: *www.umsl.edu/~biofuels/Fuelcellcarlabs.html*
- Review of fuel cell car kits: *www.umsl.edu/~biofuels/Fuelcellcarreview.html*

Biofuel YouTube videos:

- MIT algae photobioreactor: *www.youtube.com/watch?v=EnOSnJJSP5c*
- H_2 from algae powers fuel cell car: *www.youtube.com/watch?v=r9vniN54Aok*

References

Bucholtz, E. C. 2007. Biodiesel synthesis and evaluation: An organic chemistry experiment. *Journal of Chemical Education* 84 (2): 296–298.

Buckley, M., and J. Wall. 2006. *Microbial energy conversion.* American Academy of Microbiology. *http://academy.asm.org/images/stories/documents/microbialenergyconversionfull.pdf*

Energy Information Administration (EIA). 2006a. U.S. Emissions of Greenhouse Gases Report, 2006. *ftp://ftp.eia.doe.gov/pub/oiaf/1605/cdrom/pdf/ggrpt/057306.pdf*

Hudson, T. 2005. Petroleum and the environment. *The Science Teacher* 72 (9): 34–35.

Lueddecke, S. B., N. Pinter, and S. McManus. 2001. Greenhouse effect in the classroom: A project- and laboratory-based curriculum. *Journal of Geoscience Education* 49 (3): 274–279.

Pintauro, P. 2004. Ask the experts. *The Science Teacher* 71 (5): 64.

About the project

The activities described in this chapter were developed as part of a partnership between the University of Missouri–St. Louis, the Science Outreach Department at Washington University in St. Louis, and a dedicated group of high school teachers in the St. Louis area. As part of this partnership, I (author) have been working with high school teachers and students to introduce lessons on greenhouse gases, global warming, and biofuels in high school science classrooms. While most activities involve structured and guided inquiry, emphasis is placed on developing questions and hypotheses, identifying experimental variables, recording and graphing data, and drawing conclusions from results.

Chapter 19

A LIFE-CYCLE ASSESSMENT OF BIOFUELS

TRACING ENERGY AND CARBON THROUGH A FUEL-PRODUCTION SYSTEM

By Sara Krauskopf

Which is a better fuel—gasoline or ethanol? Which fuel provides better gas mileage? Produces fewer greenhouse gas emissions? Requires less energy to produce? With the exception of the gas mileage question (ethanol produces two-thirds the energy of gasoline per gallon), none of these questions has an easy answer.

Concerns about climate change, energy independence, and fossil fuel supplies are increasingly in the news. Biofuels such as biodiesel and cellulosic ethanol could help cut U.S. dependence on petroleum-based transportation fuels by up to 30% (U.S. DOE 2006). But with all of the options for alternative fuels out there, how do we determine which is the "best"?

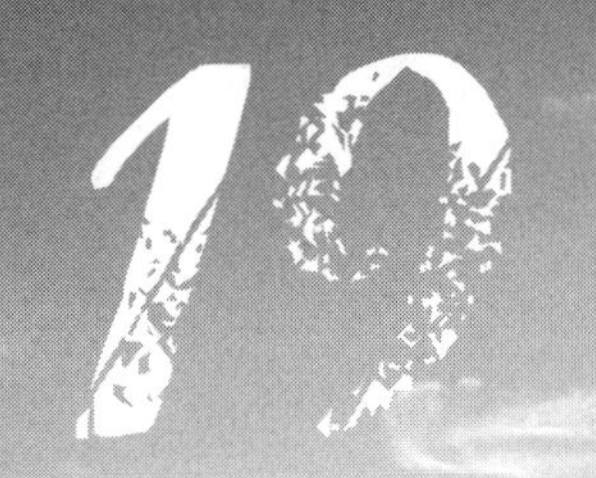

When asked to compare fuel alternatives, many of us think only about what happens in the car itself (i.e., how many miles per gallon it gets and its tailpipe emission ratings). However, a more accurate comparison would be to examine the processes from "cradle to grave," accounting for all of the steps before fuel is even put into the car—beginning, for example, with petroleum extraction, ocean transport, refinery operation, and the gasoline's transportation to the pump.

A life-cycle assessment (LCA) is a tool used by engineers to make measurements of net energy, greenhouse gas production, water consumption, and other items of concern. The Environmental Protection Agency's National Renewable Fuels Standard sets limits for net greenhouse gas emissions in biofuel manufacturing—compared to gasoline; it also requires biofuel producers to conduct LCAs that evaluate emissions during farming, transportation of raw materials, refining, and so on (EPA 2010).

As with all sustainability issues, getting people to think about the life cycle, or full system, is a challenging yet valuable approach. The first step is to develop a clear picture of all the steps in the system in which materials and energy are used or transformed.

This chapter describes an activity designed to walk students through the qualitative part of an LCA. It asks them to consider the life-cycle costs of ethanol production, in terms of both energy consumption and carbon dioxide emissions. In the process, they trace matter and energy through a complex fuel-production system.

Once students understand what a qualitative LCA looks like, they are better able to assess the quantitative results and critically consider the modes of production for different fuel types. By the end of the activity, students can suggest efficiency improvements in the biofuel production system and ways to reduce net greenhouse gas emissions. In addition, it often gets students thinking about other products they use and how these choices might affect the environment.

LEARNING OUTCOMES

I designed this lesson, which can be used in an integrated science, biology, or environmental science course, to take two 50-minute class periods. Given the complexity of the material, plenty of time should be allotted for discussion and questions. During this activity, students will

- trace energy and carbon through a transportation fuel-production system;
- describe steps in fuel processing in which carbon dioxide is sequestered and released;

- identify ways to process transportation fuels with fewer inputs of fossil fuels; and
- evaluate the sustainability of gasoline, corn ethanol, and cellulosic ethanol.

(**Note:** All of the images and handouts described in this chapter are available online, along with teacher guides, PowerPoint presentations, extensions, and more [see "On the web"]).

FORMATIVE ASSESSMENT

To begin the activity, I have students complete a brief questionnaire (Figure 1a), which asks deliberately ambiguous questions about gasoline and ethanol. After answering the questions individually, students share their responses in small groups and then discuss them as a class. This formative assessment helps gauge students' background knowledge about fuels and brings questions and misunderstandings into the open (Koba and Tweed 2009). Because many articles and advertisements advocate strongly for or against ethanol, students may have heard biased arguments—and these will often come up in the class discussion.

Figure 1

Formative assessment questions

A. Which is a better fuel—gasoline or ethanol? For each of the criteria below, place a check mark below which fuel you think is better.

	Gasoline	Ethanol
Energy per gallon/miles per gallon		
Energy required to produce a gallon		
Carbon dioxide emissions		
For the environment		
Amount available		

To create ethanol, sugars in plant biomass (i.e., the leaves, stems, and other plant parts) are harvested and converted into fuel (C_2H_5OH). Based on what you know, check the statement(s) with which you agree.

____ Carbon dioxide is released when ethanol is produced from plant biomass.
____ Creating ethanol from plant biomass is carbon neutral.
____ Creating ethanol from plant biomass contributes to climate change.

B. Create a drawing or cycle that explains the production of ethanol from plant biomass. Show the movement of carbon or carbon dioxide from location to location as best you can.

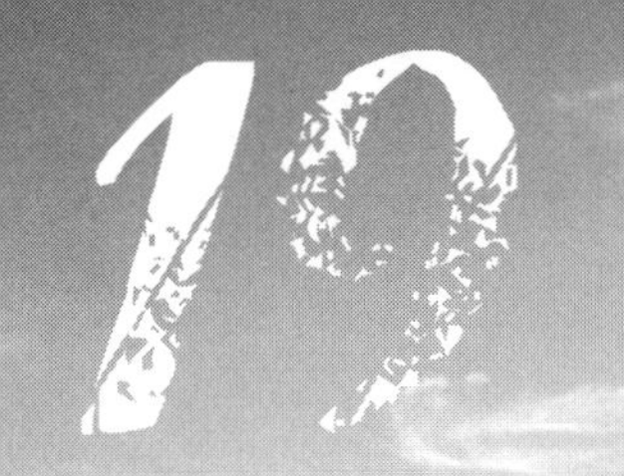

The carbon-cycle drawing students complete (Figure 1b) allows me to see whether or not they understand how these fuels are made and how carbon cycles through the fuel-production system. This drawing can be completed again at the end of the activity and compared for increased understanding.

BACKGROUND

After completing the discussion on the "better" fuel, I give students some background on how fuel-grade ethanol is produced. Ethanol can be made from different plant materials, and the starting material selected makes a difference in the total processing costs.

In the United States, most ethanol is made from corn grain, though considerable research is now being done on making ethanol from cellulosic materials. Corn ethanol is created from corn grain, a starchy material that is fairly easy to ferment; amylase (an enzyme) and heat can break the starch into glucose, which is readily fermentable.

Cellulosic ethanol, on the other hand, is created from leaves, stems, and woody materials, which are not easy to ferment. Enzymes that digest cellulose are not readily available at industrial scales, and a great deal of energy and chemicals currently go into isolating cellulose and preparing it for fermentation. Source material for cellulosic ethanol can come from farm residues, yard waste, prairie grasses, and a host of other feedstocks.

Depending on students' level of understanding, a quick review of photosynthesis, fermentation, and combustion may also be necessary to complete the activity successfully.

Figure 2

Life cycle of biofuels

(**Note:** The refining station is divided into two substations—the fermentation and plant operation phases—each with its own process tool. Thus, there are seven stations total.)

THE WALK-THROUGH

Figure 2 provides an overview of the stages necessary to create ethanol from agricultural materials. To simplify this diagram, which can be used to explain

ethanol production from corn grain or cellulosic materials, I divide the room into six stations, and post a picture on the wall for each station. These stations include

- plants,
- farming practices,
- feedstock transport,
- refining (which includes fermentation and plant operation substations),
- product transport, and
- the car.

Below each station's picture is a process tool (Figure 3). This tool was developed by Mohan and colleagues to help students grapple with the difficult concepts of conservation of energy and matter (Mohan, Chen, and Anderson 2009). Students tend to struggle with transformations of energy and changes in matter from one form to another. Often, they mistakenly describe matter turning into energy, or vice versa, when in reality, chemical changes cause matter to be *converted* into other forms of matter, and energy into other forms of energy. The process tool helps students separate these two types of transformations.

Figure 3

Process tool

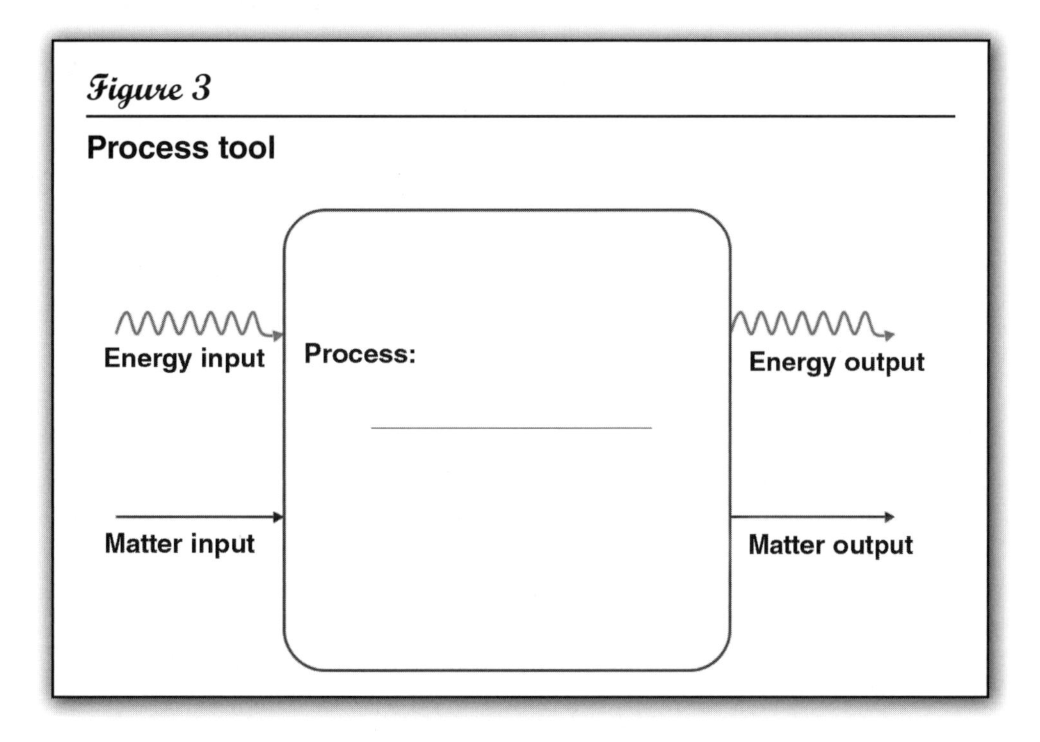

I then divide students evenly among the stations. Once at their stations, they begin to determine the energy inputs and outputs and the matter inputs and outputs for their particular part of the process. To facilitate, I project Figure 4 on a screen. This provides students with a limited number of choices for energy and matter inputs and outputs.

Student groups then discuss which choices best fit their particular station and write the appropriate energy and matter choices on the blanks in their process tool (Figure 3). This is a difficult task for many students, who often need prompting to make decisions. I ask students to restrict their choices to only those in Figure 4, and if they get stuck, I encourage them to make sure that their energy inputs match their matter inputs and that their energy outputs match their matter outputs. (For example, at the fermentation substation, glucose [matter input] is a molecule that stores chemical energy, and ethanol [matter output] is a molecule that stores chemical energy.) Students also fill in the name of the transformative "process" that occurs at their station (e.g., photosynthesis, combustion).

Once student groups have filled in their process tools, they pick up one set of "process tool cards"—cardstock with images and words—that matches their chosen inputs and outputs. These cards are cutouts of the individual choices from Figure 4. (It is a good idea to make seven cards of each choice, although not every card will be used at every station.)

Figure 4

Process tool choices

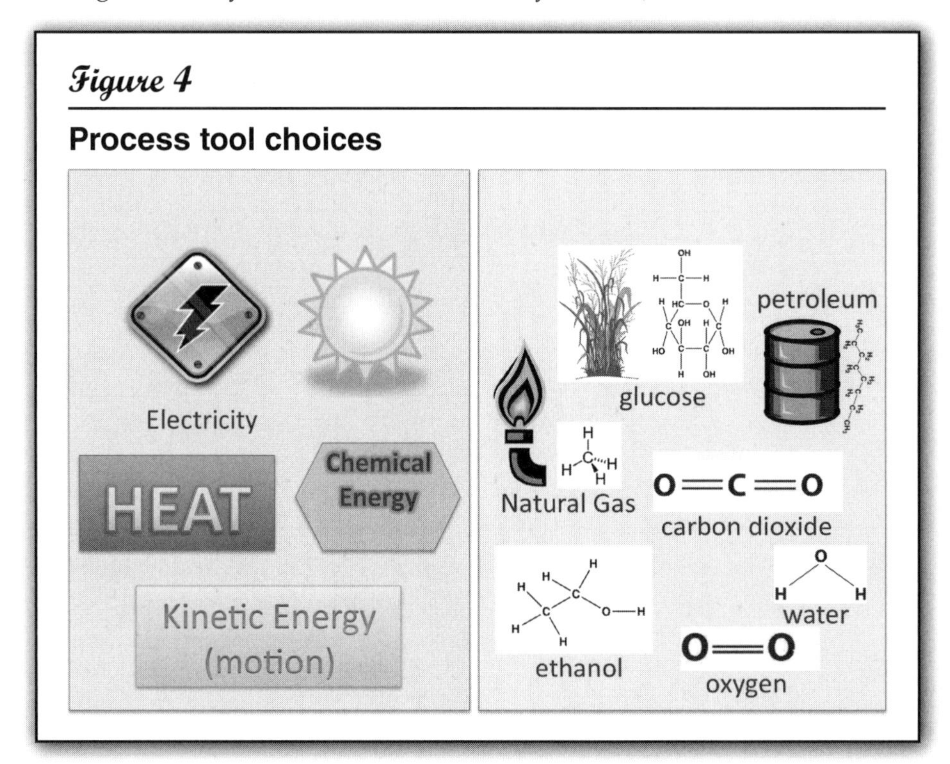

Plant station

Once each group has their process tool cards in hand, I bring the class back to attention but have them remain at their stations. Beginning with the plant group, I ask students at each station to explain their choices to the rest of the class. For example, at the plant station, solar energy is the energy input, and carbon dioxide and water are the matter inputs. The process of photosynthesis transforms solar energy into chemical energy, which is stored in the form of glucose and water. (I always make a disclaimer before beginning this step, letting students know that this is a difficult process and that I do not expect all of their answers to be correct. Any mistakes are discussed as a class.)

Next, I ask the plant group, "What form of matter is produced here that will be useful for creating ethanol (glucose)? What form of matter came off as 'waste' for the process (oxygen)? And where does the glucose go next in the process (fermentation)?"

I then have a student from the plant group walk the glucose and chemical energy card to the fermentation subgroup (part of the refinery station), thus moving the matter and stored chemical energy to the next stage in the fuel-production cycle. I skip farming practices and feedstock transportation because glucose is not transformed at those stations. First, the solar energy is traced through the system; other areas are addressed later.

Fermentation substation (part of the reifnery station)

I then move to the fermentation substation and ask these students a similar line of questions. At this substation, ethanol and carbon dioxide are the matter outputs. Ethanol, a valuable matter product, can be walked to the car station by one of the students in this group; carbon dioxide is the "waste" matter. Students often get stuck at the fermentation substation because they want to include energy inputs for heating the yeast or other processes that are separated into the "plant operation" category. Keeping a clear, traceable cycle of matter and energy is important, so these inputs should not be included at this point.

Car station

I then move to the car station, again tracing only the solar energy stored in the plants to ethanol formation and to the vehicle. In the car, combustion transforms the chemical energy in ethanol into kinetic energy and heat, and carbon dioxide and water are released. I ask students if any valuable products are created in this step. They should recognize that the carbon dioxide and water can be used by the plants, thus closing the life-cycle loop. The carbon dioxide from the fermentation substation can also be returned to the plants. Figure 5 shows the energy and matter connections between these three stations.

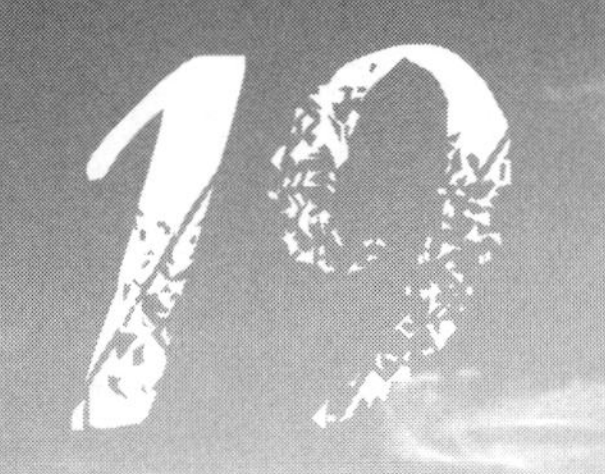

Figure 5

Energy and matter connections

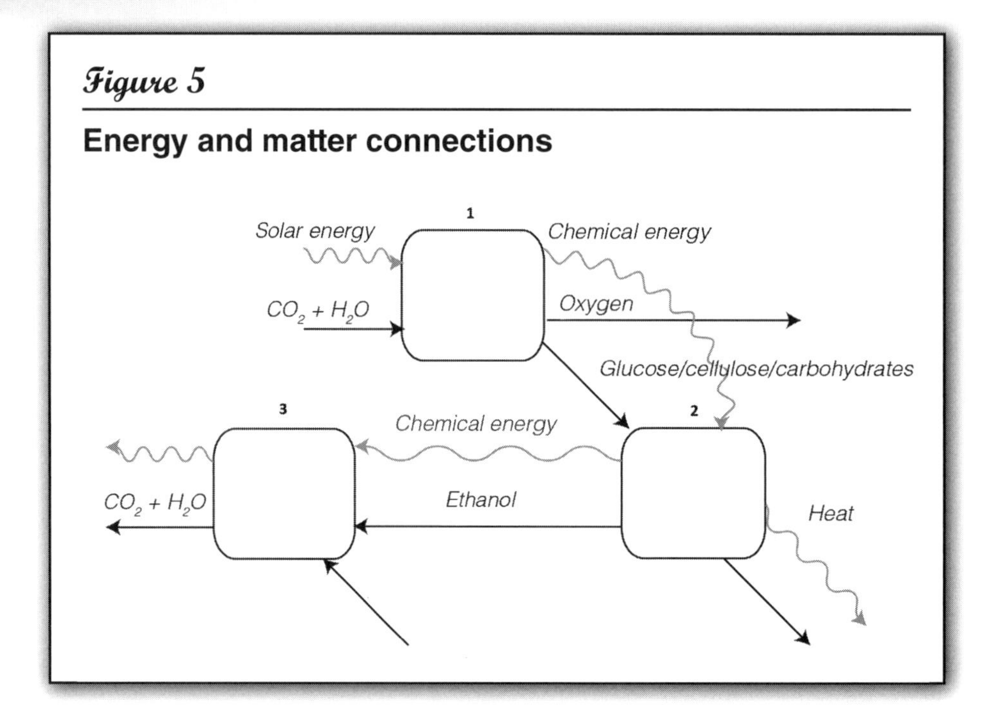

At this point, I review what happened at the stations and substations we have visited so far—plant, fermentation, and car—and ask, "Given the steps we have considered together, is the process of producing ethanol carbon-neutral?" The answer is yes. The amount of carbon dioxide absorbed to create glucose in photosynthesis is equal to the total amount released in fermentation and combustion.

Other stations

If we then return to the skipped stations (farming practices, feedstock transport, plant operations substation, and product transport), we can see where additional energy and carbon enters the system. Fossil fuels provide much of the chemical energy needed at the other stations—to create fertilizer, run tractors and fuel trucks, power refineries, and so on. The combustion of these hydrocarbons from gasoline, diesel, natural gas, and coal all release carbon dioxide. I ask students, "Is the carbon dioxide released in these products 'waste' or is it recyclable?"

In the current system, the carbon dioxide is waste that enters the atmosphere, making the system carbon-positive. However, students might suggest a number of alternative scenarios: growing more plants, increasing plant yield per acre, or growing perennial crops that store organic biomass underground. Students often realize that there are also ways to reduce fossil fuel use—by increasing vehicle efficiency or reducing driving distances from the transport stations.

At this point, I ask students to consider other opportunities to use alternative energy sources along the way. For example, cellulosic ethanol refineries could burn plant biomass to heat the fermenters and generate electricity. These inputs are carbon-neutral. Electricity used at a refinery could also be produced using wind or solar power instead of coal. This final discussion can be quite rich and synthesizes a good understanding of the system involved.

SUMMATIVE ASSESSMENT

After the class makes it through the entire cycle, I use two forms of assessment to make sure the learning outcomes were achieved. First, I present students with a handout that links blank process tools from all of the stations together and ask them to individually trace energy and matter through the system (see "On the web" for these handouts). I also have students write individual essays on which fuel is "better," using the information gained in this activity (they can also use their own research to support their arguments). In the essay, I ask students to address the following questions:

- What considerations must be included for a fair comparison of gasoline and ethanol?
- What processes are included in our system boundaries?
- Where does the greatest variability occur in the life cycles of ethanol and gasoline and where can engineering improvements be made?
- What other economic and social considerations, such as those of the farmer or world food supplies (if crops are used to produce ethanol instead of food), should be evaluated?

A fair comparison must include a similar LCA of gasoline production from drilling to shipping to refining. Students might also discuss other costs and risks of gasoline production, including oil spills, military protection of oil-rich countries, and delivery pipelines. For those living in a heavily irrigated area, an LCA of water can be added.

CONCLUSION

There is no perfect transportation fuel. The purpose of an LCA is to provide us with a way to investigate a complex production system from beginning to end—allowing for a more comprehensive comparison of our options. This activity takes students from the basic science of describing energy and matter transformations to a place where they can think about possible engineering

Suggested activity extensions (See "On the web" for more information on these activities.)

- Run a fermentation and measure carbon dioxide output or gas pressure under different conditions.
- Measure carbon dioxide released from soils on farm fields under different management practices.
- Conduct a quantitative life-cycle assessment using a Microsoft Excel model.

solutions to produce a more sustainable fuel. A logical next step after completing these materials would be to investigate the results of LCA research studies on biofuels. Now that students understand the basic system, they can do a quantitative comparison of the options they are considering.

In general, once students are introduced to this activity, they begin questioning the life cycle of other products, helping them become more informed, critical consumers and stewards of our natural resources.

Acknowledgments

Thank you to Jonathon Schramm (Michigan State University, Division of Science and Mathematics Education) for his help editing this chapter. Thank you also to John Greenler and Kristin Wratney (Great Lakes Bioenergy Research Center, University of Wisconsin–Madison) for their considerable assistance in developing this activity. This work was funded by the DOE Great Lakes Bioenergy Research Center (DOE BER Office of Science DE-FC02-07ER64494).

On the web

Educational materials (Great Lakes Bioenergy Research Center): *www.glbrc.org/education/educationalmaterials*

Process tool summative assessment handout: *www.nsta.org/highschool/connections/201012ProcessToolAssessment.pdf*

References

Environmental Protection Agency (EPA). 2010. Renewable fuels: Regulations and standards. Renewable Fuel Standard (RFS2): Program amendments. *www.epa.gov/otaq/fuels/renewablefuels/regulations.htm.*

Koba, S., and A. Tweed. 2009. *Hard-to-teach biology concepts: A framework to deepen student understanding.* Arlington, VA: NSTA Press.

Mohan, L., J. Chen, and C. W. Anderson. 2009. Developing a multi-year learning progression for carbon cycling in socio-ecological systems. *Journal of Research in Science Teaching* 46 (6): 675–698.

U.S. Department of Energy (U.S. DOE). 2006. *Breaking the biological barriers to cellulosic ethanol: A joint research agenda,* DOE/SC-0095, U.S. Department of Energy Office of Science and Office of Energy Efficiency and Renewable Energy. *www.doegenomestolife.org/biofuels.*

Chapter 20

FALL COLORS, TEMPERATURE, AND DAY LENGTH

STUDENTS USE INTERNET DATA TO EXPLORE THE RELATIONSHIP BETWEEN SEASONAL PATTERNS AND CLIMATE

By Stephen Burton, Heather Miller, and Carrie Roossinck

Along with the bright hues of orange, red, and yellow, the season of fall represents significant changes, such as day length and temperature. These changes provide excellent opportunities for students to use science process skills to examine how abiotic factors such as weather and temperature impact organisms. The activity described in this chapter uses available internet data to encourage students to explore the relationship between seasonal patterns and climate (plant phenology) as well as day-length data. In particular, students develop and test hypotheses to explain latitudinal differences in leaf color change.

IDENTIFYING PATTERNS

To engage students, we begin the lesson by asking them to examine the normal peak times of fall color across North America using a map (Figure 1) reproduced from The Weather Channel's website (see "On the web"). As a class, students make observations and describe patterns. Students easily notice that colors in the Midwest follow a latitudinal gradient. Students also observe that the latitudinal gradient does not always seem to hold true over mountainous regions, and entire regions of the United States show no evidence of color change.

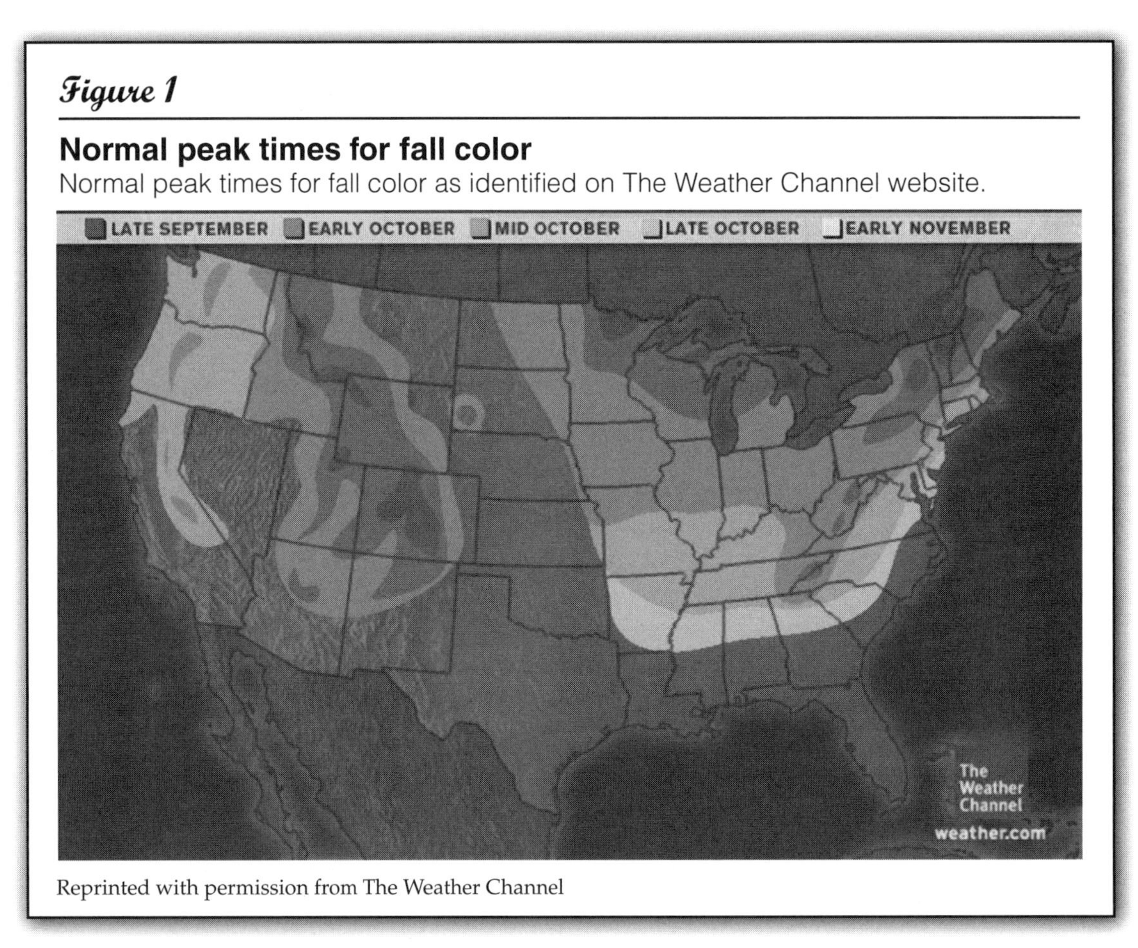

Figure 1

Normal peak times for fall color

Normal peak times for fall color as identified on The Weather Channel website.

Reprinted with permission from The Weather Channel

Students are then asked to suggest possible explanations for the observed patterns. In particular, we ask students to focus on the latitudinal variation in peak color. Students often offer temperature differences as the potential explanation for earlier peak color in northern areas. This lesson occurs well into the semester when students are already familiar with scientific inquiry, which allows us to reinforce the idea that their explanations are potential

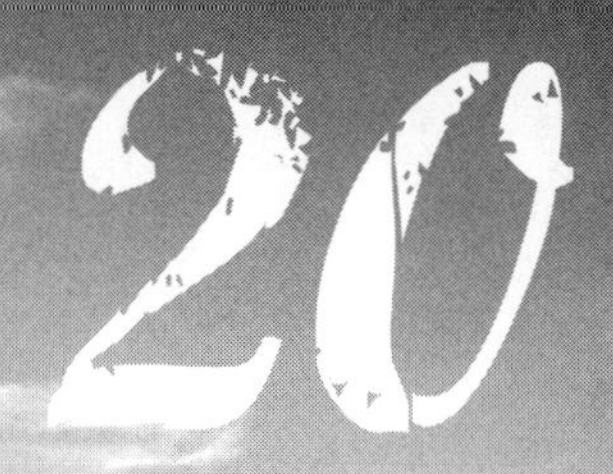

hypotheses that can and should be tested. Students are keenly aware, from previous discussions, that scientists are never satisfied with just generating hypotheses. Students understand that a possible explanation requires evidence to determine its validity. Therefore, we challenge students (usually working in groups of three to four) to design experiments to provide evidence to refute or support their hypotheses.

TESTING THE TEMPERATURE EXPLANATION

As students begin to discuss possible methods for testing suggested explanations (hypotheses), it becomes clear that the first step is to determine if latitudinal differences result in temperature differences. Students are encouraged to design experiments in which they select single cities at different latitudes in each of the peak color time regions and investigate temperature differences among them. For example, students often predict that cities at higher latitudes should be cooler.

Once the design has been established, we present students with data collected via the internet from three cities (shown as stars on Figure 1) at different latitudes. The data results support the hypothesis that temperature differences may explain the differences in peak times (Figure 2). However, we help students examine the issue of small sample size and a fair test and discuss the need for more data points. To address this shortfall, we ask each group to collect data on one to three additional cities within the region being investigated. With each group collecting data on a few cities at different peak times, it is possible to amass a data set large enough to evaluate the results. Of course, students are interested in finding out how we are going to collect this type of weather data on such short notice.

Current and past weather data for most U.S. locations are now easily accessible through the internet. One of our favorite websites that contains appropriate data is Weather Underground (see "On the web"). This website provides a vast array of weather-related data that can be used to answer a variety of questions connecting weather and biological phenomena. The weather data is reported from airports within the region. Selecting a particular city identified on a map that does not have an airport defaults to the nearest airport automatically. A key feature in collecting data is the identification of cities around the same latitude. Because cities along the edge of the peak times may overlap in temperature, we encourage students to find cities at the middle of the peak time band shown in Figure 1. A great way to help students select locations is to provide an atlas of North America showing major cities. (**Note:** Students can also use Google Earth to locate cities and towns, and obtain data for elevation, latitude, and longitude [see "On the web"].) To minimize problems of altitudinal variation, a teacher could also provide a list of cities at different latitudes with similar altitudes and allow students to select from the list. To prevent overlap, the teacher can list the cities selected on the board.

As students begin the data collection process, we encourage them to focus on the mean temperature of each location for the peak weeks identified on the original map (Figure 1). This is a great opportunity to reinforce scientific inquiry skills required to appropriately evaluate hypotheses. In particular, we ask students to determine if using the temperature from one year as a comparison is a fair test. Students often realize that averaging across multiple years is a better test as it potentially takes into consideration variability that might occur across multiple years. Fortunately, the Weather Underground website provides access to archived weather data allowing students to collect data from multiple years.

As students examine the data, they find that their results corroborate the original results showing that cities at higher latitudes (and thus earlier peak color times) are cooler during the fall (Figure 2). Students are often satisfied

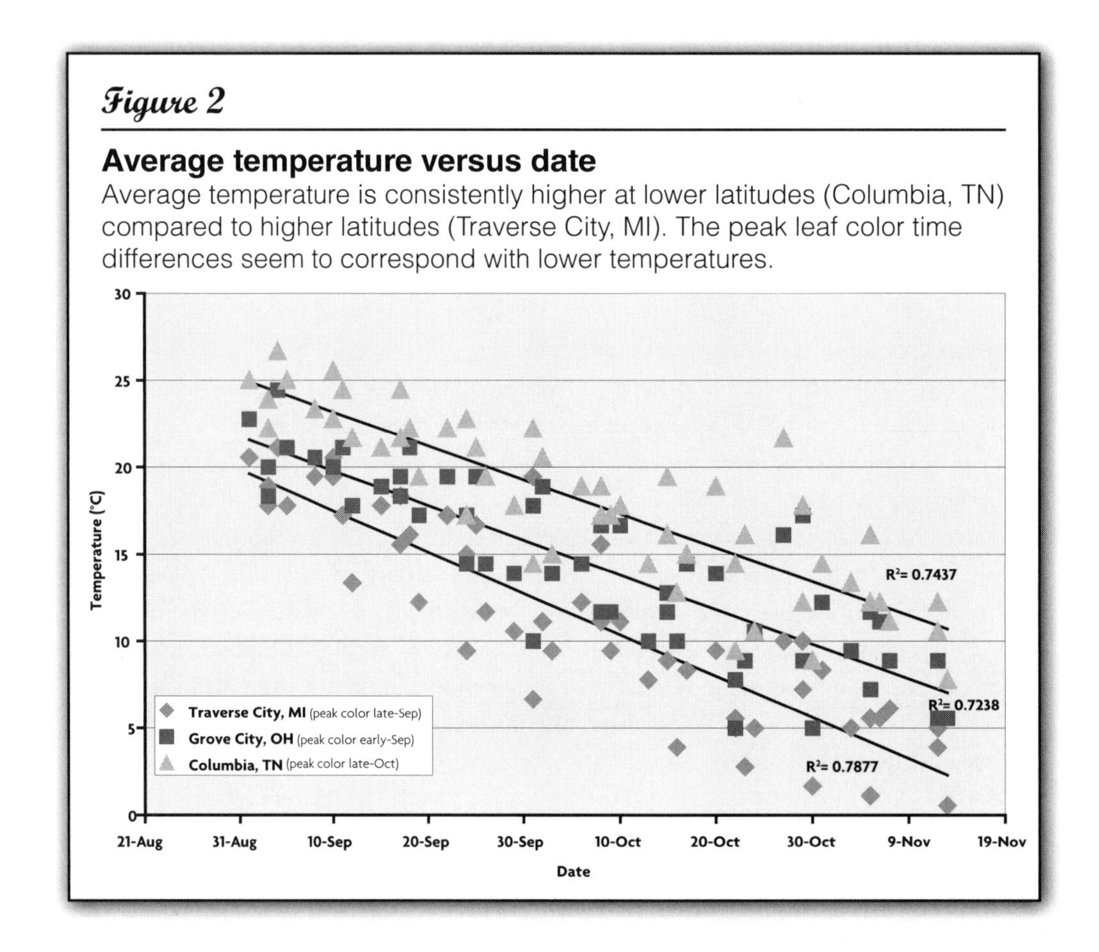

Figure 2

Average temperature versus date

Average temperature is consistently higher at lower latitudes (Columbia, TN) compared to higher latitudes (Traverse City, MI). The peak leaf color time differences seem to correspond with lower temperatures.

with the conclusion that temperature plays a significant role in leaf color change. We encourage students to consider other correlations and alternative hypotheses to determine if there are other possible explanations.

CONSIDERING OTHER VARIABLES

To reinforce the tentative nature of science, we give students a graph showing that day length (more correctly, length of darkness) is also greater in a city with an earlier peak color zone (Figure 3). This would suggest that leaf color change may be influenced by length of darkness in a day as well. Students are challenged to consider what variable—temperature or daylight hours—would explain the date of peak color.

Figure 3

Average darkness versus date

Daily darkness is consistently greater at higher latitudes (Traverse City, MI) compared to lower latitudes (Columbia, TN). The peak leaf color time differences seem to correspond with more darkness.

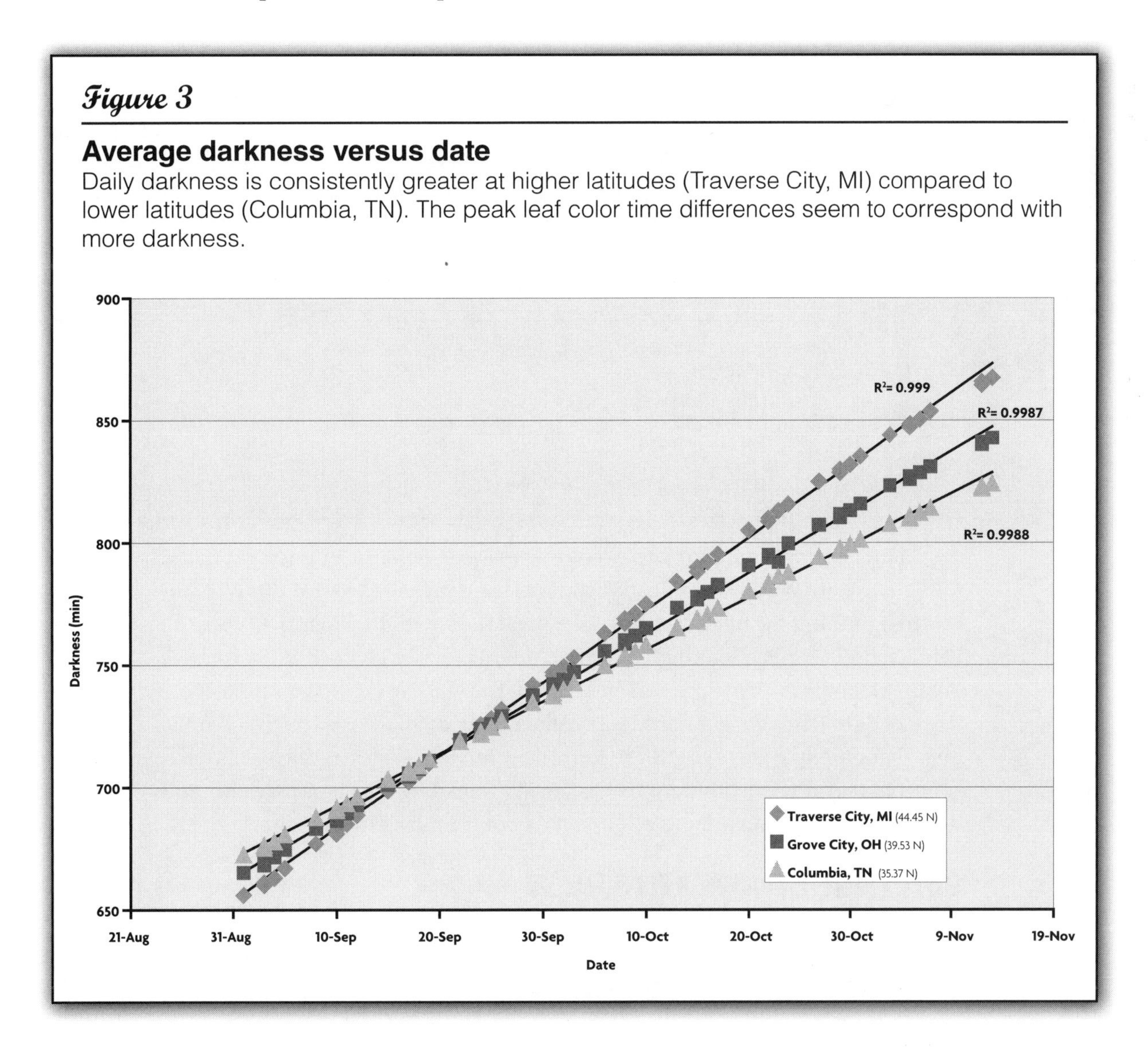

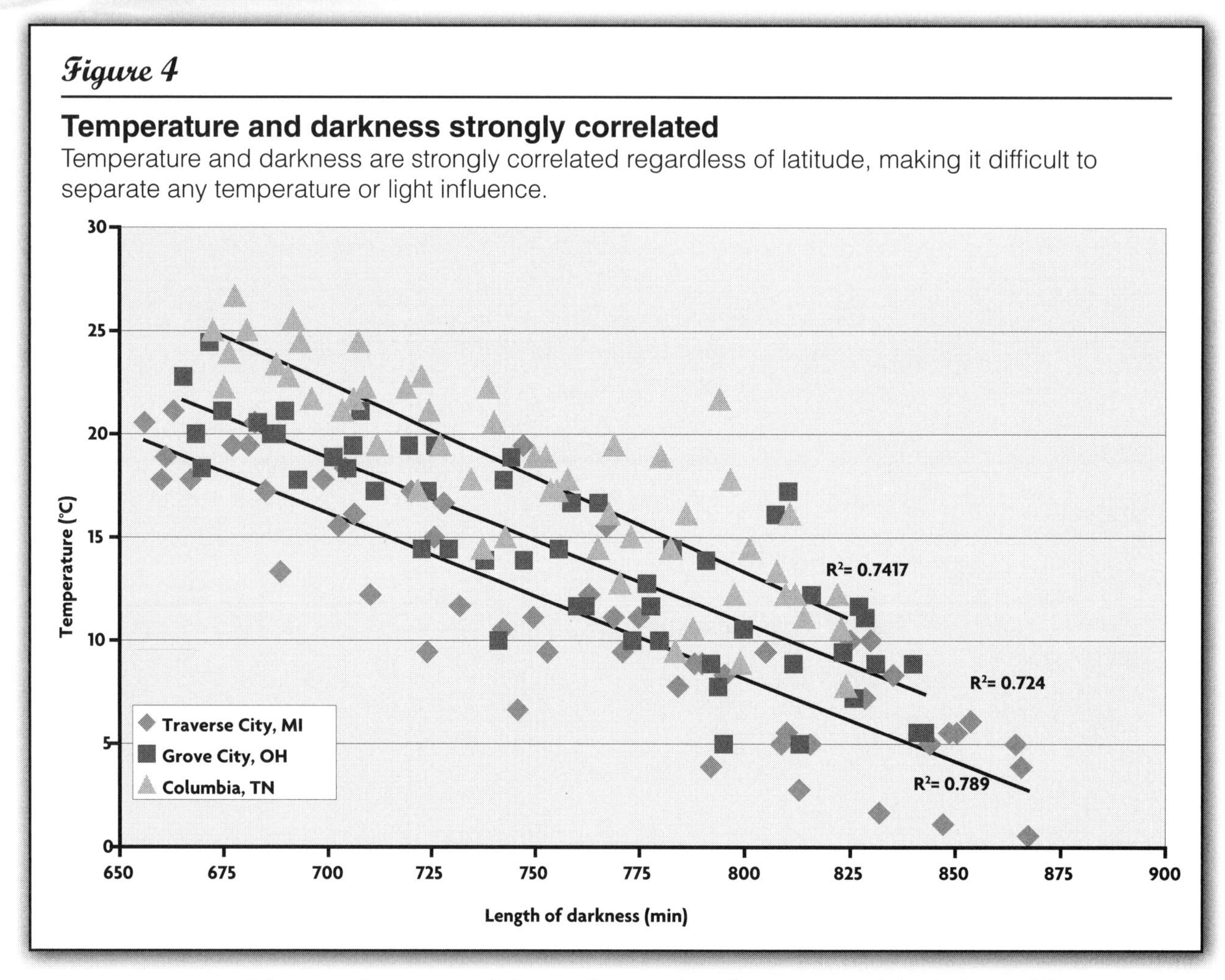

Figure 4

Temperature and darkness strongly correlated

Temperature and darkness are strongly correlated regardless of latitude, making it difficult to separate any temperature or light influence.

Many students are quick to point out that the amount of daylight could potentially influence temperature. Students are then asked to test this hypothesis by graphing the temperature of the peak time with day length (Figure 4). Day length is available at the Weather Underground website but also can be obtained for almost any location and year at the U.S. Naval Observatory website (see "On the web"). As students collect their data, we try to reinforce the idea that date and day length are highly correlated. Further, our calendar is set around changing day lengths so correlations between temperature and day length are not surprising but certainly confounding for our study.

CONTROLLING VARIABLES

How do we eliminate day length and focus on temperature? Controlling for day length is a challenge for students. We encourage students to go back to the original map (Figure 1, p. 162) and identify areas with different peak times

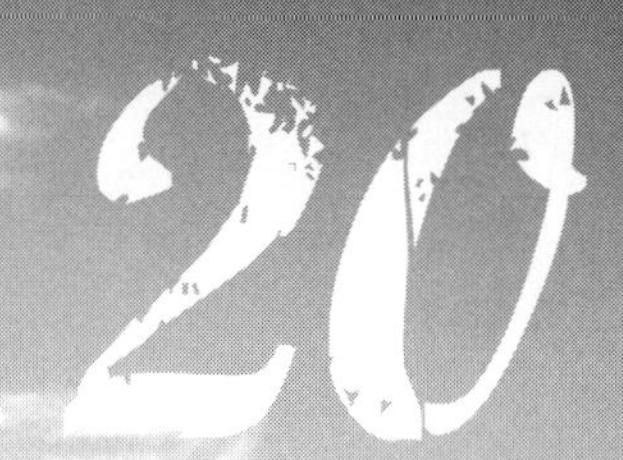

Figure 5

Map of Virginia showing peak leaf color dates

Sites selected for comparison were at the same latitude, but varied only by elevation. Higher elevations (Abingdon, VA) appear to reach peak color before lower elevations (Emporia, VA). Student-gathered data is shown below the map.

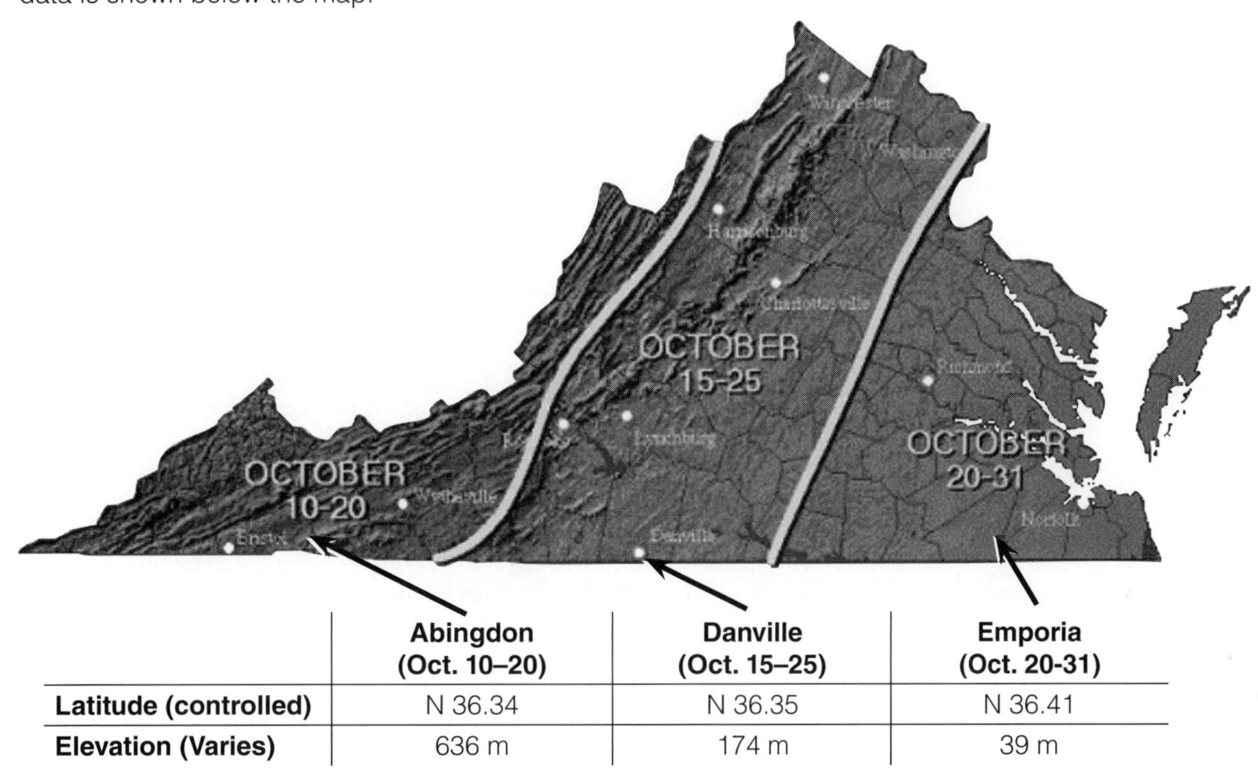

	Abingdon (Oct. 10–20)	Danville (Oct. 15–25)	Emporia (Oct. 20-31)
Latitude (controlled)	N 36.34	N 36.35	N 36.41
Elevation (Varies)	636 m	174 m	39 m

but where it is unlikely day length would be different. For instance, we have students take a closer look at Virginia using a figure from the University of Virginia Climatology Office (Figure 5), which shows that the peak times follow an east-to-west pattern where there is no difference in daylight hours (2006).

After students have identified regions showing longitudinal differences, we show students Figure 6 which provides evidence that suggests a relationship between peak time and temperature. Again, we encourage students to collect more locations to increase sample size. Not surprisingly, the results are consistent with the data in Figure 6.

While this activity certainly implies that temperature is an important variable when considering peak time, we provide a classic research paper by E.B. Matzke (1936) to discuss the fact that multiple factors (variables), outside of day length and temperature, may influence an organism. The paper describes a study that demonstrates trees close to street lights were shown to have maintained their leaves longer into the fall compared to trees farther from street lights.

EXPLORING OTHER PATTERNS

At the conclusion of the activity, we give students the opportunity to look for other variables that might explain leaf color change. We also encourage students to examine other interesting patterns—such as the start of, peak, length of, and intensity of fall colors—to look for possible explanations. To do this, students consider what data is available from the websites and identify potential tests that would have the capability of falsifying their hypotheses. With the availability of online data, most students can design and test these hypotheses in a few days. For instance, we have had students explore factors such as average low temperature, degree growing days (accumulation of heat over a particular biologically important threshold, e.g., 25°C), and precipitation.

Students have found it difficult to identify any major relationships with peak time for leaf color and other weather variables. This is not surprising considering that day length (again, more correctly length of darkness) and temperature are the two driving factors influencing the start and progression of leaf color change. Once students have completed their tests, we often emphasize that negative results showing no correlation are just as important as positive results in helping scientists gain knowledge.

Fall colors are such an attraction that tourism organizations in many states provide updates during the fall to describe how the colors are progressing (e.g., the Virginia map). This information may be in formats such as written descriptions or detailed maps. We prefer the Foliage Network (see "On the web") that contains foliage reports for the past five years for a variety of states because it provides reports in map form for much of the eastern United States. Further, the resource provides frequent updates during the peak season (i.e., more data collection points than most resources). Frequent reporting times allow students the option to examine leaf color change across broad regions as it occurs. We caution students that these foliage reports are probably not completely accurate because the website is focused on reporting for tourism not science. Therefore, when examining their hypotheses, we encourage students to account for error possibly present in the data acquired from this website.

The relevancy of leaf color change in some areas can be very beneficial in getting students engaged in inquiry and exploring relationships among organisms and abiotic environmental variables. However, this activity can also be used as a way to encourage students to think about factors that

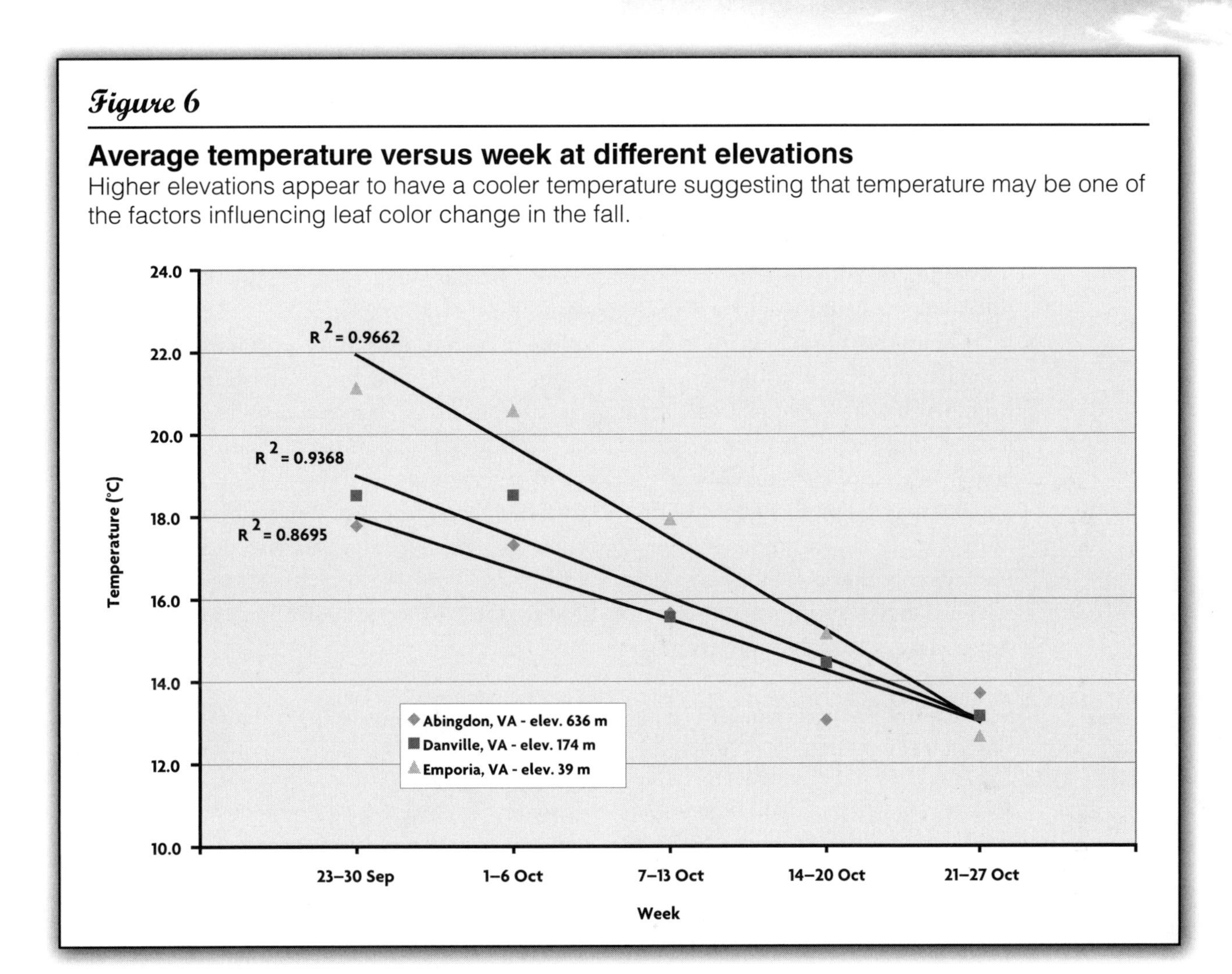

Figure 6

Average temperature versus week at different elevations

Higher elevations appear to have a cooler temperature suggesting that temperature may be one of the factors influencing leaf color change in the fall.

influence climate and weather in different areas around the world. An obvious extension of this lesson would be for students to look at factors that influence climate latitudinally, including the tilt of Earth and light from the Sun. Also, students can see from the map of Virginia that the east-west pattern appears to be related to elevation (Figures 5 and 6). Therefore, students can use the data from the internet to explore the relationship between elevation and temperature and the relationship between adiabatic cooling and rainfall (which can be used to explain the rain shadow on the east side of the Rocky Mountains).

For an ecology unit, one of the more powerful next steps is the examination of biome distributions. For instance, when looking at the peak time map from The Weather Channel (Figure 1), students often observe the lack of color in some regions of the United States. This observation lends itself to an exploration of biomes—students can look at climate in those regions by collecting the average temperature and rainfall data for several years using the internet and then creating their own climate graphs. The biome exploration provides an opportunity to discuss natural selection caused by abiotic factors such as temperature and precipitation. The study of biomes becomes more interesting to students when they design their own climate graphs to explain differences in certain regions and understand these patterns.

Using the inquiry approach to explore relevant questions provides a greater opportunity for students to become engaged and build a lasting framework of knowledge. Unfortunately, if students do not see results within a short timeframe, inquiry activities can lose appeal and impact. Inquiry activities that allow students to collect and interpret some types of data quickly may be more successful in bridging the gap. The activity described in this article allows students to see that some questions cannot be answered immediately without long-term data. At the same time, the activity allows students to ask questions and obtain long-term data quickly to assist them in developing and evaluating hypotheses.

On the web

Google Earth: *http://earth.google.com*

The Foliage Network: *www.foliagenetwork.com*

The Weather Channel (fall foliage maps): *www.weather.com/maps/fallfoliage.html*

Duration of daylight/darkness table for one year: *http://aa.usno.navy.mil/data/docs/Dur_OneYear.php*

Weather Underground: *www.wunderground.com*

References

Matzke, E. B. 1936. The effect of street lights in delaying leaf-fall in certain trees. *American Journal of Botany.* 23 (6): 446–452.

University of Virginia Climatology Office. 2006. Typical fall foliage peak color periods for Virginia. *http://climate.virginia.edu/foliage.htm*

Chapter 21

A USEFUL LABORATORY TOOL

STUDENTS BUILD AND TEST A THERMAL GRADIENT TO CONDUCT MEANINGFUL LABS

By Samuel A. Johnson and Tye Tutt

Many phenomena in nature are controlled by temperature. It is an important aspect of most species' niches, or their relational positions in their ecosystems. Living organisms' adaptations to their niches depend to some degree on their ability to carry out life processes at the temperature ranges in which they occur, a fact that is particularly relevant in a world of global climate change. It is no accident then that the biological communities in the Arctic and in the tropics have little overlap. But are the differences in these communities a function of temperature more than of other variables in the environment? Many questions such as this one—focused on temperature and living organisms—came up in our Science Club in 2008.

The club, which often consists of six to eight ambitious high school students each year, meets to ponder scientific conundrums and discuss investigations that we might undertake to satisfy our curiosity. The club is encouraged to think in utilitarian terms and find ways to produce high-visibility projects that can be used in biology and chemistry classes to both inspire interest and encourage inquiry-based approaches to lab investigations.

In 2008, the club generated a large number of questions involving temperature. Therefore, we decided to construct a thermal gradient apparatus in order to conduct a wide range of experiments beyond the standard "cookbook" labs. We felt that this apparatus could be especially useful in future ninth-grade biology classes, in which students must design and conduct individual, inquiry-based experiments as part of their training in scientific methodology.

This chapter describes our experience building and testing a thermal gradient for laboratory use. Simple in design, relatively cheap, and highly effective, the thermal gradient we constructed can be built with readily available materials and appliances.

HISTORY OF THERMAL GRADIENTS

Thermal gradients have been used in scientific research for many years to study the effects of temperature variation on living organisms. A thermal gradient is a tool that produces a range of temperatures that gradually change from cold to warm.

More than 40 years ago, a thermal gradient (0–40°C) made with aluminum stock was fitted with digital output devices that allowed automatic monitoring of the distribution and movement of small animals (Morrison and Warman 1967). Pentecost (1974) used a metal-floored box variably heated with incandescent lightbulbs to produce a thermal gradient (17–47°C) on which the behavior of lizards could be monitored. In 2003, Yamada and Ohshima used an aluminum slab with warm- and cold-water circulating pipes to produce a thermal gradient (8.5–29°C) on which they could study the movement of worms. More recent still, highly sophisticated and expensive chambers using Peltier devices and platinum sensors have been made to create highly controlled and truly linear gradients (although these only span 1–24°C) (Corbett Life Science 2008). Peltier devices are expensive semiconductor-based thermoelectric modules that respond to a low-voltage direct current and produce a heating or cooling effect.

Our Science Club wanted to create a reliable, linear thermal gradient as cheaply as possible with a temperature range that spanned most living environments (5–40°C) for students to use in conducting research. At the time, no such apparatus appeared to be on the market.

Figure 1

The cooling system

A copper coil in a plastic tub in the refrigerator compartment carries antifreeze through an ice (or dry ice) bath. Then it is pumped by the small, white RV-sink pump (outside of refrigerator) into a second, S-looped coil, which draws heat out of the specimen tray. The pump is powered by a variable-voltage DC adapter from the 120 V wall circuit. The speed of the pump can be modified by changing the voltage. A small drain tube carries melt water out of the refrigerator when ice is added.

CONSTRUCTION OF THE APPARATUS

Our temperature gradient was designed following a suggestion from ecologist Peter Marchand, a specialist in winter ecology who wrote a monthly column in *Natural History* for many years. The materials used to build the apparatus were relatively inexpensive. We constructed our gradient by suspending a slab of aluminum 6061 alloy—commercially listed as 0.5 in. × 8 in. × 8 ft. (1.3 cm × 20 cm × 2.4 m)—between the freezer box of a small refrigerator on one end and a small, single-coil hotplate on the other.

Aluminum is not inexpensive, but it is within the budget of most schools. We purchased our aluminum at a metal warehouse for $125. The small refrigerator and hot plate were salvaged from garage sales. Parents' associations are usually a good source for such materials, but be sure that donated appliances are in good (safe) working order before using them with students. Including a few additional materials—such as ethylene glycol, elastic bands, c-clamps, and small beakers for water—the total cost of constructing the thermal gradient was about $155.

We needed to cut a slot into the metal shell of the refrigerator's freezer compartment to accommodate the entrance of the aluminum plate. To accomplish this, we (the teachers) used a metal cutting blade on a 6 in. circular handsaw, while students, wearing safety goggles, stood well away from the action. A Dremel tool, which both teachers and students could use with proper protection of hands and eyes, easily cut the short sides of the rectangular slot. The interior of the freezer compartment was plastic and was also cut through as the fiberglass insulation was pushed out of the way. Students wore safety gloves and goggles to manipulate the fiberglass after the cutting was completed. (**Safety note:** Cutting metals can be dangerous, so only teachers trained in the use of power cutting tools should attempt this. Oftentimes, an industrial arts teacher can be called upon for assistance if necessary.)

The hot plate was raised off the tabletop to achieve a level surface using bricks (which are inflammable). A specimen tray for holding soil and subjects was made from a 4 × 2.5 in. (10.2 × 6.4 cm) galvanized steel U-channel. This material can be purchased at any lumberyard or hardware store that supplies construction crews with materials. Using metal shears, we cut the open ends of the U-channel, with advice from an art teacher familiar with these materials. The cut ends were folded and riveted to make neat box ends.

Across the 2.4 m span of the aluminum slab, we found an approximate 1 cm bow, due to the mass of the metal plate. To correct this and to create good contact between the floor of the specimen tray and the aluminum slab, we installed a typical photographer's tripod—with a polystyrene cap to prevent the tripod from acting as a significant heat sink (which would absorb heat from the apparatus)—that could be easily cranked up to help level the surface.

The aluminum slab was inserted into the freezer compartment slot, then gaps were filled in with expanding spray foam insulator material used for sealing cracks in houses. Bags of ice were placed above and below the plate to maintain 0°C (or lower) temperatures in the freezer.

Although some may find it counterintuitive, the freezer does not cool the cold end of the plate below about 12°C. This is because the freezer acts as a heat sink, drawing heat continuously down the plate, which tends to make the freezer run more or less continuously as long as an experiment lasts.

Figure 2

Temperature tolerance limits for germinating seeds

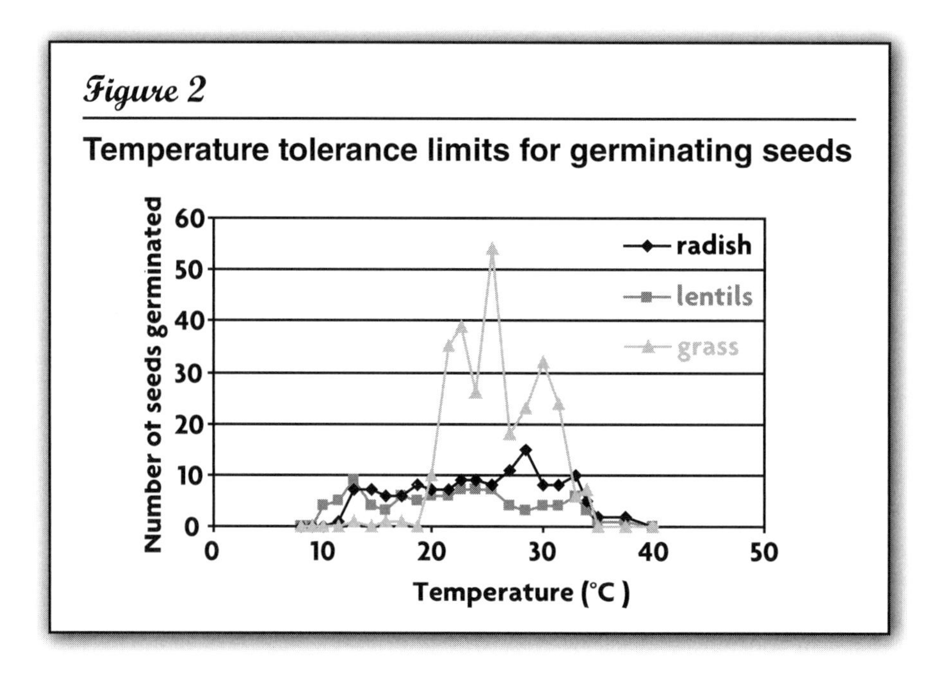

MAKING IT WORK

At this point, our device created a reliable thermal gradient that ran from 12–40°C, providing interesting experimental possibilities. For many uses, this range may prove adequate, and such a device can be built in a single day once the materials are procured.

Our club was not content with a minimum of 12°C, however, because we live in a climate in which daily temperatures are often below that mark. Therefore, we created a means to further cool the plate and thus widen the gradient. A system of two heat-transfer coils was added to the cold end—one in the refrigerator and the other in the specimen tray. The system—described in the following paragragh—was based on a relatively simple concept: antifreeze was chilled in an ice (or dry ice) bath and then circulated under the cold end of the specimen tray.

Figure 3

Temperature preferences of red worms and earthworms

The response of red worms and earthworms to the temperature gradient shows obvious preferences.

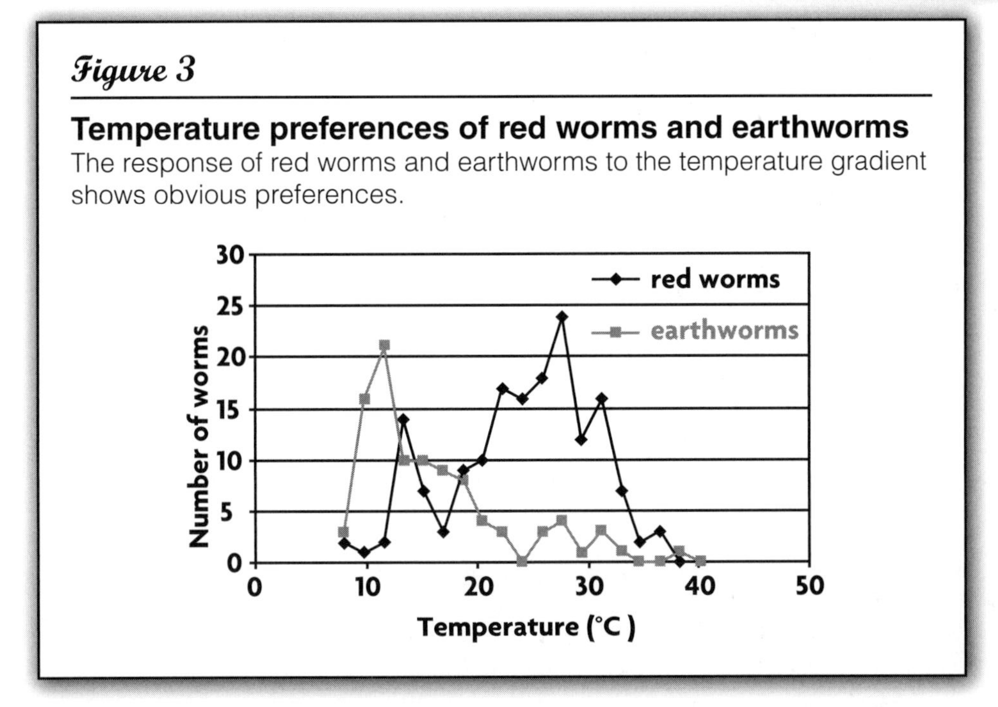

This cooling system was fashioned from two 0.25 in. (0.625 cm) copper coils connected by flexible polyethylene tubes that ran through a small pump. The primary coil was wrapped into a series of S-shaped loops, with two centrally located supply and return ports. This coil was then placed on the bottom of the specimen tray beneath the soil (or whatever medium was used to house specimens) (Figure 1, p. 172). The secondary coil was located in a dry ice container or an ice water bath inside the refrigerator. The primary and secondary coils were tied directly together with rubber tubing and a small 12 V RV-sink pump was then installed and connected between them.

The entire system was charged with a premixed (50% water, 50% ethylene glycol) antifreeze solution. (**Safety note:** Ethylene glycol is a toxic compound; although the major danger is through ingestion, it also can pass through the skin, so students were supervised and required to wear chemical-splash goggles and gloves when examining or charging the coils.) Using this system with an ice bath, a cold-end constant of about 8°C was achieved. Using dry ice on the freezer coil produced a constant cold-end temperature of about 5°C, but it was expensive to maintain this for long periods of time.

A tee joint was placed approximately 10 cm from the return on the primary coil, at a high point, so that antifreeze could be easily injected into the system with a syringe. Bubbles in the system automatically collected at this high point and removed themselves from the coil. We found that if left unused for several weeks, the pump became sticky with antifreeze and had trouble starting. However, thumping it sharply a couple of times always restarted it.

Using the ice bath, coils, and a series of computer-based laboratory temperature probes, a linear equation was experimentally derived to relate the temperature of the aluminum plate in degrees Celsius to the distance *(d)* in centimeters away from the hot end: $°C = -0.183d + 41$ ($r = 0.98$). (**Note:** $r = 0.98$ reflects a good linear relationship.) The fact that the gradient was linear was a matter of academic interest, but was not directly useful, as temperature measures during experimental tests were taken directly from the media, except in the case of the worm test (see "Experiment 2"). In that case, temperature measures were taken in the middle of each station. Variations of this equation were found using dry ice, and without the cooling coils. The largest deviation from the linear equation was at the hot end of the gradient. Very similar equations were found for damp soil and water in beakers.

TESTING THE EQUIPMENT

Our club was less invested in any specific single experiment than in demonstrating the applicability of such a device to a broad range of experiments. Students came up with several lines of inquiry, and invented three trials to test the usefulness of the apparatus, beginning with an emphasis on niche partitioning and temperature tolerance ranges of widely different species. Niche partitioning is an ecological concept whereby competing species adapt by occupying different niches (in this case, areas of different temperatures).

Experiment 1

The first demonstration involved germinating seeds. We used lentils, radishes, and lawn grass, all three of which germinate quickly and produce distinctive seedlings. We wondered if these seeds would be forced to compete, or would avoid competition by partitioning the niche.

The specimen tray was placed in the center of the aluminum plate, lined with plastic to prevent electrochemical reactions between the soil and the galvanized metal, and filled with potting soil to approximately three-quarters of its depth. The radish, lentil, and grass seeds were then sown along the back edge of the tray, evenly over the entire length of the gradient, and covered with approximately 0.4 in. of soil. The seeds were then watered continuously using cotton water wicks in containers placed beside the tray at intervals. The wicks were threaded through plastic tubes, which had been heated and bent to retain their shape. In this way, water of approximately the same temperature as the soil was constantly being applied to the treatment. We did not designate a control, although plants germinated nearby at room temperature could have filled such a need.

The gradient proved effective for all three plants in terms of limits, exceeding plant tolerances on both ends. In addition, a rough temperature optimum was determined for grass (about 25°C) (Figure 2, p. 174). Lentils and radishes appeared to occupy a wider niche, especially on the cold end of the gradient. They might be considered more generalist in this respect than our grass species.

These results suggested that growth rates of plants, in terms of height or biomass, might be measured along the gradient. One might also test plant susceptibility to herbivore damage or other selective pressures at a range of temperatures. For example, potted plants in small containers, all equally infested with aphids or scale, might be placed on the gradient to see if temperature affects the viability of the attack. Weakened plants are often cited as being vulnerable to herbivory, so one could also test whether or not stressing a plant by extreme temperatures makes it more vulnerable.

Experiment 2

The second experiment tested worms' response to temperature. We added 50 red worms and 50 earthworms, scattered evenly across the entire gradient, to the tray filled with commercial worm medium. We chose these two species because we knew that common earthworms are active in very cool soil, while, according to a gardener friend, red worms tend to occur in compost that retains a good deal of heat. We hypothesized that these species would segregate on the gradient.

The worms were allowed to move through the medium for 72 hours, after which the tray was inverted on a bench that had a calibrated liner. Worms were sorted into 10 cm stations, each correlating with a temperature along the gradient, and the numbers of each worm species were recorded and graphed. The experiment was then repeated to verify the findings.

The result (Figure 3, p. 175) provided a nice demonstration of competition avoidance by niche partitioning, as dictated by the competitive exclusion principle, and also provided evidence supporting the hypothesis of a higher temperature preference for red worms. Students suggested further experiments using isopods, centipedes, or millipedes as subjects.

Experiment 3

The third demonstration used algae from the *Spirogyra* genus in pond water. This algae was chosen because students noticed that our school pond becomes green with algae early in the spring when water temperatures are still very low. They wondered if cold water was a necessity or was merely tolerated by this species.

To investigate, 12 identical 100 ml beakers were placed at equal distances along the plate. A well-stirred pond water sample was poured equally into each, and to this we added approximately equal samples of *Spirogyra*. The beakers were loosely capped with plastic cups to retard evaporative water loss. The treatments were allowed to stand for two weeks. Every day or two they were examined and photographed. In this way, a temperature tolerance was determined for this species.

Our *Spirogyra* species survived well from 8 to 25°C, but visibly suffered at higher temperatures. Nothing survived more than a few days above 32°C. Although these observations produced only qualitative data, we then stirred the beakers after two weeks and ran samples through a spectrometer set at 350 nm. The trend was strongly suggestive ($r = 0.89$) that the brownish-yellow dead cells of the algae (or growing cultures of decomposers) could be used as a quantitative measure of survival if the water was strained at the beginning of the experiment. A scanning spectrometer might identify an optimum wavelength for absorption by dead cells. If qualitative data are sufficient, competing algae species could probably be grown this way to determine if they partition the niche thermally, as the worms do. As an extension, the pond might be sampled to see if those algae species are distributed predictably along the thermocline—a distinct layer in a body of water in which temperature changes rapidly with depth—in the summer or winter.

ENDLESS POSSIBILITIES

The possible uses of the thermal gradient as a lab tool are nearly endless. Cultures of safe strains of commercial bacteria, streaked together on petri dishes, might be set on the gradient (or parts of it) to see if competitive advantages shift at different temperatures. *Ceratopteris* ferns (C-ferns), often grown in AP biology labs, might be cultured on the gradient to see if germination success, growth rates, or sex-determination might be a function of temperature.

The 10.2 cm specimen tray perfectly accommodates petri dishes, although sweating and drying of agar at higher temperatures would have to be controlled. Aquatic species such as salamanders, tadpoles, small fish, or invertebrates could be tested similarly. In such a case, the whole specimen tray might be filled with water, with small, partial baffles—devices, such as small plates or blocks, that regulate flow or passage—at intervals to discourage thermal mixing of water, while still allowing movement of animals over the baffles to find cooler or warmer environments. Such baffles could be made about 3.5–4 cm tall, using 5 mm thick Plexiglas, cut to fit tightly within the specimen tray. Or, transpiration rates in plants could be studied in insulated chambers along the gradient.

Chemistry teachers have suggested running rate labs, solubility labs, or equilibrium reactions along the aluminum plate. Any reaction that is temperature-sensitive might be run on this sort of gradient, although extremely wide ranges of temperatures might be difficult to achieve. Turning the hot plate to a higher setting raises the whole gradient.

We tried a couple of reactions to test the effect of temperature, and although we did not take quantitative measurements, we found interesting results. For example, the reaction of limestone in hydrochloric acid will increase with temperature, and the bubbles formed by carbon-dioxide generation can be collected in a graduated cylinder as a rate measure. Our bubbles were much larger on the warmer end of the gradient than on the refrigerated end. This is much more informative than running the traditional three treatments—boiling water, ice, and room temperature baths. It would allow much finer resolution of the curve on the graph of reaction rate versus temperature. Students could then suggest additional experiments to test whether the larger bubbles were caused by an increased reaction rate, or by the expansion or decreased solubility of the gas at the higher temperature.

Teachers can also develop other experiments using the thermal gradient to address such questions as:

- Do communities on the sunny and shady sides of large trees vary because of temperature differences?
- To what degree are flowering times of plants, which are generally governed by photoperiod, modified by temperature?
- What might the threshold temperatures of invertebrate activity be?
- Might adaptation or acclimatization result from indirect responses to physical or chemical properties of materials in the environment? For example, does the heat capacity of Styrofoam, wood, or rock determine the community of organisms that live under such an object in a sunny field?
- To what degree might species expand or modify their thermal niches over time (Bennett and Lenski 1993)?

CONCLUSION

Students who helped design and construct the thermal gradient apparatus were visibly proud of its utility. We have found that designing even basic equipment can be very rewarding for students, especially if they are interested in engineering. The simplicity of dumping a bag of ice and turning on a single switch (on the power strip) to activate the gradient, which equilibrates within about two hours, has proven a pleasure well worth the investment of time and money.

Since the gradient's installation, grow lights have been added to accommodate longer-term plant growth experiments and to provide light-and-shadow effects for invertebrates. With climate change so often in the news, investigations of how organisms respond and adapt to changing temperature are particularly relevant and interesting.

References

Bennett, A. F., and R. E. Lenski. 1993. Evolutionary adaptation to temperature II: Thermal niches of experimental lines of *Escherichia coli. Evolution* 47 (1): 1–12.

Morrison, P., and N. Warman. 1967. A thermal gradient chamber for small animals, with digital output. *Medical and Biological Engineering* 5: 41–45.

Pentecost, E. D. 1974. Behavior of *Eumeces laticeps* exposed to a thermal gradient. *Journal of Herpetology* 8 (2): 169–173.

Yamada, Y., and Y. Ohshima. 2003. Distribution and movement of *Caenorhabditis elegans* on a thermal gradient. *Journal of Experimental Biology* 206: 2581–2593.

Chapter 22

FIRE AND ECOLOGICAL DISTURBANCE

A 5E LESSON TO ADDRESS AN IMPORTANT MISCONCEPTION

By Michael Dentzau and Victor Sampson

"The most important single factor influencing learning is what the learner already knows."

—Ausubel 1968 (p. iv)

When considering this statement through the lens of constructivism, the significance of misconceptions takes center stage. Misconceptions are not simply factual errors or a lack of understanding, but rather explanations that are constructed based on past experiences (Hewson and Hewson 1988). If students' misconceptions are not directly engaged in the learning process, they may persist—even when faced with instruction to the contrary (Bransford, Brown, and Cocking 1999).

This chapter presents a 5E instructional model to help students overcome misconceptions about fire and its role in ecological succession. Through this activity, students learn that fire plays an important—and beneficial—role in ecology.

ECOLOGY MISCONCEPTIONS

Much of the literature on misconceptions has centered on the physical sciences, with much less emphasis on biology, and even less on ecology (Driver, Guesne, and Tiberghien 1985; Munson 1994; McComas 2002). Marek (1986), however, found that even after instruction in an environmental unit, 33% of students still had misconceptions about ecosystems. Consider this in light of the goal of scientific literacy, which aims to equip students with the tools needed "to participate thoughtfully with fellow citizens in building and protecting a society that is open, decent, and vital" (AAAS 1989), and the relevance of ecological misconceptions to science education is unmistakable.

Based on the expert opinion of professional ecologists, Cherrett (1989) generated a list of 50 concepts integral to understanding ecology. Though agreement on all 50 concepts is not universal, it seems reasonable to consider the top 10 as essential to an operational understanding of the discipline (Munson 1994). These 10 concepts include

- the ecosystem,
- succession,
- energy flow,
- conservation of resources,
- competition,
- niche,
- materials cycling,
- the community,
- life-history strategies, and
- ecosystem fragility.

Understanding that these concepts are interconnected, we have selected an aspect of ecology that touches on many of them: the role of fire in the ecology of natural communities. Fire, or lack thereof, is directly related to an understanding of succession, but also involves energy flow, materials cycling, life-history strategies, and competition. Gibson (1996) found that succession is often oversimplified in textbooks, especially those for nonscience majors—leading to the assumption that climax communities are the deterministic "end product" in ecology.

The notion of a climax community, however, is more widely viewed as the exception, and not the rule (Simberloff 1982; Noss and Cooperrider 1994).

If students view succession on the traditional continuum of grasses to shrubs to forest, we would also expect them to view fire as an ecological negative.

But fire can be beneficial, even though it disrupts the process of succession. For example, southeastern coastal plain fire—driven by a combination of lightning strikes, Native Americans, and early European settlers—was once a consistent disturbance that shaped the plant and animal composition of natural communities in these areas (Wade, Ewel, and Hofstetter 1980). Frost (1995) suggests that, historically, less than 5% of the coastal plain landscape was protected from fire, with fire-return intervals ranging from 1 to 300 years. This lies in stark contrast to the popular conception that fire is negative and destructive, and has a predominant anthropogenic genesis (Hanson 2010).

The idea that naturally occurring fire has substantially shaped the form and function of wetland ecosystems can be difficult for students to understand. This association is not intuitive (Lugo 1995), largely because of another misconception that the hydrology of a wetland is constant (i.e., it provides "wet" land all of the time). In reality, the hydrology of wetlands fluctuates greatly, and even outside of a drought period, wetlands will burn.

To address this, we developed a unit that relies on the 5E—Engage, Explore, Explain, Elaborate, and Evaluate—instructional model (Bybee 1993; Bybee et al. 2006). Figure 1 provides a summary of the lesson. Each phase challenges the notion of an inevitable climax successional stage, advances fire as an integral part of the natural ecosystem, and reinforces the impacts of ecological disturbances.

Figure 1

Summary of 5E lesson

Engage
What can you tell me about fire in southeastern wetland communities? Is it good or bad? Natural or unnatural?
Explore
Use vegetation-monitoring data over time to reflect changes in the community with periodic fire. Analyze data.
Explain
Introduce material on fire ecology in the southeast, including publications and short video demonstrations.
Elaborate
Extend the discussion with prompts that require additional research and assimilation of information provided.
Evaluate
Implement a refutational writing exercise convincing a reader of the value of fire in southeastern ecosystems.

ENGAGE

The Engage phase is designed to pique students' interest in the topic and determine their prior knowledge and misconceptions. Though this phase is often enacted using a demonstration in physics and chemistry, many ecological concepts—and subsequent misconceptions—are not easily portrayed inside the classroom. We engage students by posing questions to the class, such as:

- Tell me what you know about the impacts of fire in southeastern wetland ecosystems in states such as Florida and Mississippi.

- What happens when fire occurs in these systems?
- Is fire good or bad for ecosystems?
- Does fire naturally occur in a wetland?
- Does fire occur in all wetlands?

Responses to these questions might include common misconceptions, such as:

- Fire is an unnatural occurrence in wetlands, caused only by humans.
- Fire kills all animals and reduces biodiversity.
- Fire always destroys property, causes smoke, and leads to traffic accidents.
- Fire kills all plants and sterilizes soil.

Alternatively, some students may correctly see fire as a mechanism for ecosystem change, a natural occurrence, or something that is beneficial to biodiversity or important for some species' survival.

At this stage, we continue prompting students until they have voiced a variety of responses that can be documented for later reference by the class. We also gauge prior knowledge with a true-or-false assessment (Figure 2). This is not a graded test, but rather a tool to determine what the class already brings to the discussion.

EXPLORE

In the Explore phase, students have a chance to build their understanding of the phenomenon. In our example, they are provided with two sets of data generated by an actual vegetation-monitoring program in northwest Florida. The first data set (Table 1, p. 187) represents the "frequency of occurrence," or the number of times a species occurred within 1 ft. intervals along a 100 ft. transect. The only other information provided is that a fire occurred in March of Year 1 and May of Year 4, after the annual monitoring data was collected.

The second data set (Table 2, p. 188) provides information about the number of flowering stalks of a rare grass, *Calamovilfa curtissii,* or Florida sandreed (Dentzau 2002). These data are complementary to the data collected in Table 1 and demonstrate fire's effect on the flowering response of a rare target species. We ask students to review the data in both Tables 1 and 2, graph any trends as appropriate, analyze the data, and prepare a plausible explanation.

Figure 2

True-or-false assessment

Fire in Pinelands	Circle one
1. Fires kill all of the wildlife in the wetlands.	T F
2. Some plants have adapted to survive and thrive with fire.	T F
3. Prior to the settlement of North America, lightning caused most fires.	T F
4. High-intensity fires are always ecologically destructive and bad.	T F
5. When fires occur regularly, the heat sterilizes the soil and kills most of the plants.	T F
6. Some wetlands historically burned as often as every one to three years.	T F
7. Many of our wetlands are less healthy because of reduced amounts of fire.	T F
8. Fires can increase the diversity of the vegetative landscape and foster a mosaic of habitats.	T F
9. Ecological communities always succeed from grasses to shrubs to trees.	T F
10. Fire recycles nutrients.	T F

Answer key: 1-F, 2-T, 3-T, 4-F, 5-F, 6-T, 7-T, 8-T, 9-F, 10-T. An answer guide is available online (see "On the web").

EXPLAIN

In the Explain phase, students discuss their findings and begin incorporating the appropriate framework for the ecological concept: the effect of fire on a wetland ecosystem. Working in groups, students discuss the following questions:

- What trends did you observe in the frequency of occurrence data provided in Table 1?
- Which species exhibits a strong response to fire, based on frequency of occurrence data?
- What is the impact of fire on the flowering response of Florida sandreed?

- Which species increased in frequency after the fires?
- Which species decreased in frequency after the fires?
- Which species seemed to be little affected by the fires?

Once these ideas, analyses, and inferences are shared in class, we introduce students to the concepts surrounding fire in the landscape. Potential resources are provided in the "On the web" section at the end of this chapter. In addition, we recommend consulting Wade, Ewel, and Hofstetter (1980), which is available online (see "References"). The essential ideas for understanding the ecological role of fire are provided in Figure 3.

In our sample data sets, there are some clear indications of fire's effects, and some that are more subtle. In Table 1, *Lachnanthes caroliniana,* or redroot, an herbaceous species, shows a clear positive response in the sampling period after the fire, with increased numbers in the year following the fire and subsequent decreases until the next disturbance. With respect to woody species, fire appears to facilitate a decline in some species, but not all. In Table 2, the Florida sandreed flowering data show a dramatic response to fire.

ELABORATE

In the Elaborate stage, we push students to extend their learning to new situations. The initial lesson questions and the true-or-false assessment (Figure 2) are a good place to start. These help students see how their conceptions have changed.

Next, we assign different prompts to student groups, and after some time for research and collaboration, have each share its information with the class. Sample prompts that require knowledge application and additional resources include the following:

1. Why would you think a beech or magnolia hardwood forest on a very dry, upland ridge would burn less frequently than a dense coastal grass marsh? (An appropriate response focuses on the types of fuels available in each community and an understanding that grasses and pine needles [fine fuels] readily carry fire, whereas hardwood leaves do not.)
2. Do you think a catastrophic wetland fire can be beneficial? If so, how? (An appropriate response includes the conditions needed for a catastrophic wetland fire [i.e., drought] and how such fires can consume organic material and lower the elevation of the substrate, allowing different species to recolonize [i.e., reset succession].)

3. Identify one plant species and one animal species that are considered fire-dependent, and explain their responses to and requirement for periodic fire. (There are several appropriate responses to this question, but students should base their answers on available data.)
4. Why did we see different species responding differently to fire? (There are many acceptable responses for this item, including the plants' health, the amount of soil moisture present, fire intensity, weather conditions during the burn, standing water in some locations, and reduced fuel in some spots.)

Table 1

Frequency of occurrence

Species' frequency of occurrence was measured along a 100 ft. transect in a Northwest Florida wetland.

Species	Type	Year 1	Year 2	Year 3	Year 4	Year 5	Year 6
Andropogon glomeratus (broomsedge)	Herbaceous	13	5	8	4	4	2
Calamovilfa curtissii (Florida sandreed)	Herbaceous	52	67	61	66	65	63
Drosera intermedia (spoonleaf sundew)	Herbaceous	28	28	23	35	56	45
Gaylussacia mosieri (woolly huckleberry)	Woody	69	70	78	71	78	76
Ilex coriacea (sweet gallberry)	Woody	23	6	11	4	5	6
Ilex glabra (inkberry)	Woody	86	99	97	93	98	99
Ilex myrtifolia (myrtle-leaved holly)	Woody	2	0	0	0	0	0
Lachnanthes caroliniana (redroot)	Herbaceous	40	100	87	55	99	78
Myrica inodora (odorless bayberry)	Woody	24	11	14	14	3	2
Pinus elliottii (slash pine)	Woody	2	0	0	0	0	0
Rhexia alifanus (meadow beauty)	Herbaceous	0	11	0	0	0	10
Smilax laurifolia (bamboo vine)	Vine	3	8	0	5	2	2

Table 2

Number of flower stalks: Florida sandreed

Measurements of the number of flower stalks of *Calamovilfa curtissii* (Florida sandreed) in ten 3.28 ft.^{2} plots were systematically placed along each of two 100 ft. transects in a North Florida wetland. One transect was within the area of prescribed fire and the second was in a control area, where no fire was implemented.

These additional data on flowering response were collected to supplement the frequency and cover data collected for all species represented in Figure 3. Even though fire's impact on this species was not clear in the frequency of occurrence data, there was a dramatic effect on the flowering response.

Plot number	Within the area that burned the prior year	Within the area that did not burn the prior year (control)
1	129	0
2	1	0
3	32	0
4	0	0
5	21	0
6	0	0
7	15	0
8	0	0
9	160	0
10	72	0

EVALUATE

The Evaluate phase brings the lesson back to the original concept—the ecological role of naturally occurring fire in community composition and succession—and students' misconceptions about it. We assess students' understanding of the target concept with a refutational writing assignment, which requires them to introduce a common misconception related to a phenomenon, refute it, describe the scientific concept, and then show that the scientific way of thinking is more valid or acceptable (Dlugokienski and Sampson 2008). A sample refutational writing prompt, which connects the fire ecology lesson and the fluctuating status of succession in natural systems, is available online (see "On the web").

SO WHAT?

Fire is an essential ecosystem component of many communities in the southeastern United States (Figure 3). Yet, when students are first introduced

Figure 3

The ecological role of fire

Adapted from Wade, Ewel, and Hofstetter 1980

Fire influences the physical-chemical environment by
• directly releasing mineral elements such as ash.
• indirectly releasing elements by increasing decomposition rates.
• volatilizing some nutrients (e.g., nitrogen, sulfur).
• reducing plant cover and thereby increasing insolation.
• changing soil temperatures because of increased insolation.
Fire regulates dry-matter production and accumulation by
• recycling the stems, foliage, bark, and wood of plants.
• consuming litter, humus layers, and, occasionally, increments of organic soil.
• creating a large reservoir of dead organic matter by killing but not consuming vegetation.
• usually stimulating increased net primary production, at least on short time scales.
Fire controls plant species and communities by
• triggering the release of seeds.
• altering seedbeds.
• temporarily eliminating or reducing competition for moisture, nutrients, heat, and light.
• stimulating vegetative reproduction of top-killed plants.
• stimulating the flowering and fruiting of many shrubs and herbs.
• selectively eliminating components of a plant community.
• influencing community composition and successional stage through its frequency or intensity.
Fire determines wildlife habitat patterns and populations by
• usually increasing the amount, availability, and palatability of foods for herbivores.
• regulating yields of nut- and berry-producing plants.
• regulating insect populations, which are important food sources for many birds.
• controlling the scale of the total vegetative mosaic through fire size, intensity, and frequency.
• regulating macroinvertebrate and small-fish populations.

(continued)

Figure 3 *(continued)*

Fire influences insects, parasites, fungi, and so on by
• regulating the total vegetative mosaic and the age structure of individual stands with it.
• sanitizing plants against pathogens such as brown spot on longleaf pine.
• producing charcoal, which can stimulate ectomycorrhizae.
Fire also regulates the number and kinds of soil organisms, affects evapotranspiration patterns and surface water flow, changes the accessibility through and aesthetic appeal of an area, and releases combustion products into the atmosphere.

to this counterintuitive concept, many ask, "So why do we suppress them?"

The answer to this question is relatively straightforward and explains why this lesson is so important: We have forever altered the historic landscape through our occupation of this country and, as a result, have changed the frequency and intensity of fire in the southeastern United States. Historically, fires in this area occurred frequently and burned for a long time at a low intensity, but today, naturally occurring fires start in areas where the amount of fuel is exceedingly high due to fragmented natural habitats. Though land managers actively seek to use controlled or prescribed fire to achieve restoration goals, there will likely always be a need for suppression in some areas, such as national parks, because of the hazardous conditions created by increased fuel loads and encroaching human development.

When considering scientific literacy and the subsumed goal of fostering productive citizens, the ability to understand ecological issues—such as the role of fire in ecosystems—is paramount. Students develop critical-thinking skills when they analyze authentic data sets and use them to construct evidence-based conclusions. The lesson we have provided here serves to actively engage student misconceptions of important ecological concepts and replace them with appropriate scientific knowledge. Such lessons assist the development of scientific literacy and bring students one step closer to becoming the citizens our society needs.

Ecological terms

Ecosystem: A dynamic complex of plant, animal, fungal, and microorganism communities and their associated nonliving environment interacting as an ecological unit.

Succession: The more-or-less predictable change in the composition of communities following a natural or human disturbance.

Energy flow: The movement of energy (the capacity to do work) through an ecosystem. The source of virtually all of the energy useful to organisms is the Sun and the basic way in which life captures this energy is through photosynthesis. Available energy is lost continuously as it moves through the ecosystem trophic levels.

Conservation of resources: The wise use of Earth's natural resources by humanity.

Competition: The struggle among organisms, both of the same and different species, for limited food, space, and other requirements needed for survival.

Niche: The place occupied by a species in its ecosystem—where it lives, what it eats, its foraging route, the season of its activity, and so on. In a more abstract sense, a niche is a potential place or role within a given ecosystem into which species may or may not have evolved.

Materials cycling: The movement of minerals, elements, and nutrients through ecosystems in a cycle of production and consumption.

Community: All of the organisms—plants, animals, and microorganisms—that live in a particular habitat and affect one another as part of the food web or through their various influences on the physical environment.

Life-history strategies: A species pattern of growth, development, age at sexual maturity, level of parental involvement, number of offspring, and senescence that are shaped by natural selection to produce the largest possible number of surviving offspring.

Ecosystem fragility: The degree to which an ecosystem can be easily damaged or changed, in part or in whole, in unexpected and undesirable ways.

On the web

Sample refutational writing prompt: *http://www.nsta.org/highschool/connections/201104SampleRefutationalWritingPrompt.pdf*

Guide for true-and-false assessment: *http://www.nsta.org/highschool/connections/201104GuideForTrueFalseAssessment.pdf*

Fire ecology: *http://en.wikipedia.org/wiki/Fire_ecology*

Benefits of prescribed burning: *http://edis.ifas.ufl.edu/fr061*

Dome swamp: *www.fnai.org/PDF/NC/Dome_Swamp.pdf*
Wet flatwoods: *www.fnai.org/PDF/NC/Wet_Flatwds.pdf*
Fire ecology in the southeastern United States: *www.nwrc.usgs.gov/factshts/018-00.pdf*
Prescribed fire training: *www.youtube.com/watch?v=dZybHZwmgZ0*
Restoring fire to native landscapes: *www.nature.org/ourinitiatives/regions/northamerica/unitedstates/florida/howwework/florida-restoring-fire-to-native-landscapes.xml*
Fire as a Habitat Restoration Tool: *www.youtube.com/watch?v= YCC6C5U60wA*
Florida's fire-dependent plants and animals: *www.nature.org/photosmultimedia/floridas-fire-dependent-plants-and-animals.xml*

References

American Association for the Advancement of Science (AAAS). 1989. *Science for all Americans: A Project 2061 report on literacy goals in science, mathematics, and technology.* Washington, DC: AAAS.

Ausubel, D. 1968. *Educational psychology.* New York: Holt, Reinhart, and Winston.

Bransford, J., A. L. Brown, and R. R. Cocking. 1999. *How people learn: Brain, mind, experience, and school.* Washington, DC: National Academies Press.

Bybee, R. W. 1993. An instructional model for science education. In *Developing biological literacy: A guide to developing secondary and post-secondary biology curricula.* Colorado Springs, CO: Biological Sciences Curriculum Study.

Bybee, R. W., J. A. Taylor, A. Gardner, P. Van Scotter, J. Powell, A. Westbrook, and N. Landes. 2006. *The BSCS 5E instructional model: Origins, effectiveness, and applications.* Colorado Springs, CO: Biological Sciences Curriculum Study.

Cherrett, J. M. 1989. Key concepts: The results of a survey of our members' opinions. In *Ecological concepts*, ed. J. M. Cherrett, 29. Oxford, England: Blackwell Scientific.

Dentzau, M. W. 2002. Prescribed burn enhances flowering of Florida sandreed (Florida). *Ecological Restoration* 20 (4): 295–296.

Dlugokienski, A., and V. Sampson. 2008. Learning to write and writing to learn in science: Refutational texts and analytical rubrics. *Science Scope* 32 (3): 14–19.

Driver, R., E. Guesne, and A. Tiberghien. 1985. *Children's ideas in science.* Philadelphia: Open University Press.

Frost, C. C. 1995. Presettlement fire regimes in southeastern marshes, peatlands, and swamps. In Proceedings, 19th Tall Timbers Fire Ecology Conference, ed. S. Cerulean and R. T. Engstrom, 39–60. Tallahassee, FL: Tall Timbers Research Station.

Gibson, D. J. 1996. Textbook misconceptions: The climax concept of succession. *The American Biology Teacher* 58 (3): 135–140.

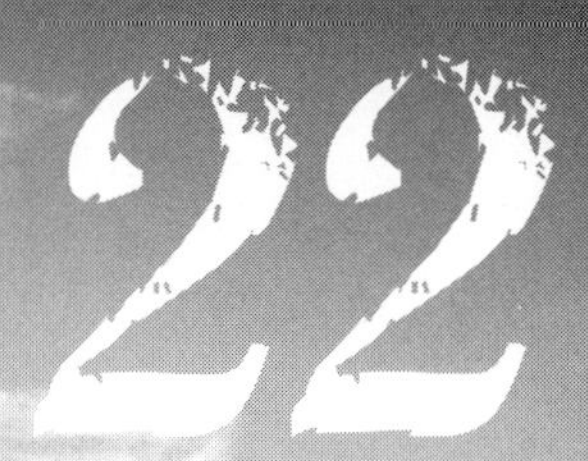

Hanson, C. T. 2010. *The myth of "catastrophic" wildfire: A new ecological paradigm of forest health.* Cedar Ridge, CA: John Muir Project of Earth Island Institute.

Hewson, P., and M. Hewson. 1988. An appropriate conception of teaching science: A view for studies of science learning. *Science Education* 72 (5): 597–614.

Lugo, A. E. 1995. Fire and wetland management. In Proceedings, 19th Tall Timbers Fire Ecology Conference, ed. S. Cerulean and R.T. Engstrom, 1–9. Tallahassee, FL: Tall Timbers Research Station.

Marek, E. A. 1986. They misunderstand, but they'll pass. *The Science Teacher* 53 (9): 32–35.

McComas, W. F. 2002. The ideal environmental science curriculum: I. History, rationales, misconceptions, and standards. *The American Biology Teacher* 64 (9): 665–672.

Munson, B. H. 1994. Ecological misconceptions. *Journal of Environmental Education* 25 (4): 30–34.

Noss, R. F., and A. Y. Cooperrider. 1994. *Saving nature's legacy: Protecting and restoring biodiversity.* Washington, DC: Island Press.

Simberloff, D. 1982. A succession of paradigms in ecology: Essentialism to materialism and probalism. In *Conceptual issues in ecology*, ed. E. Saarinen, and D. Reidel, 63–102. A Dordrecht: Pallas.

Wade, D. D., J. J. Ewel, and R. H. Hofstetter. 1980. Fire in south Florida ecosystems. Forest Service general technical report SE, 17. Asheville, NC: Southeastern Forest Experiment Station. *www.srs.fs.usda.gov/pubs/206.*

Chapter 23

A COOPERATIVE CLASSROOM INVESTIGATION OF CLIMATE CHANGE

STUDENTS INVESTIGATE ENVIRONMENTAL CHANGES AND THEIR IMPACT ON PENGUIN COMMUNITIES

By Juanita Constible, Luke Sandro, and Richard E. Lee Jr.

At the global level, strong evidence suggests that observed changes in Earth's climate are largely due to human activities (IPCC 2007). At the regional level, the evidence for human-dominated change is sometimes less clear. Scientists have a particularly difficult time explaining warming trends in Antarctica—a region with a relatively short history of scientific observation and a highly variable climate (Clarke et al. 2007). Regardless of the mechanism of warming, however, climate change is having a dramatic impact on Antarctic ecosystems. In this article, we describe a standards-based, directed inquiry we have used in 10th-grade biology classes to highlight the ecosystem-level changes observed on the western Antarctic Peninsula. This activity stresses the importance of evidence in scientific explanations and demonstrates the cooperative nature of science.

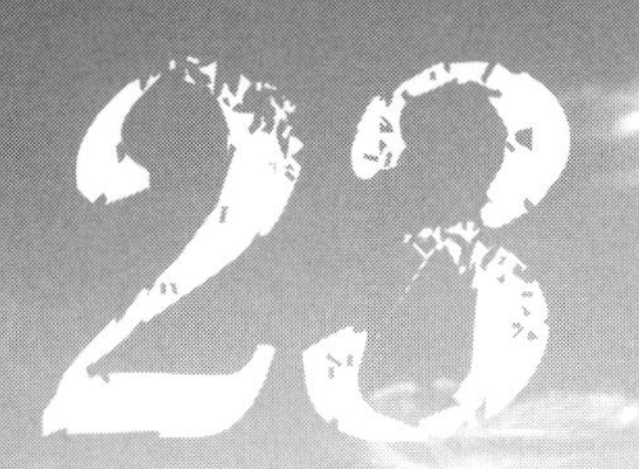

WARMING CLIMATE, WANING SEA ICE

Air temperature data indicate that the western Antarctic Peninsula has warmed by about 3°C in the last century (Clarke et al. 2007). Although this relatively short-term record is only from a few research stations, other indirect lines of evidence confirm the trend. The most striking of these proxies is a shift in penguin communities. Adélie penguins, which are dependent on sea ice for their survival, are rapidly declining on the Antarctic Peninsula despite a 600-year colonization history. In contrast, chinstrap penguins, which prefer open water, are increasing dramatically. (**Note:** See "Chinstrap and Adélie penguins," p. 198, for additional information on the penguins.) These shifts in penguin populations appear to be the result of a decrease in the amount, timing, and duration of sea ice (Figure 1; Smith, Fraser, and Stammerjohn 2003).

Figure 1

Effects of regional warming on sea ice, krill, and penguin communities of the Antarctic Peninsula

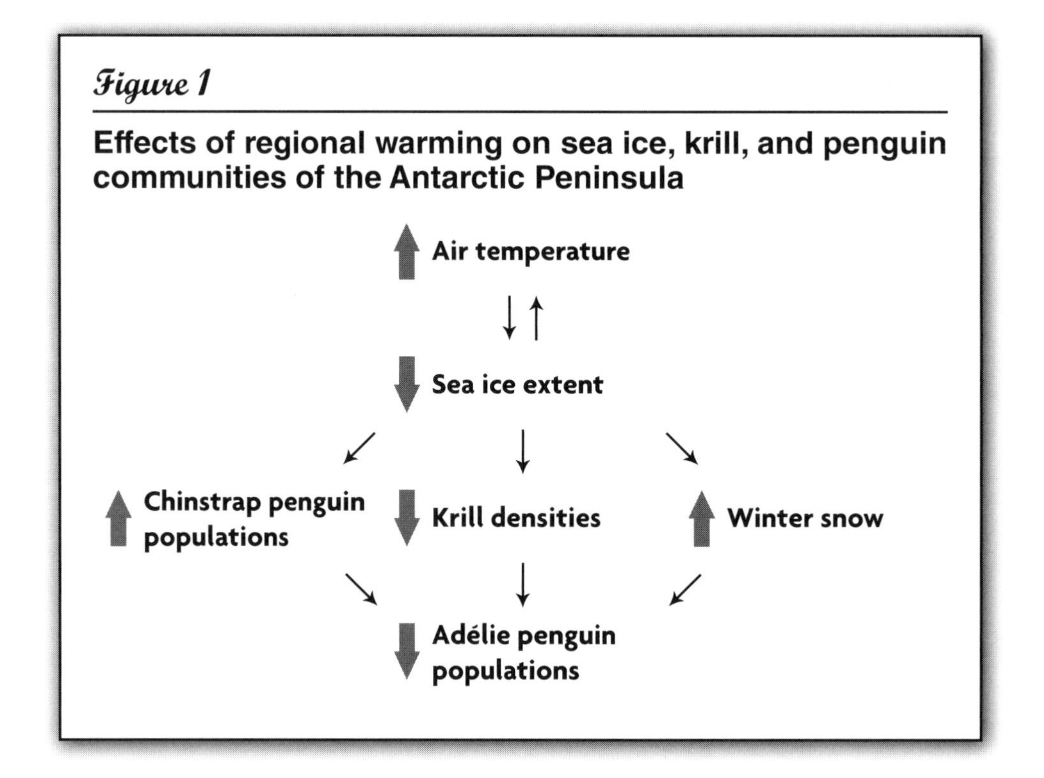

Why is sea ice so important to Adélie penguins? First, sea ice is a feeding platform for Adélies. Krill, the primary prey of Adélies on the peninsula, feed on microorganisms growing on the underside of the ice (Atkinson et al. 2004). For Adélie penguins, which are relatively slow swimmers, it is easier to find food under the ice than in large stretches of open water (Ainley 2002). Second, sea ice helps control the local climate. Ice keeps the peninsula cool by reflecting solar radiation back to space. As air temperatures increase and sea ice melts,

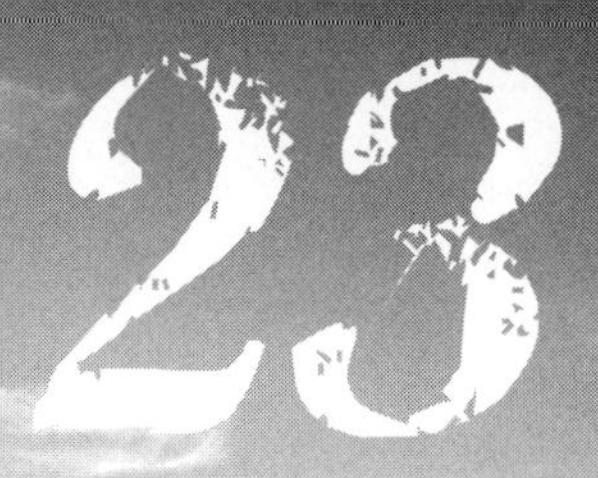

open water converts radiation into heat and amplifies the upward trend in local air temperatures (Figure 2; Wadhams 2000). Third, ice acts as a giant cap on the ocean, limiting evaporation. As sea ice declines, cloud condensation nuclei and moisture are released into the atmosphere, leading to more snow. This extra snow often does not melt until Adélies have already started nesting; the resulting meltwater can kill their eggs (Fraser and Patterson 1997).

Figure 2

Melting sea ice amplifies the effects of atmospheric warming.

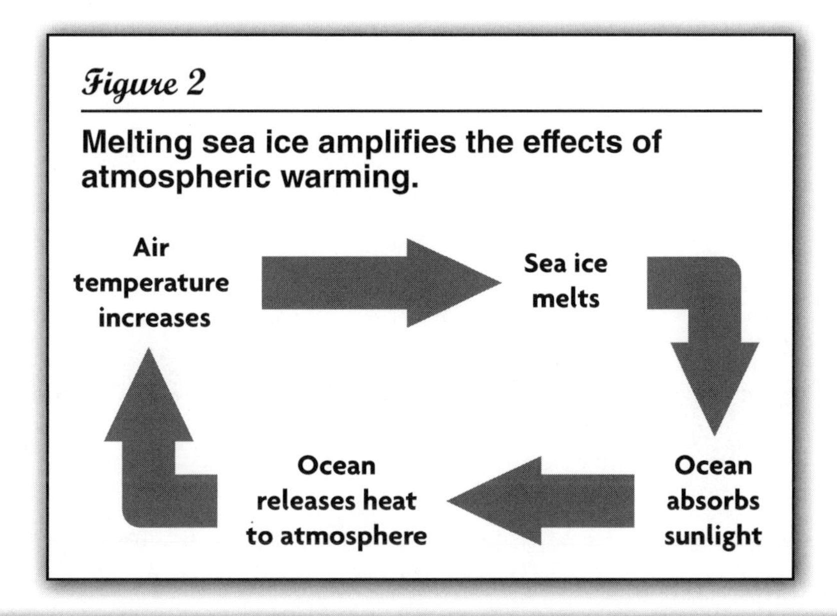

Figure 3

Specialist fact sheet

Each Home Group contains five different specialists.

Ornithologist: A scientist who studies birds. Uses visual surveys (from ship or on land), diet analysis, and satellite tracking to collect data on penguins.

Oceanographer: A scientist who studies the ocean. Uses satellite imagery, underwater sensors, and manual measurements of sea ice thickness to collect data on sea ice conditions and ocean temperature.

Meteorologist: A scientist who studies the weather. Uses automatic weather stations and visual observations of the skies to collect data on precipitation, temperature, and cloud cover.

Marine ecologist: A scientist who studies the relationship between organisms and their ocean environment. Uses visual surveys, diet analysis, and satellite tracking to collect data on a variety of organisms, including penguins.

Fisheries biologist: A scientist who studies fish and their prey. Collects data on krill during research vessel cruises.

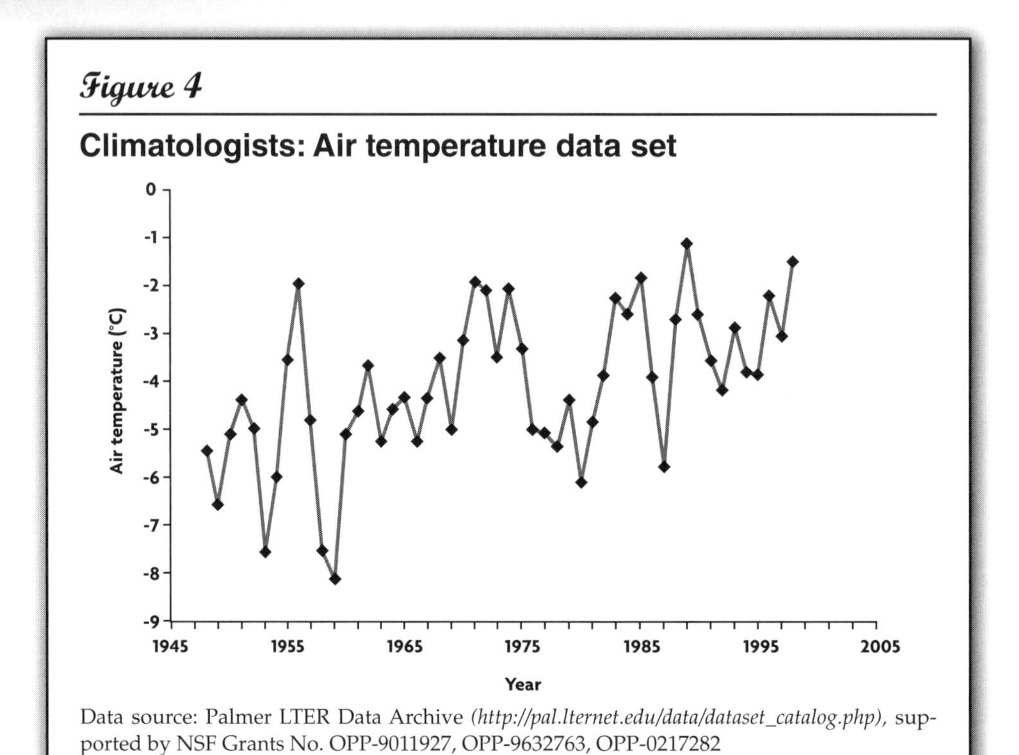

Figure 4

Climatologists: Air temperature data set

Data source: Palmer LTER Data Archive *(http://pal.lternet.edu/data/dataset_catalog.php)*, supported by NSF Grants No. OPP-9011927, OPP-9632763, OPP-0217282

Chinstrap and Adélie penguins

Chinstrap penguins *(Pygoscelis antarctica)* are primarily found on the Antarctic Peninsula and in the Scotia Arc, a chain of islands between the tip of South America and the peninsula. Their name comes from the black band running across their chins. Adult chinstraps stand 71–76 cm tall and weigh up to 5 kg.

Adélie penguins *(Pygoscelis adeliae)* breed on the coast of Antarctica and surrounding islands. They are named after the wife of French explorer Jules Sébastien Dumont d'Urville. Adult Adélies stand 70–75 cm tall and weigh up to 5 kg.

ACTIVITY OVERVIEW

For our directed inquiry, we use the *jigsaw technique*, which requires every student within a group to be an active and equal participant for the rest of the group to succeed (Colburn 2003). To begin, students are organized into "Home Groups" composed of five different specialists. (**Note:** See "Procedure" in the next section for specific instructions on how each student assumes the identity of a different specialist.) Specialists from each Home Group then reorganize into "Specialist Groups" that contain only one type of scientist (e.g., Group 1 could include all of the ornithologists and

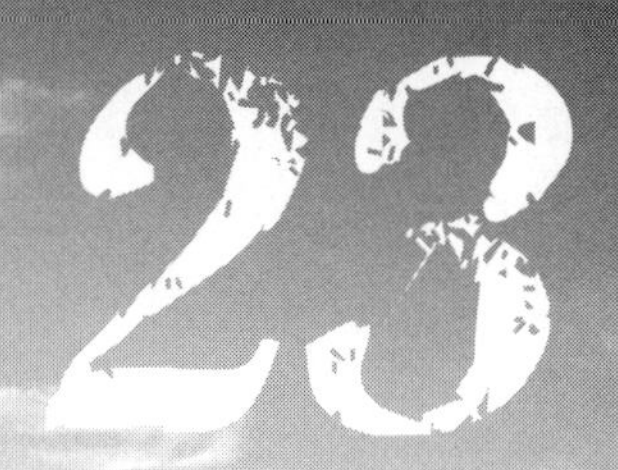

Group 2 all of the oceanographers). Each Specialist Group receives a piece of the flowchart in Figure 1, in the form of a data table. With only a few facts to guide them, the Specialist Groups create graphs from the data tables, brainstorm explanations for patterns in their data, and report results back to their Home Groups. Finally, Home Groups use the expertise of each specialist to reconstruct the entire flowchart (Figure 1).

Figure 5

Year	# Breeding pairs of Adélie penguins
1975	15,202
1979	13,788
1983	13,515
1986	13,180
1987	10,150
1989	12,983
1990	11,554
1991	12,359
1992	12,055
1993	11,964
1994	11,052
1995	11,052
1996	9,228
1997	8,817
1998	8,315
1999	7,707
2000	7,160
2001	6,887
2002	4,059

Ornithologists: Adélie penguin data set

Michael Elnitsky

- Adélie penguins spend their summers on land, where they breed. They spend winters on the outer extent of the sea ice surrounding Antarctica, where they molt their feathers and fatten up.
- Adélies are visual predators, meaning they need enough light to see their prey. Near the outer part of the pack ice, there are only a few hours of daylight in the middle of the winter. There is less sunlight as one moves further south (closer to land).
- On the western Antarctic Peninsula, Adélie penguins mostly eat krill, a shrimplike crustacean.
- Several countries have been harvesting krill since the mid-1960s.
- Adélie penguins need dry, snow-free places to lay their eggs. They use the same nest sites each year and at about the same time every year. Heavy snowfalls during the nesting season can bury adult Adélies and kill their eggs.
- Female Adélies lay two eggs, but usually only one of those eggs result in a fledged chick (fledged chicks have a good chance of maturing into adults). The two most common causes of death of eggs and chicks are abandonment by the parents (if they cannot find enough food) and predation by skuas (hawklike birds).
- In the water, Adélies are eaten mostly by leopard seals and killer whales.
- Adélies can look for food under sea ice because they can hold their breath for a long time. They are not as good at foraging in the open ocean because they cannot swim very fast.
- Adélie penguins have lived in the western Antarctic Peninsula for at least 644 years.

Data source: Smith, Fraser, and Stammerjohn 2003

Before starting this activity, students should have at least a rudimentary knowledge of Antarctica. Teachers can find a collection of links to our favorite Antarctic websites at *www.units.muohio.edu/cryolab/education/AntarcticLinks.htm.* Teachers also can engage student interest in this inquiry by showing video clips of penguins, which are naturally appealing to students of all ages. We also have short movies of Adélies feeding their young and battling predators on our website at *www.units.muohio.edu/cryolab/education/antarcticbestiary.htm.*

THE ACTIVITY

Materials

- specialist fact sheet (Figure 3; 1 for each student or 1 overhead for the entire class)
- temperature data (Figure 4; 1 overhead for the entire class)
- data sets for each Specialist Group (Adélie penguins [Figure 5], sea ice [Figure 6], winter snow [Figure 7], chinstrap penguins [Figure 8], and krill [Figure 9])
- Specialist Group report sheets (found online at *http://www.nsta.org/highschool/connections/200709SpecialistGroupReportSheet.pdf*; 1 for each student)
- sheets of graph paper (1 for each student), or computers connected to a printer (1 for each Specialist Group)
- sets of 6 flowchart cards (1 complete set for each Home Group); Before the inquiry, teachers can make flowchart cards by photocopying Figure 1 and cutting out each box (e.g., "Air Temperature," "Sea Ice Extent," etc.)
- paper, markers, and tape for constructing flowcharts

Procedure: Class period 1 (45–60 minutes)

1. Split the class into Home Groups of five students each. Assign the name of a different real-life research agency to each group (see *www.units.muohio.edu/cryolab/education/AntarcticLinks.htm#NtnlProg* for examples).
2. Instruct students to read the Specialist fact sheet (Figure 3). Within a Home Group, each student should assume the identity of a different specialist from the list.

3. Introduce yourself: "Welcome! I'm a climatologist with the Intergovernmental Panel on Climate Change in Geneva, Switzerland. In other words, I study long-term patterns in climate. My colleagues and I have researched changes in air temperatures on the Antarctic Peninsula since 1947. We have observed that although air temperatures on

Figure 6

Oceanographers: Sea ice data set

Year	Area of sea ice extending from the Antarctic Peninsula (km^2)
1980	146,298
1981	136,511
1982	118,676
1983	88,229
1984	85,686
1985	78,792
1986	118,333
1987	142,480
1988	90,310
1989	44,082
1990	79,391
1991	111,959
1992	110,471
1993	94,374
1994	103,485
1995	95,544
1996	86,398
1997	100,784
1998	73,598
1999	79,223
2000	79,200
2001	69,914

- In August (midwinter), sea ice covers more than 18 × 106 km^2, or 40%, of the Southern Ocean (an area larger than Europe). In February (midsummer), only 3 × 106 km^2 (about 7%) of the ocean is covered by sea ice.
- Sea ice keeps the air of the Antarctic region cool by reflecting most of the solar radiation back into space.
- Open water absorbs solar radiation instead of reflecting it and converts it to heat. This heat warms up the atmosphere.
- Sea ice reduces evaporation of the ocean, thus reducing the amount of moisture that is released to the atmosphere.
- As sea ice melts, bacteria and other particles are released into the atmosphere. These particles form condensation or freezing nuclei, which grow into rain or snow.
- Rain helps to stabilize the sea ice by freezing on the surface.
- Sea ice can be broken up by strong winds that last a week or more.
- An icebreaker is a ship used to break up ice and keep channels open for navigation. Icebreakers were first used in the Antarctic in 1947.

Data source: Palmer LTER Data Archive (*http://pal.lternet.edu/data/dataset_catalog.php*)

the Peninsula cycle up and down, they have increased overall (show Figure 4). We think this might be occurring because of an increase in greenhouse gases, but we are unsure of the impacts on the Antarctic ecosystem. Your team's job is to describe the interconnected effects of warming on Antarctica's living and nonliving systems."

4. Direct the specialists to meet with their respective Specialist Groups. Specialist Groups should *not* interact with one another.
5. Distribute the data sets and Specialist Group report sheets to each Specialist Group. The specialists should graph their dataset and interpret the graph.

Procedure: Class period 2 (45–60 minutes)

6. Hand out a complete set of flowchart cards to the reconvened Home Groups. Each specialist should make a brief presentation to his or her Home Group approximating the format on the Specialist Group report sheet. Home Groups should then construct their own flowchart using all the flowchart cards. Remind the students throughout this process that they should use the *weight of evidence* to construct the flowcharts. In other words, each idea should be accepted or rejected based on the amount of support it has.
7. Consider these discussion questions at the end of the period as a class, by Home Group, or as homework for each student:
   - How has the ecosystem of the Antarctic Peninsula changed in the last 50 years? What are the most likely explanations for these changes?
   - Is there sufficient evidence to support these explanations? Why or why not? What further questions are left unanswered?
   - Did your Specialist Group come up with any explanations that you think are not very likely (or not even possible!), based on the complete story presented by your Home Group?

ASSESSMENT

To assess student learning, we use a simple performance rubric that can be found online at *http://www.nsta.org/highschool/connections/200709PerformanceRubric.pdf*. The rubric focuses on group work and the nature of science. Depending on the unit of study in which this inquiry is used, a variety of specific content standards also may be assessed. For example, in an ecology unit,

Figure 7

Meteorologists: Winter snow data set

Year	% precipitation events that are snow
1982	49
1983	67
1984	72
1985	67
1986	81
1987	80
1988	69
1989	69
1990	68
1991	72
1992	70
1993	70
1994	83
1995	77
1996	74
1997	81
1998	81
1999	83
2000	77
2001	90
2002	82
2003	76

- In the winter, most of the precipitation in the western Antarctic Peninsula occurs as snow. There is an even mix of snow and rain the rest of the year.
- It is difficult to accurately measure the amount of snowfall in the Antarctic because strong winds blow the snow around.
- The Antarctic Peninsula has a relatively warm maritime climate, so gets more rain and snow than the rest of the Antarctic continent.
- Most of the rain and snow on the peninsula is generated by cyclones from outside the Southern Ocean. Cyclones are areas of low atmospheric pressure and rotating winds.
- When there is less sea ice covering the ocean, there is more evaporation of the ocean and therefore more moisture in the atmosphere.
- As sea ice melts, bacteria and other particles are released into the atmosphere. These particles form condensation or freezing nuclei, which grow into rain or snow.

Data source: Antarctic Meteorology Online, British Antarctic Survey *(www.antarctica.ac.uk/met/metlog/)*

teachers may wish to determine student knowledge of interactions among populations and environments; in an Earth science unit, teachers could check student understandings about weather and climate.

Figure 8

Marine ecologists: Chinstrap penguin data set

Year	# breeding pairs of chinstrap penguins
1976	10
1977	42
1983	100
1984	109
1985	150
1989	205
1990	223
1991	164
1992	180
1993	216
1994	205
1995	255
1996	234
1997	250
1998	186
1999	220
2000	325
2001	325
2002	250

- Chinstrap penguins breed on land in the spring and summer and spend the rest of the year in open water north of the sea ice. The number of chinstraps that successfully breed is much lower in years when the sea ice does not melt until late spring.
- Chinstraps mostly eat krill, a shrimplike crustacean.
- Whalers and sealers overhunted seals and whales, which also eat krill, until the late 1960s.
- Chinstraps primarily hunt in open water, because they cannot hold their breath for very long.
- The main predators of chinstraps are skuas (hawklike birds), leopard seals, and killer whales.
- Chinstraps will aggressively displace Adélie penguins from nest sites in order to start their own nests, and may compete with Adélies for feeding areas.
- Although chinstrap penguins have occupied the western Antarctic Peninsula for over 600 years, they have become numerous near Palmer Station only in the last 35 years.

Data source: Smith, Fraser, and Stammerjohn 2003.

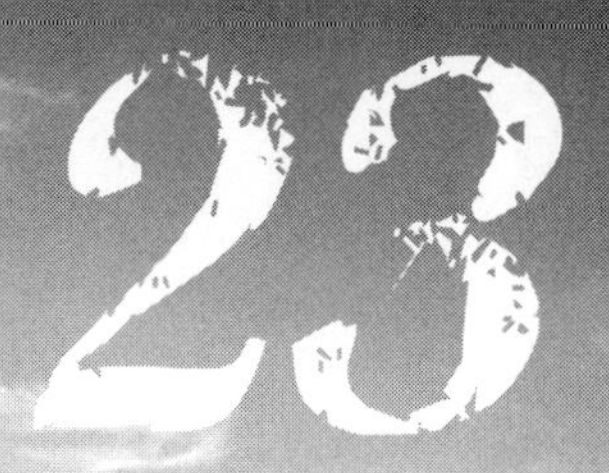

MODIFICATIONS

Some students have initial difficulties with the construction and interpretation of flowcharts. Once students have connected their flowchart cards with arrows, it may be useful to have them label each arrow with a verb. For instance:

 sea ice extent **causes** →  winter snow

Figure 9

Fisheries biologists: Krill data set

Year	Density of krill in the Southern Ocean (no./m^2)
1982	91
1984	50
1985	41
1987	36
1988	57
1989	15
1990	8
1992	7
1993	22
1994	6
1995	9
1996	31
1997	53
1998	46
1999	4
2000	8
2001	31
2002	8
2003	3

Data source: Atkinson et al. 2004

- Krill are a keystone species, meaning they are one of the most important links in the Antarctic food web. All the vertebrate animals in the Antarctic eat either krill or another animal that eats krill.
- Krill eat mostly algae. In the winter, the only place algae can grow is on the underside of sea ice.
- Several countries have been harvesting krill since the mid-1960s.
- Ultraviolet radiation is harmful to krill and can even kill them. Worldwide, ozone depletion is highest over Antarctica.
- Salps, which are small, marine animals that look like blobs of jelly, may compete with krill for food resources. As the salt content of the ocean decreases, salp populations increase and krill populations decrease.

Richard E. Lee Jr.

Teachers can shorten this lesson by starting immediately with Specialist Groups, rather than with Home Groups. Another option is to provide graphs of the data rather than having Specialist Groups create their own.

To make this lesson more open-ended, students may do additional research on the connections between sea ice, krill, and penguins. Instructors should keep in mind, however, that the majority of resources on this topic are articles in primary scientific journals. If teachers have access to a university library, they might wish to make a classroom file of related journal articles. A more engaging extension would be for students to generate ideas for new research studies that would address questions left unanswered by the current inquiry. This type of activity could range from asking students to formulate new hypotheses to asking students to write short proposals that include specific research questions and plans to answer those hypotheses.

Many students have trouble comprehending how just a few degrees of atmospheric warming (in this case, 3°C) could make a difference in their lives. The decline of a charismatic species such as the Adélie penguin is an example of how a seemingly minor change in climate can pose a major threat to plants and animals. Beyond the effects of climate change, however, the activity illustrates the multidisciplinary, international, and above all cooperative nature of science. We want social teenagers to realize that they do not have to sit alone in a lab to do science.

Acknowledgments

The authors would like to thank personnel at Palmer Station, especially W. Fraser, for valuable background information related to this article. M. Kaput, S. Bugg, S. Waits, S. Metz, S. Nuttall, A. Carlson, and an anonymous review panel provided constructive criticism on previous drafts of this chapter. This project was supported by NSF grants No. OPP-0337656 and No. IOB-0416720.

References

Ainley, D. G. 2002. *The Adélie penguin: Bellwether of climate change.* New York: Columbia University Press.

Atkinson, A., V. Siegel, E. Pakhomov, and P. Rothery. 2004. Long-term decline in krill stock and increase in salps within the Southern Ocean. *Nature* 432:100–103.

Clarke, A., E. J. Murphy, M. P. Meredith, J. C. King, L. S. Peck, D. K. A. Barnes, and R. C. Smith. 2007. Climate change and the marine ecosystem of the western Antarctic Peninsula. *Philosophical Transactions of the Royal Society B* 362: 149–166.

Colburn, A. 2003. *The lingo of learning: 88 education terms every science teacher should know.* Arlington, VA: NSTA Press.

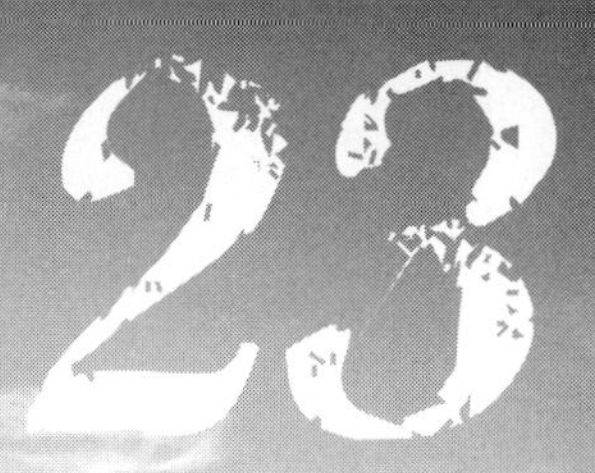

Fraser, W. R., and D. L. Patterson. 1997. Human disturbance and long-term changes in Adélie penguin populations: A natural experiment at Palmer Station, Antarctic Peninsula. In *Antarctic communities: Species, structure and survival*, eds. B. Battaglia, J. Valencia, and D. W. H. Walton, 445–452. Cambridge, UK: Cambridge University Press.

Intergovernmental Panel on Climate Change (IPCC). 2007. *The physical science basis: Summary for policy makers*. Working Group I Report. *www.ipcc.ch.*

Smith, R. C., W. R. Fraser, and S. E. Stammerjohn. 2003. Climate variability and ecological response of the marine ecosystem in the Western Antarctic Peninsula (WAP) region. In *Climate variability and ecosystem response at Long-Term Ecological Research (LTER) sites*, eds. D. Greenland, D. Goodin, and R. C. Smith, 158–173. New York: Oxford Press.

Wadhams, P. 2000. *Ice in the ocean.* Amsterdam: Gordon and Breach Science Publishers.

Chapter 24

CLIMATE PHYSICS

USING BASIC PHYSICS CONCEPTS TO TEACH ABOUT CLIMATE CHANGE

By William Space

Numerous connections exist between climate science and topics normally covered in physics and physical science courses. For instance, lessons on heat and light can be used to introduce basic climate science, and the study of electric circuits provides a context for studying the relationship between electricity consumption and carbon pollution. To highlight some of these connections, this chapter describes a series of lessons and activities that use basic physics concepts to teach about climate change. The examples may help teachers notice and emphasize additional correlations between their own existing physics programs and climate science.

In my high school physics classes, climate change provides a unifying theme for the second semester. The curriculum outline in Figure 1 shows the full range of topics I have been able to connect to climate science in recent years. I spread these lessons across the entire semester, but they could also be taught consecutively as a separate unit in a physics or environmental science class.

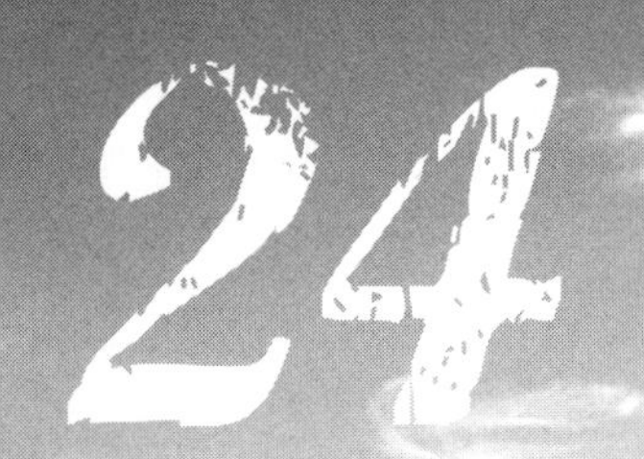

Figure 1

Sample climate physics curriculum outline

Thermal physics (February, about two weeks)

- Thermodynamics and heat flow
- Thermal expansion of seawater
- Heat engines and efficiency

Electromagnetism (March and April, about six weeks)

- Electric circuits and consumption in homes
- Resistive heating and transmission losses
- Electric motors and generators

Waves and light (April and May, about six weeks)

- Absorption spectra and the greenhouse effect
- Blackbody radiation and temperature
- The photoelectric effect and photovoltaic cells

Nuclear physics (May and June, about two weeks)

- Nuclear fission and nuclear power plants
- Nuclear fusion as a potential energy source

This chapter first explains how I introduce and integrate basic climate topics into units on heat and electromagnetic radiation. The second part of this chapter explains how a unit on electric circuits provides a useful context for discussing ways to solve the problem of climate change. Explanations of specific activities are kept brief to allow for a relatively comprehensive survey of the possibilities.

HEAT AND ELECTROMAGNETIC RADIATION

The second law of thermodynamics tells us that energy naturally flows from hot things to cold things, and the Stefan-Boltzmann law states that hotter objects radiate more energy than cold ones. These two laws form the basis for a very simple model of the interaction between the Sun, Earth, and empty space (Figure 2). I generally complete the following sequence as a class demonstration to facilitate discussion. However, the experiments described could easily be conducted by small groups if safety concerns posed by hot objects are adequately addressed.

To begin, place a beaker full of water on a hot plate and set the plate to low or medium. The water will warm at first but eventually will reach an equilibrium temperature. This happens because as the temperature of the water increases, the corresponding rate at which the water loses energy to the cooler surrounding air also increases. Eventually the increasing rate at which the water loses energy to its environment equals the rate at which the water absorbs energy from the hot plate, and the system reaches equilibrium. At this point, I explain that Earth and the Sun are in a similar state of equilibrium, in which the amount of energy Earth receives from the Sun is equal to the amount radiated back into space. Careful application of the Stefan-Boltzmann law, however, predicts a much lower temperature on Earth than is actually observed, so the model is incomplete.

Figure 2

Climate physics summarized

The Sun radiates energy, partly in the form of visible light, because it is hot. The Earth also radiates energy, but because it is not as hot as the Sun, most of Earth's radiation takes the form of an invisible form of radiation called *infrared*. Visible light from the Sun passes through the atmosphere and warms Earth's surface, but infrared light radiated from Earth is absorbed by carbon dioxide molecules in the atmosphere. The carbon dioxide in the atmosphere thus forms a blanket that keeps our planet warm enough for us to survive. Adding carbon dioxide to Earth's atmosphere, which we do when we burn fossil fuels, is somewhat like throwing an additional blanket on top of a contented sleeper.

Now place a loose lid on top of the beaker. The rate at which heat leaves the beaker will decrease and the temperature will begin to increase. This temperature increase will cause the rate of heat loss to once again increase until a new, higher-temperature equilibrium is reached. I explain to my students that carbon dioxide plays the role of the lid in Earth's atmosphere, and the carbon dioxide that exists naturally in the atmosphere keeps Earth's surface warm enough to support the various life forms that students study in their biology classes. Adding another lid to better insulate the top of the beaker simulates the consequences of adding carbon dioxide to the atmosphere by burning fossil fuels—a new higher equilibrium temperature will be reached.

Of course, this model is far from complete. Earth's atmosphere, surface, and oceans form a complex, interacting system. Some of these complexities are relevant, such as the fact that warming will melt reflective white ice caps or may lead to the formation of more reflective white clouds. Because some of the more poorly understood of these processes have the potential to partially counteract the warming effects of additional carbon dioxide, the processes have been much discussed by people who argue against strong government responses to climate change (Figure 3). Students should be encouraged to

Figure 3

Climate skeptics

There is no scientific debate about the role that carbon dioxide plays in keeping us warm, or about the fact that the concentration of carbon dioxide in the atmosphere is increasing significantly as we burn fossil fuels. Also, there is a broad and growing consensus that some warming has already been observed. However, skeptics of global warming argue that the media consistently exaggerates the need for preventative action. These people make the following three points:

- Current understanding of Earth's climate is not complete enough to support specific quantitative predictions of future warming. For example, the role of clouds is not well understood.
- Climate changes, if they occur, will have some positive effects, such as lengthened growing seasons.
- It would be very difficult and expensive to significantly limit carbon emissions.

One powerful skeptic is Senator James Inhofe of Oklahoma. His statements on the subject can be downloaded from the U.S. Senate website and used to stimulate classroom discussion. Inhofe is the ranking member of the Senate Environment and Public Works Committee; an interesting debate can be started by comparing Inhofe's views with those of the chairman of the committee, Barbara Boxer. A good place to start would be the March 21, 2007 hearing of the committee that featured former Vice President Al Gore *(http://epw.senate.gov/public/index.cfm)*. Will science ever resolve this controversy?

discuss these physical processes and political controversies, but teachers should also stress the degree to which most climate scientists expect the underlying effect, captured in the model just described, to predominate.

A more serious problem with the hot plate model is that it does not show how Earth's carbon dioxide "lid" keeps heat in without keeping sunlight out. In the same way walls are transparent to radio waves but opaque to visible light, carbon dioxide is transparent to visible sunlight but opaque to the infrared light emitted from the much cooler Earth. Infrared radiation causes the chemical bonds in carbon dioxide to vibrate and thus absorb the infrared radiation, trapping it in the atmosphere.

To show that radiation from hot and cold objects differs in wavelength, use a Bunsen burner to heat an iron rod until it is orange or even white hot. Then watch the color change as it cools and explain that the iron rod stops glowing not because it stops emitting electromagnetic radiation, but because our eyes are insensitive to the infrared radiation emitted by cooler objects. I have found the computer simulations *Blackbody Spectrum* and *The Greenhouse Effect*—available for free download from the Physics Education Technology project at the University of Colorado—to be invaluable in teaching about this concept (Figure 4).

In general, these problems with the hot plate model can be approached in two very different ways. I have generally chosen to stress the fact that all

Figure 4

Blackbody spectrum

How does the blackbody spectrum of the Sun actually compare to that of Earth? Students can compare the spectra using a simulation available at *http://phet.colorado.edu/en/simulation/blackbody-spectrum.*

Students adjust the temperature (using the interactive slider pictured) to see that the Earth and Sun spectral curves peak at very different wavelengths, and teachers explain that carbon dioxide absorbs the longer wavelength radiation emitted by the cooler Earth. This "greenhouse effect" is more explicitly modeled in another simulation available from the same source.

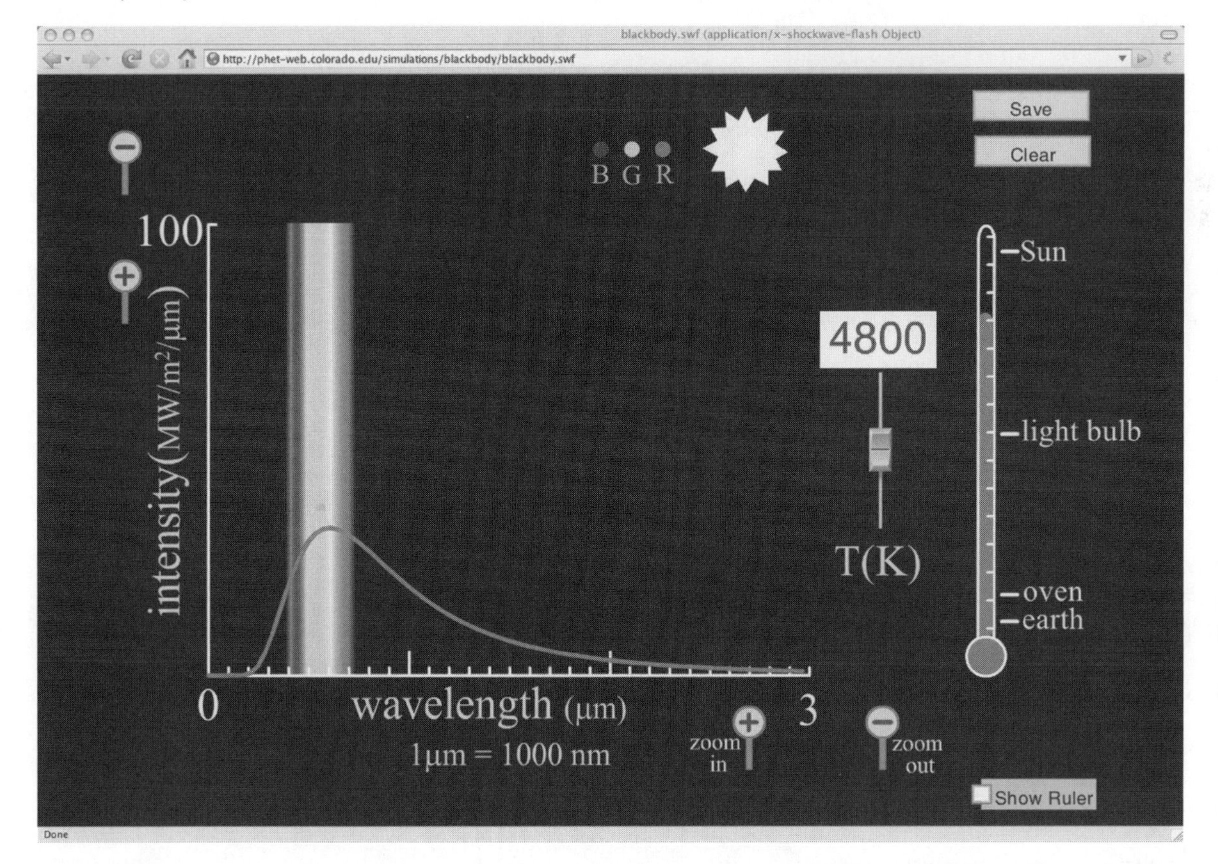

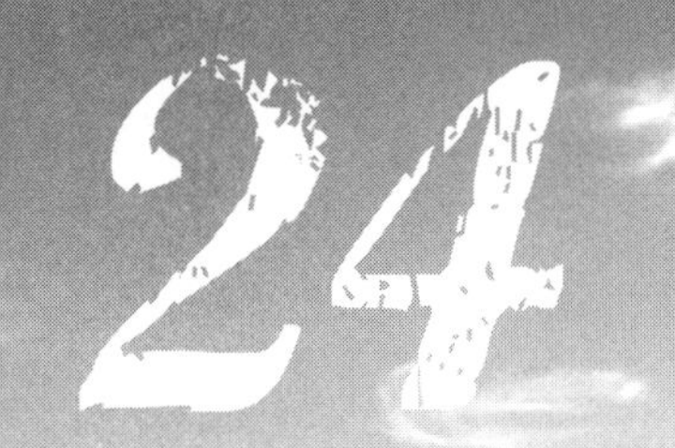

models are incomplete. While the hot plate/lid/beaker system is different from the corresponding Sun/atmosphere/Earth system in many fundamental ways, it has enough similarities to be useful and students can benefit from a discussion of its strengths and weaknesses. Another approach would be to challenge students to improve the model, for example, by using a heat lamp instead of the hot plate and using a glass lid.

ELECTRIC CIRCUITS

Some students are aware that carbon dioxide is created and released into the atmosphere when gasoline is burned. But few students may realize that when they turn on a light switch, a power plant somewhere burns a little more coal or natural gas and emits a corresponding amount of carbon dioxide. While the United States does generate some of its electricity from nonfossil sources, including nuclear reactors, production from these power plants is constrained by factors other than short-term variations in demand (Figure 5). A unit on electric circuits provides a useful context for introducing the idea that, by changing the way electricity is generated, transmitted, and consumed, carbon dioxide emissions can be decreased (Figure 6).

For students, the generation of electricity is the most interesting and least understood part of the sequence, so I discuss the operation of a traditional, fossil fuel–fired power plant at the beginning of the semester. These power plants convert chemical potential energy into heat and then electrical energy, so this provides a good opportunity to explain the relevant chemistry and review the law of conservation of energy. At this point each student in my class chooses an alternative, carbon-free energy source that they will research throughout the semester and present to the class at the end of the year.

My students also study two things to familiarize themselves with electricity consumption in their homes. First, students write down information about voltage, current, resistance, and power that is printed on electrical devices in their homes and use the information to estimate monthly electric consumption for different devices. Second, students examine their families' electric bills to determine the

Figure 5

Nuclear power

Growing concern about climate change may revive the debate about nuclear power. Nuclear plants are currently the largest source of carbon-free electricity in the United States, and some people have begun to argue for new reactors to combat climate change. However, others, including many of the environmentalists who worry most about climate change, oppose the construction of new reactors because of concerns about reactor safety, nuclear waste, and weapons proliferation.

After a lesson about nuclear fission, students can debate or discuss the following hypothetical scenario: A company proposes the construction of a new nuclear power plant in your state, asserting that the new plant is necessary if efforts to limit carbon emissions are to succeed. Would you support such a proposal? (**Note:** Both sides of this debate produce propaganda that can be misleading. For this to be a learning experience, the teacher must have some familiarity with the various arguments and be prepared to show impartiality for both sides.)

Figure 6

Motors, circuits, and energy efficiency

Efficiency is a measure of how little energy is required to accomplish a specific task. Students can conduct the following activity to develop a mathematical understanding of efficiency. Connect a small electric motor to a power source, ammeter, and voltmeter. Clamp the motor to a stand and wrap a string around the shaft. Use the motor to lift a small weight.

- Calculate the amount of gravitational potential energy gained by the weight. ($E_g = mgh$; m = mass, g = 9.8 m/s^2, and h = height.)
- Calculate the amount of electrical energy provided by the motor. ($E_e = VIt$; V = voltage, I = current, and t = time.)
- To calculate the efficiency, divide the gravitational energy by the electrical energy and write the answer as a percentage. The answer should be much less than one.
- Repeat the experiment with different weights and different speeds. Do these changes affect the efficiency of the motor? How? Why?

The discussion of efficiency can be extended by asking students to research how cars can be made to use gasoline more efficiently. An entire class should be able to compile a very long list. Students can also calculate savings that result from replacing incandescent lightbulbs with much more efficient compact fluorescent bulbs. In each case, using less energy cuts carbon emissions.

cost of each kilowatt-hour of electricity. These activities are included here not because they are new or creative, but rather to suggest that teachers should use the additional opportunities to point out the relationship between electrical consumption and carbon emissions.

GETTING STARTED WITH CLIMATE PHYSICS

There was no single point in time when I decided to organize the second semester of my physics classes around the theme of climate change. Instead, the more I learned about the topic, the more I noticed connections with my physics curriculum. Hopefully teachers who read this chapter will also realize these connections and, ideally, find additional connections not identified here.

Biology teachers have, in recent decades, been able to use a constant stream of new discoveries and very relevant news stories to instill appreciation for

science as an ongoing, relevant, and often controversial process (e.g., the public discussion about evolution).

In contrast, the standard physics curriculum includes relatively few such opportunities. By teaching about climate change, physics teachers can help students see physics as useful, relevant, and interesting. Teachers will find that students begin to ask questions about discussions heard at home, things seen on television, and stories read in newspapers. These student-centered discussions are important because they excite the interest of students in a way that no amount of isolated direct instruction can.

Interwoven instruction about climate change makes lessons about heat, light, electricity, and nuclear physics more relevant and interesting for both students and teachers. Perhaps more importantly, knowledge of basic physics can help students better understand and eventually influence political debates about climate change.

Resources

Boeker, E., and R. Grondelle. 2001. *Environmental science: Physical principles and applications*. United Kingdom: Wiley.

Houghton, J. 2004. *Global warming: The complete briefing*. United Kingdom: Cambridge University Press.

Sweet, W. 2006. *Kicking the carbon habit: Global warming and the case for renewable and nuclear energy*. New York: Columbia University Press.

Chapter 25
SEEING THE CARBON CYCLE

By Pamela Drouin, David J. Welty, Daniel Repeta, Cheryl A. Engle-Belknap, Catherine Cramer, Kim Frashure, and Robert Chen

The most important biochemical reactions for life in the ocean and on Earth are cellular respiration and photosynthesis. These two reactions play a central role in the carbon cycle. The ocean-based carbon cycle is highly relevant to students because of its key role in global warming. The Earth's atmosphere maintains the temperature of the Earth within a relatively narrow range that can support life. The atmosphere is made up of gases such as carbon dioxide, methane, water vapor, and others that allow radiant energy to pass through, but prevent heat loss back into space. As the composition of the atmosphere changes due to more carbon dioxide emissions produced by humans generating power, the insulating capacity of the atmosphere increases and more heat is trapped. As carbon dioxide levels increase, so does the average temperature of Earth. Besides the obvious solution of cutting carbon dioxide emissions, which does have

economic drawbacks, the possibility of sequestering carbon dioxide in long-term storage stages of the carbon cycle is an experimental option (Figure 1). The following lessons outline a classroom experiment that was developed to introduce students to the carbon cycle. The experiment deals with transfer of CO_2 between liquid reservoirs and the effect CO_2 has on algae growth (Figure 2). It allows students to observe the influence of the carbon cycle on algae growth, explore experimental design, collect data, and draw a conclusion.

TEACHING THE CARBON CYCLE

In this activity, students observe how different levels of carbon dioxide in the atmosphere affect the growth of algae. Before getting started, students review the steps in an experiment and write an "If…, then…" prediction with a "because" clause on how the carbon cycle might influence growth of algae. For example, "If there is more CO_2, then there will be less algae growth because too much CO_2 will kill the algae." Or, "If CO_2 increases, then algae growth increases because it is used by algae during photosynthesis to make sugar."

Figure 1

Causes of increased CO_2 emissions and effects/results

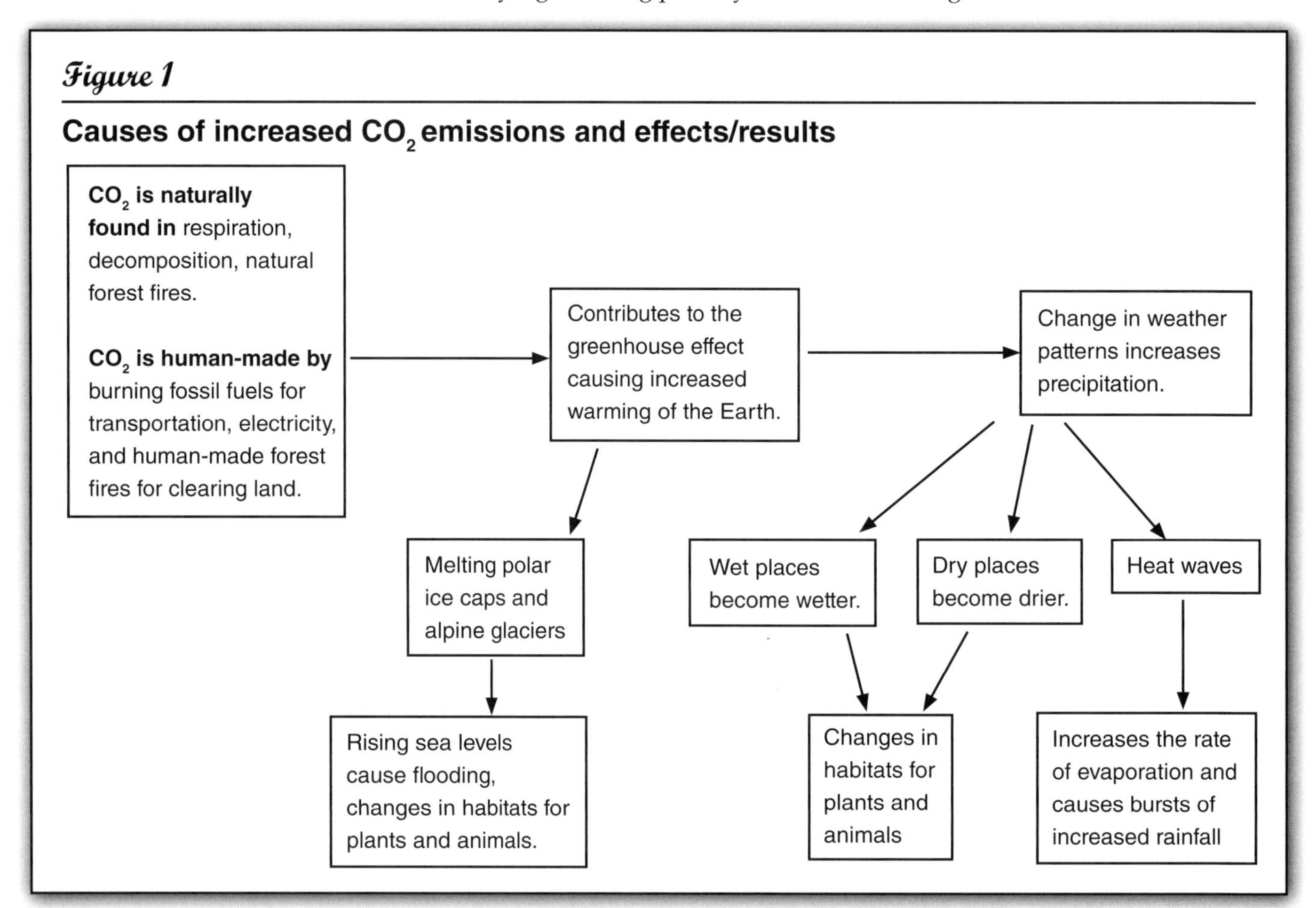

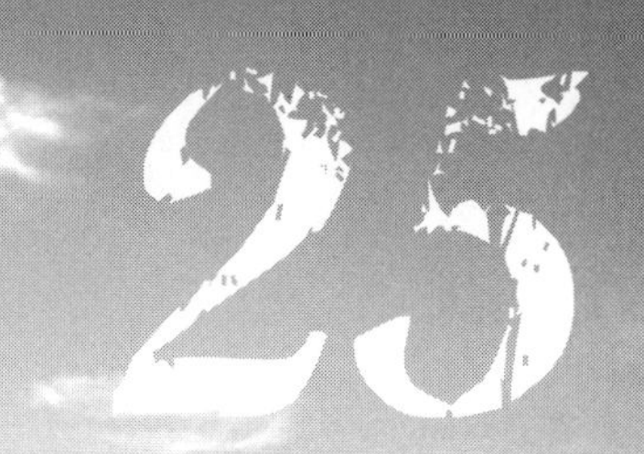

Figure 2

Experimental design's relationship to photosynthesis

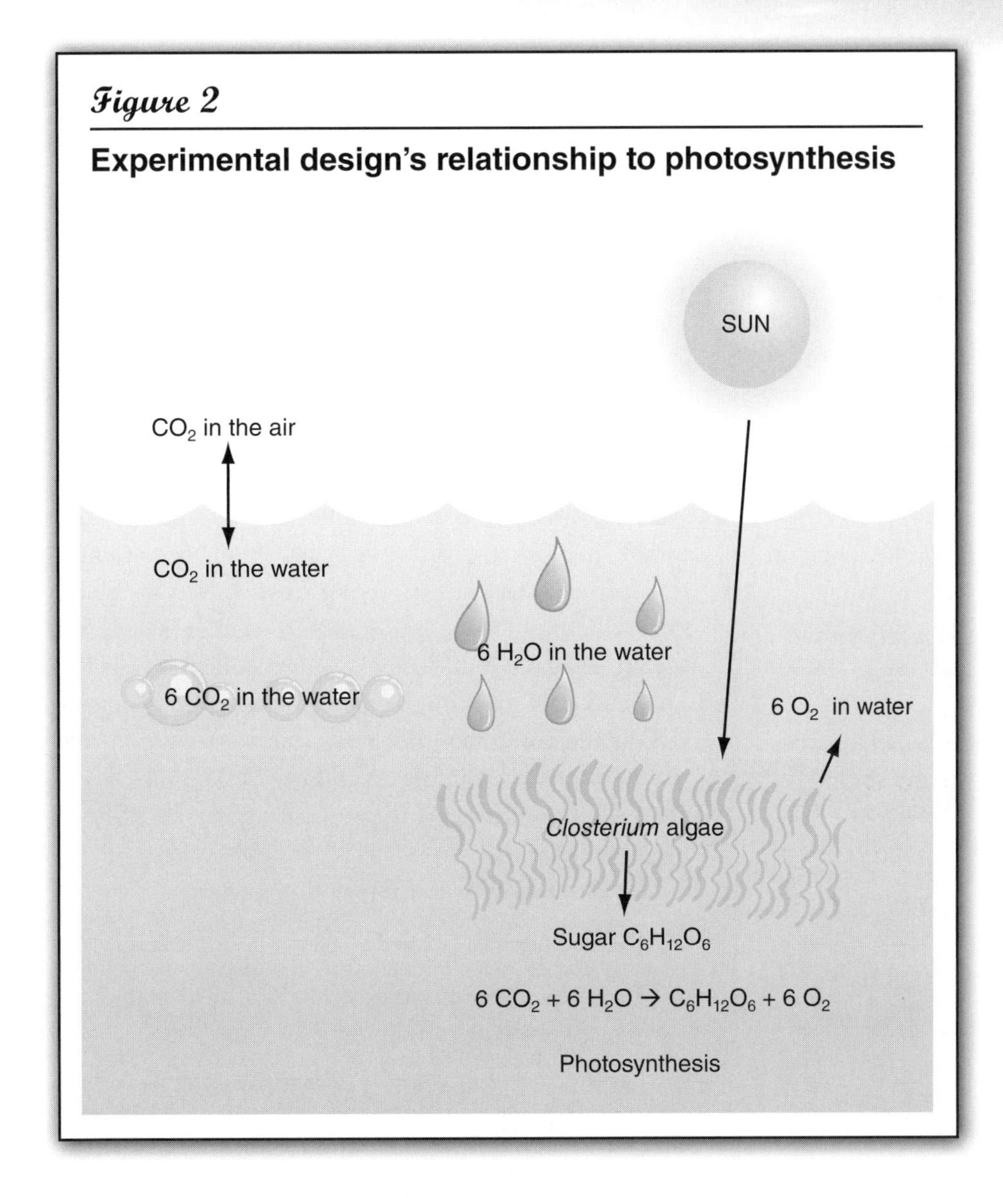

EXPERIMENTAL DESIGN

This single experiment is set up by the teacher for multiple classes to observe and analyze because

- the amount of equipment (Figure 3, p. 220) required is impractical for multiple setups;
- the experiment is simple enough to set up for a science teacher, but too complex for a student to handle;
- there is a 0.4 molar potassium hydroxide (KOH) solution in the negative control that presents a safety hazard to students; and

Figure 3

Materials needed for the experiment

- 2 L of distilled water
- 1 culture of actively growing *Closterium* algae (Carolina Biological Supply #HT-15-2115, $4.95)
- 1 tube of 50X Alga-Gro (Carolina Biological Supply #HT-15-3751, $27.60)
- Schultz, 10-15-10 Plant Food Plus ($2.95)
- 3 half-gal. Rubbermaid screw-cap containers
- 4 600 ml Pyrex beakers (Ball canning jars work as a substitute, but must be able to fit inside the half-gallon container; plastic containers may have impurities that inhibit algae growth; beakers are cleaned and rinsed twice with distilled or bottled water)
- 0.5 L of carbonated water (seltzer water)
- 0.6 L 0.4 molar potassium hydroxide solution (obtain from a chemistry teacher), wear safety glasses when handling
- 500 ml graduated cylinder
- 50 ml graduated cylinder
- rubber band
- plastic wrap or parafilm
- glass-marking pen
- 2 fluorescent lights
- electric on-off timer

- the amount of algae transferred to each beaker needs to be consistent, as large differences in algae settling time could influence how much algae is in each group.

The students' responsibilities during the experiment are to make observations, collect data, reach a conclusion on how different levels of CO_2 affect algae growth, and then apply that knowledge to determine how greenhouse gases might be tied to global warming.

The experiment consists of a set of three half-gallon containers partially filled with solutions that establish different CO_2 levels. Inside each container is a culture of actively growing freshwater *Closterium* algae (Figure 4). Each half-gallon container is a closed system that prevents gas exchange with the outside. The beaker of growing algae sits in a liquid reservoir within the container. The reservoir solution controls the amount of CO_2 in the air (Figure 5). The *Closterium* algae is grown under a fluorescent light on a 12-hour light/dark cycle.

The experiment consists of three groups:

(1) the normal control with a tap water reservoir;

(2) the experimental group with a carbonated water reservoir; and

(3) the negative control with a 0.4 M KOH reservoir.

The tap water represents the natural condition of dissolved gases in the environment. The experimental group is a 1:2 mixture of bottled carbonated water to tap water, which is enriched with CO_2 gas. CO_2 reacts with water to form carbonic acid (H_2CO_3).

$$CO_2\,(g) + H_2O\,(l) \rightarrow H_2CO_3\,(aq)$$

Because 0.4 M KOH reacts with CO_2 to form K_2CO_3 (s), KOH effectively depletes CO_2 from the atmosphere and water of the closed system.

$$2\,KOH\,(aq) + CO_2\,(g) \rightarrow K_2CO_3\,(s) + H_2O\,(l)$$

Depending on the level of your students, you can explain the reactions occurring inside each container, or simply label them as Depleted CO_2, Regular CO_2, and High CO_2. Students should be instructed that as the algae culture grows the cells will become more crowded and the culture will become a darker green. This is called *cell density* and indicates a higher algae number from greater growth.

After the teacher has set up the experiment, students observe the systems over the next two weeks and record their observations in a data table (Table 1).

At this point in the lesson, students start making connections between carbon dioxide levels and algae growth. Specifically, they begin to realize that algae depend on photosynthesis for survival, which is why algae live within the sunlight zone in the ocean. To further student understanding, the following equations for photosynthesis and cell respiration are introduced and explained with the use of guided questions.

$$6\,CO_2 + 6\,H_2O \text{—Photosynthesis} \rightarrow C_6H_{12}O_6 + 6\,O_2$$

$$C_6H_{12}O_6 + 6\,O_2 \text{—Cell respiration} \rightarrow 6\,CO_2 + 6\,H_2O$$

- What type of gas will collect in the container as photosynthesis occurs?
- How would photosynthesis be affected if too little or too much carbon dioxide were present?
- How would respiration be affected if too little carbon dioxide were present?
- In which container would you expect to find the greatest amount of algae growth? Why?
- In which container would you expect to find the smallest amount of algae growth? Why?

More advanced students can be introduced to the role of carbon, hydrogen, and oxygen in the carbon cycle. They can also study the atomic structure, electron orbitals, valence electrons, and the molecules of cellular respiration and photosynthesis: carbon dioxide, glucose, oxygen, and water. The relationship between the products and the reactants can also be discussed and the balancing of the equation examined.

RESULTS

During this experiment, students will observe that the experimental group with carbonated water grows better than the normal control. For this reason, the experimental group has more algal cells and is a darker green. The negative control grows less well than the normal control; consequently, the resulting culture has less algal cells and is a lighter green. These results demonstrate what would be predicted from the photosynthesis reaction: When there is more carbon dioxide, the algae grow better. When there is less carbon dioxide, the algae grow poorly. The negative control of 0.4 M KOH will show little detectable algae growth. Students should compare these observations to their predictions about how the CO_2 levels would affect the algae growth.

Figure 4

Growing *Closterium* algae

1. Add 50X Alga-Gro to 1 L of distilled water to make 1X Alga-Gro.
2. Transfer to the glass beaker 400 ml of 1X Alga-Gro.
3. Add 1 drop of plant food.
4. Transfer half the volume of the *Closterium* algae culture to the beaker.
5. Cover with plastic wrap and secure with a rubber band.
6. Punch 6 holes in the plastic wrap.
7. Place in a sunny window or under a fluorescent light for about a week. As the culture grows, the green color will intensify.

As a follow-up, ask students the following questions to assess what they learned about the process:

- What is the reaction that this experiment demonstrates?
- What part of the reaction does the experiment test?
- What would you expect to be the outcome for the algae in the experimental group?
- What would you expect to be the outcome for the algae in the negative control?
- What is going on in each chamber of the experiment?

- Identify ecological/environmental conditions that are similar to the conditions found in each container.

CONCLUSION

Through this experiment students learn that CO_2 has a positive effect on algae growth and, in fact, is essential to the growth of algae. At the end of the activity, students should be able to make the connection that algae growth is dependent on photosynthesis. Through a discussion of deforestation and consumption of fossil fuels, students should be led to the concept of long-term carbon reservoirs being depleted and an increased amount of carbon being put into the atmosphere. (One way to remove more carbon dioxide from the atmosphere is to stimulate algae and plant growth.) Finally, the concept of food chains built upon photoautotrophic organisms that use sunlight, water, and carbon dioxide to make sugar should be introduced. The sugars then support the energy demands of the herbivore heterotrophs and carnivore heterotrophs. Students should be able to grasp the concept that since most life is directly or indirectly dependent on photosynthesis, if photosynthesis stopped due to sunlight being blocked, then all life on Earth would be in jeopardy. However, if animal respiration stopped, which is only one of several sources of carbon dioxide for plants, plants could survive from carbon dioxide released by volcanoes and the ocean.

EXTENSIONS

You can supplement this activity by measuring the amount of oxygen released by the algae and performing serial dilutions and filtrations to further quantify the amount of algae growth. By performing serial dilutions, it is possible to determine the algae concentration per ml, so

Figure 5

Experimental setup

1. Transfer into each half-gallon plastic container 600 ml of one type of reservoir liquid:
   - normal: 600 ml of tap water
   - experiment: 400 ml of tap water and 200 ml of carbonated water
   - negative: 600 ml of 0.4 M potassium hydroxide
2. Label beakers *Normal*, *Experiment*, or *Negative* with a marker.
3. Transfer 350 ml of 1X Alga-Gro to each beaker.
4. Transfer 50 ml of actively growing algae into each beaker.
5. Resuspend the algae after each transfer.
6. Carefully insert beakers into half-gallon containers.
7. Close and wrap tightly with plastic wrap or parafilm.

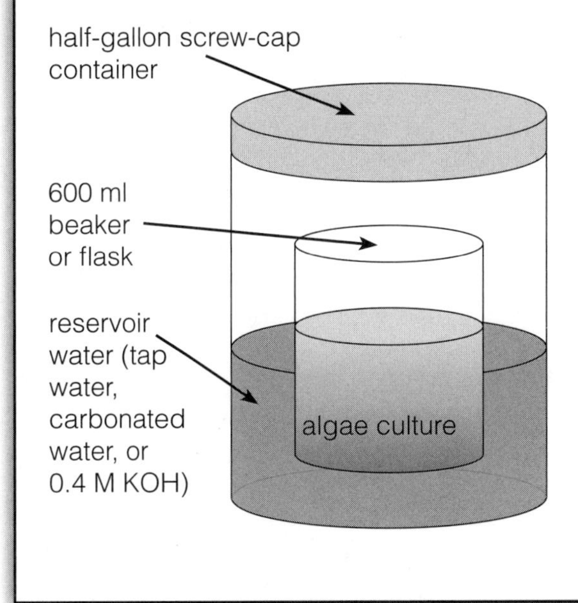

Table 1

Student data table

Day	Observation	Depleted CO_2	Regular CO_2	High CO_2
1	Darkness (1–10)			
	Growth (1–10)			
3	Darkness (1–10)			
	Growth (1–10)			
7	Darkness (1–10)			
	Growth (1–10)			
10	Darkness (1–10)			
	Growth (1–10)			
14	Darkness (1–10)			
	Growth (1–10)			

students can compare 1×10^7 cells per ml to 1×10^9 cells per ml. This will allow them to find the dilution where the algae cells limit out following dilution. Filtration allows all of the algae to be captured on a solid substrate for densitometry or better photographic documentation.

Acknowledgments

This experiment grew out of collaborative work in the Ocean Science Education Institute (OSEI), which develops and implements high-quality ocean science education for middle school students through projects that connect with existing district curricula and effective science educational practices. OSEI is a project of the Center for Ocean Science Education Excellence–New England (COSEE-NE), an NSF-funded partnership between the New England Aquarium, the University of Massachusetts/Boston, and the Woods Hole Oceanographic Institution (WHOI). The OSEI format includes a five-day workshop, numerous classroom visits, and two follow-up days. During the 2004–2005 school year, researchers and Massachusetts

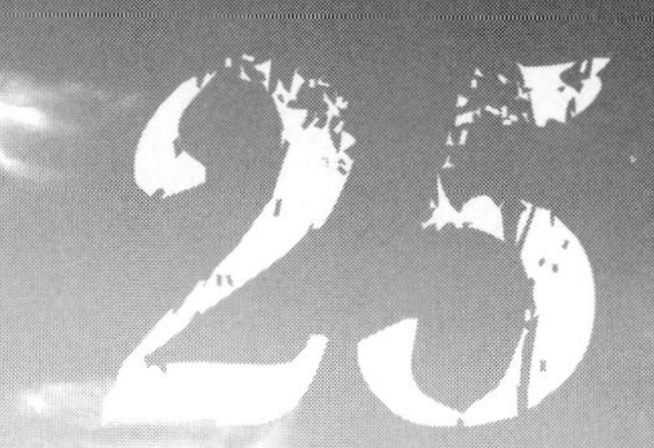

middle school teachers, district science coordinators, and facilitators teamed up to produce districtwide, inquiry-based science curricula for middle school students based on current ocean science research. To find out more about OSEI and other COSEE-NE programs, please visit our website at *www.cosee-ne.net*.

Chapter 26

HOW MUCH CARBON IS IN THE FOREST?

A PROJECT-BASED SCIENCE INVESTIGATION OF TREES' ROLE IN OFFSETTING GLOBAL WARMING

By Leah Penniman

At the start of class, my coteacher and I presented our students with the following challenge: "How much carbon is stored in the Normanskill Preserve?" We then told them they had one month to investigate and present their results, and asked, "What do you need to begin?"

This hook served to introduce our integrated Algebra I and Environmental Science class to the next project in our 100% project-based curriculum. More information on designing Standards-based projects is available online (see "On the web").

This chapter describes the project in more detail and what students learned along the way.

THE PROJECT

The Normanskill Preserve is a forest in upstate New York, where our school is located. After we had presented the class with the driving question—"How much carbon is stored in the Normanskill Preserve?"—students, working in teams of four, began compiling their "need-to-know" lists. Most lists included some of the following:

- How big is the forest?
- Is carbon found in trees?
- Why is anyone interested in carbon?
- How can you weigh a forest?
- What is the best way to present the results?
- Do we get to go outside?

Carbon sequestration
Carbon sequestration is the process by which growing trees and plants capture carbon dioxide from the atmosphere and store it in the form of biomass. A carbon sink has a faster carbon dioxide removal rate than its rate of release. Young, fast-growing trees are more effective at sequestering carbon than mature trees. Enhancing carbon sinks is one way in which society can combat global warming (EPA 2010).

Using these need-to-know lists as living documents and guides for classroom scaffolding, we then gathered the information resources, physical materials, and outside experts needed to support students in this project.

To begin, students conducted background research to better understand the concept of carbon sink (see "Carbon sequestration") and the important role that forests play in the environment. Using online resources provided by the New York Department of Environmental Conservation and the Environmental Protection Agency (EPA) (see "On the web"), students created "carbon sink" concept maps on butcher block paper. They then shared their work through whole-class discussion, which built excitement around the role of forests in offsetting global warming.

SAMPLE PLOTS

To determine the amount of carbon stored in their study site—the Normanskill Preserve—students began by determining its area. Most teams used Google Earth (see "On the web") to view satellite imagery of the forest and the "polygon" tool to outline the forest's boundaries. They then broke the irregular-shaped polygons into rectangles and triangles and used the "ruler" tool to measure the dimensions. With the appropriate formulas, students

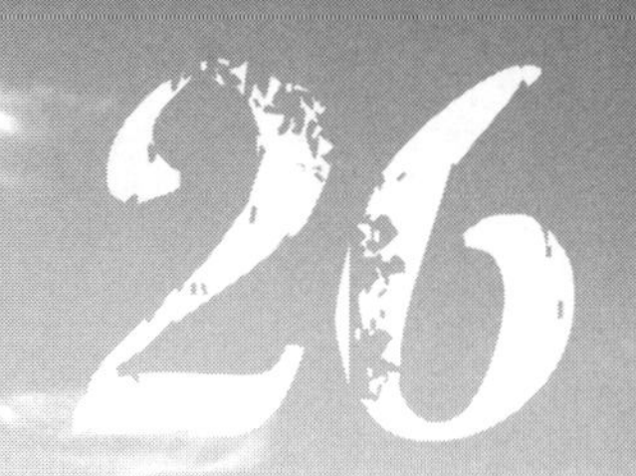

were able to find the area of each individual shape and the sum of these shapes for the forest's total area.

Other teams chose to work manually. Using a paper map, these students overlaid a transparent grid or divided the forest into small triangles and rectangles and calculated the sum of their areas.

After determining the total area of the forest—approximately 1.3 million m^2—students realized there was no way to "measure" the entire forest in just one month.

We then had a guest speaker from the United States Geological Service (USGS) visit the classroom to answer students' questions about methods of unbiased sampling and study design. After hearing the speaker, students were energized with new ideas.

They then set to work devising methods to randomly identify sample plot locations within the forest. Their suggested methods varied, from using an online random-number generator and corresponding Global Positioning System (GPS) coordinates to tossing rice onto their paper map. Ultimately, students selected a systematic sampling method in which plots were laid out uniformly along an imaginary grid overlaying the forest. They decided to use a uniform

A student measures the distance between the observer and the tree.

All photos courtesy of the author

Safety note

Teachers should always visit outdoor areas to review potential safety hazards prior to conducting an activity. They need to check with the school nurse about student medical issues (e.g., allergies, asthma) and be prepared for medical emergencies. Students should have a means of communication (e.g., cell phones, radios) in case of emergency. Parents should also be informed of the on-site field trip.

Proper footwear (i.e., closed-toe shoes), clothing (i.e., long-sleeve shirts, pants, hats), and eye protection (i.e., safety glasses or goggles) are required for fieldwork. Students should also be cautioned about poisonous plants (e.g., ivy, sumac), insects (e.g., bees, ticks, mosquitoes), and hazardous debris (e.g., broken glass or other sharps). Students should wash their hands with soap and water after returning to the classroom.

plot size for each team so they could share data easily and have a larger sample size. Each team was assigned two, 4.6 × 4.6 m (15 × 15 ft.) plots. They would then take the results from their sample plots and combine them to cover the entire forest.

METHODOLOGY

Through online research on carbon and information gleaned from a guest speaker from the Department of Environmental Conservation, students discovered that the Intergovernmental Panel on Climate Change (IPCC) estimates that carbon makes up about 50% of the mass of living trees but is also present in the detritus, soil, and underground biomass of the forest (Pearson, Brown, and Birdsey 2007). They also learned that scientists have methods for estimating biomass that do not involve actually "massing" a tree. By determining volume, and using known densities, mass can be calculated.

Student teams then worked to determine how they would find the volume of trees in their plot. Most approximated the shape of a tree as a cylinder; others used a more complex formula (Pearson, Brown, and Birdsey 2007):

$$v = 1/3\pi \times h \times [r1 + r2 + r1 \times r2]$$

In this equation, h is the height of the tree, $r1$ is the radius at the base of the tree, and $r2$ is the estimated radius at the top of the tree; students then multiplied the total volume by 1.2 to account for the approximate volume of the tree's branches and leaves (Pearson, Brown, and Birdsey 2007).

To hone their estimation skills, students placed objects at the far end of the classroom and estimated their radii. They then measured the actual radii of each object to evaluate their estimation skills and make improvements.

Ultimately, students decided to measure tree height by triangulation, using the tangent function. In a given right triangle, the tangent of an acute angle is equal to the length of the opposite side divided by the adjacent side. To calculate tree height, the observer stands at a distance from the tree so that the angle sighted to the top is between 30° and 60°. The angle is then measured with a clinometer.

The adjacent side of the triangle is the distance between the observer and the base of the tree, which can be measured using a meterstick or measuring tape. The opposite side of the triangle is the unknown and represents the height of the tree minus the height of the observer's eye.

The Global Learning and Observations to Benefit the Environment (GLOBE) program has an excellent description of this method for determining tree height (see "On the web"). Students practiced this method before going into the forest

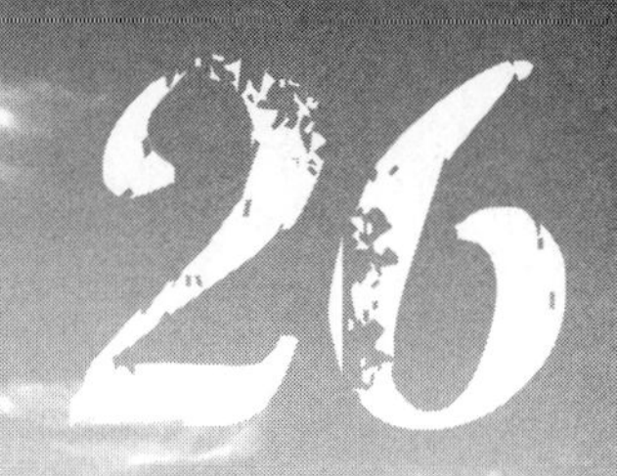

by conducting a lab in which they estimated their own heights using the tangent, then measured their actual heights and determined the percent error.

Students decided they would find the tree trunk's radius by measuring its circumference at chest height and working with the formula for a circle's circumference ($\pi \times 2r$). They then found the tree's volume using the formula for the volume of a cylinder ($\pi \times r^2 \times h$). Given that volume multiplied by density equals mass, students simply had to look up the density of the trees in their plot to calculate mass. They spent time familiarizing themselves with the Peterson field guides for tree species identification and locating an online lumber density table (see "On the web"). In accordance with the IPCC's recommendation, they estimated that 50% of their tree's mass was made up of carbon.

A student uses a clinometer to measure the angle made by the imaginary line between the observer's eye and the top of the tree.

FIELDWORK

The process of researching and developing a methodology (described in the previous sections) took about two weeks of class time. Students were then ready to transition into the implementation phase. To prepare for fieldwork, my coteacher and I conducted safety workshops and assessments (see "Safety note" on p. 229). Students need to stay with their assigned teams and at their assigned plots, and proper care of equipment should be stressed. Each student team had a 100 m measuring tape, clinometer, tree identification guide, data-collection form, compass, and flagging tape. They also carried a cell phone in case of an emergency. (All teams were within ear and eye shot, but we wanted to be prepared in case a team went out of bounds.) The fieldwork took less than two, 90-minute class periods, and students worked efficiently since they had detailed and practiced their methods in advance.

RESULTS

Back in the classroom, teams worked furiously to convert their raw data (i.e., angle to the top of the tree, observer's distance from the tree, height of the

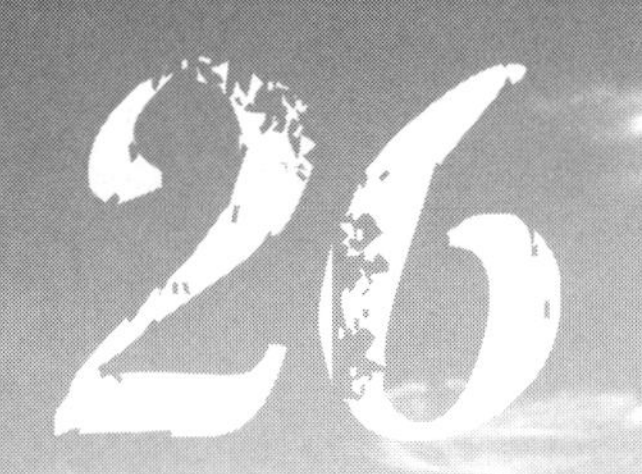

observer's eye, the tree trunk's circumference at chest height, the tree species) into calculations of the mass of carbon in their trees. We displayed a master chart for the class's data on the interactive white board, and as teams completed their analysis, they entered their data into the chart. (If a computer and projector are available, entering data into an Excel spreadsheet facilitates data analysis; for example, calculating class averages.)

Outliers were quickly identified by student comments such as, "That number looks way too big!" and "What units were you using?" Some teams went back to their original data and made adjustments. The entire class had ownership over the quality of the numbers because each team's numbers affected the class results. In this way, students acted as peer reviewers for their classmates' data and calculations.

Once the master chart for the amount of carbon in each plot was complete, assuming representative sampling, students were able to determine the amount of carbon in the entire forest with a simple proportion. For the sake of time and simplicity, most teams ignored the carbon stored in nonliving and underground biomass.

Teams reported their final results in kilograms, but also converted to other units to enhance the meaning of the numbers. For example, 1,000,000 kg of carbon (the rounded class average) sounds impressive, but using the EPA statistic that an average car uses 581 gallons of gasoline annually and a gallon of gas contains 2.4 kg of carbon, students can calculate a more illuminating equivalency (Coe 2005). The amount of carbon stored in the living trees of the Normanskill Preserve is approximately equal to the annual emissions of 720 cars:

$$\frac{1{,}000{,}000 \text{ kg}}{581 \frac{\text{gal}}{\text{car}} \times 2.4 \frac{\text{kg}}{\text{gal}}} - \frac{1{,}000{,}000 \text{ kg}}{1{,}394 \frac{\text{kg}}{\text{car}}} = \sim 720 \text{ cars}$$

PRESENTATION

Student teams presented their results in a formal scientific paper and presentation, both of which were scored using detailed rubrics (see "On the web"). Students also completed individual written assessments of their content acquisition. This revealed that student strengths were in understanding scientific study design and bias, the importance of carbon sinks, and basic trigonometry. Students' major weakness was in algebraic proportions, which we attribute to the fact that the project only had one application of this concept—scaling the data collected in the sample plots up to the entire forest. The quality of student reports varied depending on the team's time-management

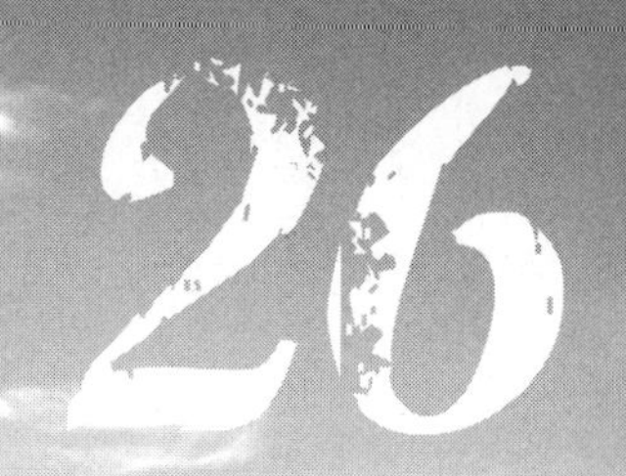

skills and the extent to which the report was revised.

CONCLUSION

Students demonstrated a high level of engagement and critical thinking throughout the project. Even reluctant learners found themselves compelled to squeeze through thickets and mud to get to their plot and report the data their classmates were depending on. Students scored each other highly on peer-to-peer collaboration evaluations, including comments such as "Everyone on the team had a different strength, and we all worked together."

In the postproject reflection, students ranked this project among their favorites, noting its contribution to improving society. As one student reflected,

> *"It is very important to communicate the importance of the trees and the forests, and how much carbon they store for us. Since forests are being cut down and state parks are closing, it is our duty to show people why we need to conserve and save the forests by calculating how much carbon their trees actually hold."*

A student measures a tree's circumference so that she can calculate its volume.

On the web

Designing Standards-based projects: *http://pbl-online.org/pathway2.html*

Environmental Protection Agency's carbon sources and sinks information: *www.epa.gov/climatechange/emissions/co2_human.html*

GLOBE protocol for measuring tree height: *http://classic.globe.gov/tctg/land_prot_biometry.pdf?sectionId=210&rg=n&lang=EN*

Google Earth: *http://earth.google.com*

Lumber density table: *www.simetric.co.uk/si_wood.htm*

N.Y. Department of Environmental Conservation's Trees: The Carbon Storage Experts: *www.dec.ny.gov/lands/47481.html*

Project rubric: *www.nsta.org/highschool/connections.aspx*

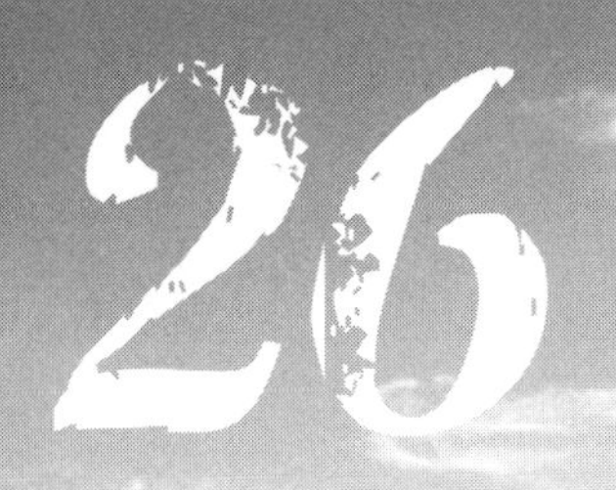

References

Coe, E. 2005. Emission facts: Average carbon dioxide emissions resulting from gasoline and diesel fuel (EPA420-F-05-001). U.S. Environmental Protection Agency, Office of Transportation and Air Quality. *www.epa.gov/oms/climate/420f05001.htm*

National Research Council (NRC). 1996. *National science education standards.* Washington, DC: National Academies Press.

Pearson, T. R. H., S. L. Brown, and R. A. Birdsey. 2007. *Measurement guidelines for the sequestration of forest carbon.* Newtown Square, PA: United States Department of Agriculture Forest Service. *www.nrs.fs.fed.us/pubs/3292*

The University of the State of New York. 1996. *Learning standards for math, science, and technology.* Albany, NY: The University of the State of New York. *www.p12.nysed.gov/ciai/mst/pub/mststa1_2.pdf*

U.S. Environmental Protection Agency (EPA). 2010. Human-related sources and sinks of carbon dioxide. *www.epa.gov/climatechange/emissions/co2_human.html*

Chapter 27

THINKING LIKE AN ECOLOGIST

STUDENTS MAKE CONNECTIONS BETWEEN THEIR INFLUENCE ON GLOBAL CHANGE AND CURRENT FIELD RESEARCH

By Jenn Carlson

Students make connections with what they know. Unfortunately, many students have never been formally introduced to human impact issues and hold a number of misconceptions. Correcting these misconceptions is imperative if students are to be cognizant of their everyday effects on the environment and make educated, ecologically conscious decisions regarding their actions in the future.

So where do we start as science teachers? This chapter presents a lesson in which students examine current field research on global change. In particular, students investigate the effect of carbon dioxide and tropospheric ozone on ecosystems by applying their knowledge of scientific inquiry and photosynthesis. The goal of the activity is for students to think like ecologists and draw connections between the data and their everyday energy choices.

CURRENT GLOBAL CHANGE RESEARCH

Since the Industrial Revolution, the documented increase in the atmospheric concentration of carbon dioxide (CO_2) has increased by over one-third, from about 280 ppm in 1850 to about 380 ppm in 2006. This number is expected to continue to rise as fossil fuel use persists and land is cleared of vegetation for development and agriculture. Beyond predictions such as rising sea levels, changes in weather patterns, and regional climate shifts, increased CO_2 is known to directly affect plant photosynthesis and water use (Karnosky et al. 2003). This could potentially increase the plant growth in both agricultural and natural ecosystems. CO_2, however, is not the only gas in our atmosphere that influences plant growth. Ozone (O_3), when found in the troposphere near ground level, is considered a pollutant that is detrimental to plant growth and also poses human health problems. The majority of tropospheric O_3 is formed by the reaction of sunlight on air containing carbon monoxide, nitrogen oxides (NOx), and volatile organic compounds (VOCs). The major anthropogenic sources of these O_3 precursors include motor vehicle exhaust, industrial emissions, and chemical solvents.

I became interested in the impacts of increased O_3 and CO_2 when I participated in a graduate class, Global Change for Teachers, offered by Michigan Technological University in Houghton, Michigan. The most memorable experience of the class was visiting a current global change research facility—the Aspen FACE research site in Wisconsin. Aspen FACE is a "multidisciplinary study to assess the effects of increasing tropospheric O_3 and CO_2 levels on the structure and function of northern forest ecosystems," specifically, on the growth of aspen trees (Facts II: The Aspen FACE Experiment 2005). FACE is the acronym for Free-Air Carbon dioxide Enrichment experiment. More than 25 FACE research sites are scattered across the globe, each with almost 100 scientists conducting long-term research on the ecological effects of expected CO_2 gas in a natural environment. The experimental design at FACE involves twelve 30 m tree rings where scientists control the concentrations of CO_2 and tropospheric O_3 to simulate the levels expected in 50 years. The effects of the gases on ecosystem balance, including changes in plant growth and soil carbon, can then be assessed (Facts II: The Aspen FACE Experiment 2005).

Unlike a greenhouse, FACE's open-air design allows the ecosystems to develop as naturally as possible; including allowing the growth of tall trees. The data collected at the FACE research sites are helping scientists anticipate the possible impacts of global change—both positive and negative.

BRINGING ASPEN FACE TO THE CLASSROOM

After spending a day in the field among the researchers at Aspen FACE, I wanted to bring this experience back to the classroom. The case studies examined in biology classes are often decades old and conducted in a lab, with corresponding activities that ask students to verify what has already been discovered. Aspen FACE is an example of current field research where the systems are complex and all of the possible variables are not known; the researchers are finding many trends, but there is still much uncertainty for the

Figure 1

Student responses to a free-write exercise

This exercise assesses student understanding of the issues of greenhouse gases, global climate change, and tropospheric ozone.

Student A: *Aerosols, we use aerosol hairspray that deplete the ozone. One big negative effect is skin cancer, this damages your skin and can hurt you a lot. On plants, it helps them grow a lot and sometimes overgrow.*
(**Misconception:** CFCs and depleting ozone relates to the stratospheric ozone layer, not tropospheric ozone which would harm, not help, the plants. In addition, aerosols no longer contain CFCs in the United States.)

Student B: *CO_2 is the greenhouse gas that is of most concern. Burning fossil fuels, automobile exhaust, cutting and burning forests worldwide, and factory pollution are human actions that have contributed to the emission of CO_2.*
(**Correct:** Although other gases like water vapor and methane also function as important greenhouse gases; in fact, the warming effect of water vapor is greater than that of CO_2. Ozone in the upper troposphere also acts as a greenhouse gas.)

Student C: *Bad ozone found near the ground is a health hazard. It's known to cause coughing, congestion, and chest pains. It's also known to worsen conditions such as asthma and bronchitis. Humans aren't the only organisms damaged by negative ozone, however. Plants are also damaged; bad ozone stops the plant from photosynthesizing.*
(**Correct.**)

Student D: *In recent years CO_2 and ozone have played a big part in the environment. The increase in CO_2 is depleting the ozone and killing plants.*
(**Misconception:** CO_2 does not deplete the ozone layer and tropospheric ozone is detrimental to plant growth [Karnosky et al. 2003]. CO_2 actually increases plant growth.)

future. Analyzing an experiment such as Aspen FACE would allow students to think like ecologists. In addition, students would

- reason through the scientific inquiry used at Aspen FACE;
- analyze and interpret the data from the Aspen FACE site;
- make connections to the effects humans (specifically themselves) have on ecosystems;
- propose solutions to global climate issues; and
- understand the uncertainties of research investigating current environmental issues.

UNDERSTANDING THE INDEPENDENT VARIABLES: CO_2 AND O_3

Before introducing students to Aspen FACE, they must understand the independent variables in the experiment: CO_2 and O_3. I begin by gauging my students' prior knowledge of the subject with a free write about what they know about ozone, greenhouse gases, and global atmospheric changes (Figure 1, p. 237). One of the biggest misconceptions I have encountered in my students is the difference between "good ozone" and "bad ozone." Good ozone is found in the stratosphere as the ozone layer that protects us from harmful ultraviolet rays. Bad ozone is found in the troposphere at ground level and can be created by a chemical reaction when NOx reacts with VOCs emitted mainly from automobiles and industry in the presence of sunlight. Students have heard *ozone action* days announced on news programs but may think that it refers to the stratospheric O_3 layer rather than ground-level ozone. Students typically understand that daily activities, such as driving automobiles, emit CO_2 and contribute to the greenhouse effect. Fewer students also understand that these activities also can contribute to low-level O_3 production. Students do not, however, always make the link between increased low-level O_3 and CO_2 and impacts on plant growth.

In order for students to understand the FACE experiment and the effects of CO_2 and O_3 on plants, I needed them to correct their misconceptions and successfully distinguish between greenhouse gases, good ozone, and bad ozone. I have found that students are most successful when they are driving the research; therefore, I schedule my first day of this activity in the computer lab. During this time, students use the internet (documenting their sources), along with their textbook and other supplementals I have provided, to research and record their findings to the questions on the worksheet found in Figure 2.

Understanding CO_2 and O_3

Your goal: Paint a complete picture of the environmental issues surrounding CO_2 and O_3 by researching and answering the questions below. Organize and record your findings in your science notebook. Be sure to keep a complete list of the resources you have used.

The greenhouse effect and global climate change

What is the greenhouse effect?

- List three greenhouse gases.

What is global climate change?

- What has been the trend over geologic time? Recently?
- What greenhouse gas is of most concern? Why?
- What human actions contribute to the emission of this gas?
- Sketch a graph of how the concentration of this gas has changed over time.
- What are the predicted consequences?
- How might it impact plants?
- Can it be stopped? Slowed? Reversed? What are some possible solutions?

Ozone

You most likely know about "good ozone."

- Where is it found?
- What is its importance?

There is also "bad ozone."

- Where is it found?
- What two main types of compounds are involved in the chemical reaction that forms it?
- What weather condition contributes to the formation?
- What human actions contribute to its formation?
- What are the negative effects overall? On plants? Find a photo showing the effects of ozone on leaves and describe.
- What are possible solutions?

Contrasting good and bad ozone

- Which type is found in the stratospheric ozone layer? Which type is found in the troposphere?
- What is an ozone action day? Which type of ozone does it address?

Resources (please list what resources you used)

-
-

The first supplemental I provide is a copy of the brochure *Ozone: Good Up High, Bad Nearby*, published by the Environmental Protection Agency (2003). The brochure compares and contrasts good and bad ozone in a straightforward and succinct manner. Another resource I provide students with is a presentation by Bill Holmes, an Aspen FACE researcher from the University of Michigan. Even though the presentation, *Elevated CO_2 and Ozone: Causes and Consequences,* was designed for teachers, many of my students found the diagrams and descriptions (in the slides and in the speaker notes) very useful. This presentation is available on the NSTA website at *www.nsta.org/highschool/connections/200801CarlsonOnlineII.ppt.*

I guide the research of my lower-level students by providing a list of websites that address the issues at an appropriate level (see "On the web"). The following day, we engage in an educated discussion regarding students' findings. I encourage students to ask questions and correct any misconceptions they had by recording corrections in a different color in their journal.

APPLYING SCIENTIFIC INQUIRY IN THE FIELD

Often in high school, scientific experiments are conducted in a laboratory and have definite, predetermined conclusions. Aspen FACE breaks both of these traditions; it is in-progress field research with no final results. The goal of the activity described in Figure 3 is for students to think like ecologists conducting research in the field. I begin by presenting students with the very basics of the Aspen FACE experiment using a color overhead of Figure 4 and asking them to brainstorm ideas that a scientist might study at the site. We then compare our ideas with the long list of interconnected ecological research that is being carried out at Aspen FACE, online at *http://aspenface.mtu.edu/investigators.htm.*

Next, I focus the class on one specific question: What are the effects of increased CO_2 and O_3 on plant growth? This question is a simplified form of what many of the FACE researchers are studying. Students take on the role of ecologists working in the field as they complete an inquiry into the Aspen FACE experiment. The inquiry is designed to lead students through designing a procedure, posing hypotheses, analyzing data, drawing conclusions, and identifying sources of error and uncertainty. In Figure 3, a guided lab report, students use the photos and descriptions in Figures 4, 5, and 6 to discover the Aspen FACE experiment by applying their knowledge of scientific investigations. For the inquiry to be effective, I am very careful not to tell students too much about the FACE research findings during the introduction. The idea is for students to analyze and interpret data (Figure 7) from a large-scale research project; some students draw conclusions similar to the experts' and others use the data to support their own hypotheses.

Figure 3

Exploring the Aspen FACE experiment

Your goal: To think like an ecologist. You will be investigating a current field research project. You must use your expertise in scientific investigations, photosynthesis, carbon dioxide, and tropospheric ozone to help analyze this real-life experiment. Use the following sections to help you organize and record your findings in your science notebook.

Problem: What are the effects of increased CO_2 and ozone on plant growth?

- **Background information:** What is needed for a plant to grow? What is the equation for photosynthesis?
- **Procedure:** Design a simple experiment to test the problem. Then, compare and contrast your design with the FACE experimental design shown in Figure 5. In the Aspen FACE experimental design:
  - What is the dependent variable?
  - How many of the treatments are experimental groups? What is the independent variable in each of the treatments?
  - What does ppm stand for? ppb? What do they both measure?
  - Which treatment is the control? Why is a control needed?
  - What are the constants?
  - Why is the "free-air" part of the design so important?
- **Hypothesis:** Keeping in mind the equation for photosynthesis and your research, create three hypotheses; one for each of the experimental groups.
- **Data:** Growth can be measured in a number of ways. We will be analyzing growth above ground, measured with a volume growth index. To calculate the volume growth index, scientists took the tree's diameter squared and multiplied it by the height of the tree. Data Table 1 shows results from three years.
  - Create a multiple line graph comparing the growth in each experimental group and the control from 1998 to 2000.
  - The photos in Figure 6 were not taken systematically. Look closely at the pictures to find a way to compare and contrast the three plots.
  - Use the trends in your graph to label the three pictures in Figure 6 as elevated CO_2, elevated O_3, or elevated $CO_2 + O_3$.
- **Conclusions:** Now, write a conclusion for each of your hypotheses.
  - Restate your hypothesis.
  - Tell whether it was correct using supporting data from your graph.
  - Explain why, using your research from the activity in Figure 2.
- **Sources of uncertainty:** How would each of the following affect the results?
  - How growth was measured.
  - What plant was used.
  - The time period.

(continued)

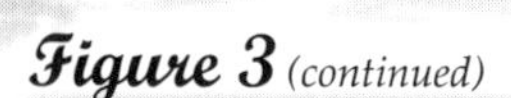
Figure 3 (continued)

Data Table 1: Volume Growth Index (diameter^2 x height in cm) of Aspen trees
(The data in this table is from Karnosky et al. 2003)

Year	Control CO_2/O_3 Concentrations (360/32 ppm)	Elevated CO_2 Concentration (560 pm)	Elevated O_3 Concentration (360 ppm)	Elevated CO_2 + O_3 Concentrations (560/360 ppm)
1998 (Volume Growth Index)	1050 cm^3	1100 cm^3	1000 cm^3	1020 cm^3
1999 (Volume Growth Index)	4500 cm^3	5600 cm^3	3300 cm^3	3800 cm^3
2000 (Volume Growth Index)	7000 cm^3	9200 cm^3	5600 cm^3	6700 cm^3

Figure 4

Introduction to Aspen FACE

One of the major tools for investigating effects of elevated CO_2 and ozone on plants and ecosystems is the Free Air CO_2 Enrichment (FACE) experimental design. The photo on the bottom shows the 12 rings at Aspen FACE in Wisconsin; a close-up of a ring is shown in the top photo. The Aspen FACE site is focused primarily on the effect of global change on the Trembling Aspen *(Populus tremuloides)*, the most widely distributed tree species in North America, but the site also contains white paper birch and sugar maple. There are over 25 FACE sites scattered over the globe representing the diverse ecosystems of the world.

Aspen FACE experimental design

	Treatment 1: Control (normal air)	**Treatment 2:** Elevated CO_2	**Treatment 3:** Elevated O_3	**Treatment 4:** Elevated $CO_2 + O_3$
Concentration of CO_2 (ppm)	360	560	360	560
Concentration of O_3 (ppb)	32	32	56	56
Number of 30 m rings	3	3	3	3

At Aspen FACE, each 30 m ring of plants is surrounded on the perimeter by a series of vertical vent pipes (see photo) which push CO_2, O_3, CO_2 + O_3, or normal air into the center of the ring. The concentrations of these gases are measured in parts per million (ppm) or parts per billion (ppb). For example, in the control, there are 360 parts of CO_2 for every million parts of air. The system that controls the concentration of gases in the air is computer controlled and adjusts the amount of gas released every second to maintain a stable, elevated concentration of CO_2 and/or O_3 throughout the experimental plot. In the past, most studies were designed on a small scale with groups of plants enclosed in open-top chambers with controlled atmospheres and were limited because as minigreenhouses, they provided unnatural protection from wind exposure and other natural occurrences. The FACE design is unique because it has open-air control of atmosphere conditions, is fairly large in scale, is being carried out for a longer period of time, and involves a large team of scientists and researchers.

MAKING CONNECTIONS

The activity in Figure 3 requires students to carefully analyze the FACE data and understand the impact of CO_2 and ground-level O_3 on plant growth. It also leads students to begin thinking about larger global issues and making connections between Aspen FACE and global change issues of CO_2 and O_3. In the activity described in Figure 2, students learned about the negative effects of increased CO_2 on ecosystems, but the FACE data analyzed in Figure 3's activity show that adding more CO_2 (the reactant) increases photosynthesis and therefore tree growth, while O_3 reduces tree growth. When asked about effects of O_3, one student says "Tropospheric ozone smothers plants and they will not be able to use CO_2 and make O_2. The human race would suffer." After analyzing the FACE tree growth data, some students make the hypothesis that "The positive growth effects of CO_2 will be cancelled out by the negative effects of O_3." Students are asked to support statements such as this with *evidence* from the FACE data. We also discuss what these opposite effects mean and if scientists really know what is going to happen as a result of global

Figure 6

Growth of trees in experimental groups at Aspen FACE

(1) Top left is elevated CO_2—notice height of trees is above top rail; (2) bottom left is elevated O_3—notice height of trees is below bottom rail; and (3) right is elevated CO_2 + O_3—notice height of trees is mostly between the rails.

change. They see that research often leads to new findings and more questions and investigations.

As their final assessment, students are asked to write a five-paragraph essay conveying their understanding of the connections between global change and the Aspen FACE experiment, and their proposals for solutions. Any remaining misconceptions become strikingly apparent in the final essays and can be remediated before their ecology assessment (Figure 8).

As students move through the activities, it is gratifying as a teacher and an environmentally conscious citizen to watch students connect their knowledge of photosynthesis to ongoing field research in such

Figure 7

Student-generated graph

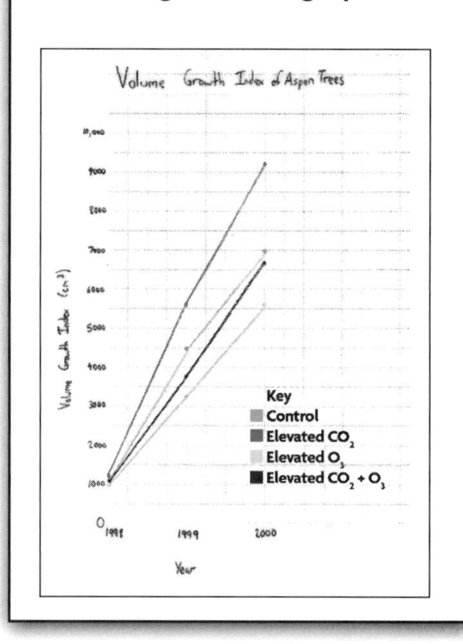

Figure 8

Student activity C: Assessment

Now that you understand the issues, it is your job as an ecologist to educate others by writing an essay explaining the environmental issues of CO_2, O_3, and how the research at Aspen FACE is contributing to an understanding of these issues. Use your science notebook; you have already collected most of the information you need. The idea now is to make connections!

	Criteria: 1	2	3
Paragraph 1: Introduction	Thesis is inadequate or not present.	Thesis is adequate, CO_2 and O_3 and consequences are introduced.	Thesis is clear, establishes connection between CO_2, O_3, and Aspen FACE.
Paragraph 2: Rising CO_2	Contributing human activities are identified.	Human activities and consequences are listed, but not explained.	Clear connections are made between multiple human activities and multiple consequences.
Paragraph 3: O_3	O_3 is defined.	Good O_3 and bad O_3 are defined.	Good O_3 and bad O_3 are defined, effects of tropospheric O_3 on plants and humans are described.
Paragraph 4: Aspen FACE: What are the effects of CO_2 and O_3 on plant growth?	Conclusions from Aspen FACE are stated, but not supported with data.	Conclusions from Aspen FACE are stated and supported with data.	Conclusions from Aspen FACE are supported with data, and uncertainties are identified.
Paragraph 5: Conclusion	Student gives one or two suggestions each for reduction of CO_2 and O_3, no mention of Aspen FACE.	Student summarizes Aspen FACE and gives two suggestions each for reduction of CO_2 and O_3.	Student relates Aspen FACE conclusions to the environmental issues of CO_2 and O_3 as previously discussed, and gives three suggestions each for reduction of CO_2 and O_3.

a prominent environmental issue such as global change and relate it to their daily lives and the energy choices they make.

On the web

Greenhouse effect: *www.epa.gov/climatechange/kids/basics/today/greenhouse-effect.html*

Global warming is hot stuff: *www.dnr.state.wi.us/org/caer/ce/eek/earth/air/global.htm*

SunWise kids ozone layer: *www.epa.gov/sunwise/kids/kids_ozone.html*

Ozone action: *www.semcog.org/OzoneAction_Kids.aspx*

What's Ozone? (Smog City): *www.smogcity.com/welcome.htm*

References

EPA Office of Air and Radiation (EPA). 2003. Ozone: good up high bad nearby. EPA. *www.epa.gov/oar/oaqps/gooduphigh.*

The Aspen Free-Air Carbon Dioxide Enrichment Experiment (FACE). Facts II: The Aspen FACE Experiment. 2005. Northern Forest Ecosystem Experiment (NFEE). *http://aspenface.mtu.edu.*

Karnosksy, D., et al. 2003. Trospospheric O_3 moderates responses of temperate hardwood forests to elevated CO_2: A synthesis of molecular to ecosystem results from the Aspen FACE project. *Functional Ecology* 17: 289–304.

Chapter 28

TEACHING ABOUT ENERGY

By Amanda Beckrich

The U.S. Energy Information Administration's website (see "On the web") offers energy basics, lesson plans, and low-tech lab activities for teachers. To start, have your students collect energy-use data from classroom equipment and perform simple calculations. Many energy monitors are commercially available—they connect to appliances and read their electrical consumption by kilowatt-hour on an LCD (liquid crystal display).

Using these meters, students can calculate energy unit conversions or determine school electrical expenses. An interesting extension is to compare an appliance's electrical consumption in different modes. For example, have students compare a computer's energy usage when it is on, off, and asleep. Multiply the results to determine daily, monthly, or even yearly energy consumption. You may be surprised by the results!

Students can also perform a National Earth Science Teachers Association lab (see "On the web") that connects energy consumption to carbon-dioxide production. If students use an energy meter that measures voltage and current thousands of times a second, they can "see the surge" of power when an appliance is first turned on.

ENERGY AUDIT

Perhaps you want to move beyond studying classroom appliances and have students perform home or school energy audits. Consultants perform audits like these at great cost to families, businesses, and schools around the world, so having your students perform at least part of this task makes curricular, environmental, and financial sense. An energy audit extends beyond monitoring electricity usage and allows students to design and implement a specific conservation plan to reduce the amount of electricity used and money spent. The Environmental Literacy Council offers a well-planned home energy audit, modifiable for use in school (see "On the web").

INTERSCHOOL ENERGY CONSERVATION CHALLENGE

How about competing against other schools to reduce your energy consumption? Although many schools have done this informally, the biggest interschool conservation challenge in the United States is the Green Cup Challenge (GCC). The first student-driven interschool energy challenge in the nation, the GCC takes place each February in hundreds of public, charter, and independent schools to call attention to peak (winter) energy use. During the 2010 GCC, participating schools saved 1,254,000 kWh of electricity—an average savings of $836 per school!

The GCC is organized by the Green Schools Alliance (GSA), a global alliance of more than 2,000 public and independent schools united to solve environmental problems through sustainable and energy-smart solutions (see "On the web").

Regardless of the amount of time and resources you are able to commit, you can teach students about their energy use during winter. Once students know about their personal energy consumption, they will be more likely to change their behavior, educate others about energy, and become more sustainable world citizens. How's that for the power of education?

On the web

Environmental Literacy Council "Home Energy Audit" lab: *www.enviroliteracy.org/article.php/1149.html*

Green Cup Challenge: *www.greencupchallenge.net*

Green Schools Alliance: *www.greenschoolsalliance.org*

National Earth Science Teachers Association "Plugged in to CO_2" lab: *www.windows2universe.org/teacher_resources/teach_pluggedCO2.html*

U.S. Energy Information Administration—Energy kids: *www.eia.doe.gov/kids/index.cfm*

Chapter 29

SOLAR RADIATION—HARNESSING THE POWER

USING NASA DATA TO STUDY ALTERNATIVE ENERGY SOURCES

By Teri Rowland, Lin Chambers, Missy Holzer, and Susan Moore

Finding real-world scientific data for use in the science classroom can be a challenge. Oftentimes, this data is too complex for students to really use. Although lab exercises do have a place in the classroom, the use of appropriate, authentic data sets can provide a global picture while helping students build both conceptual understanding and understanding of the data-analysis process.

My NASA Data (Chambers et al. 2008; see "On the web") is a teaching tool available on NASA's website that offers microsets of real data in an easily accessible, user-friendly format—and there are over 60 high school–level lesson plans available. The data sets are created at the NASA Langley Research Center from atmosphere, biosphere, cryosphere, land surface, and ocean data gathered by near-Earth satellites and ground-based sources. Teachers can retrieve raw data to place into spreadsheets, graphs, or grids. Available

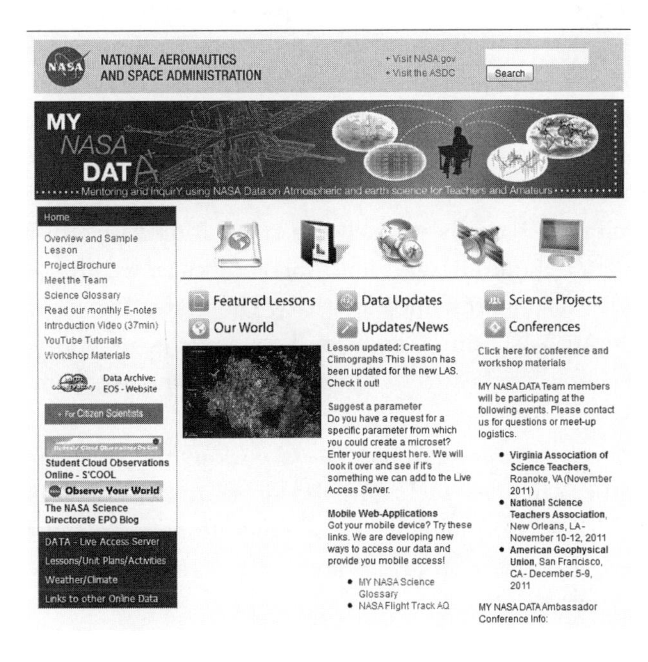

data sets include information about air quality, atmospheric radiation, clouds, precipitation, surface conditions, and surface cover—to name a few. A good way to begin is to visit the website and browse the data sets available. By first visiting "Science Basics" and then the "Learn More About Our Data" pages, teachers can identify the temporal and spatial limits of these data sets.

In this chapter, we describe a lesson plan based on an activity from My NASA Data, in which students explore parts of the United States that they would want to live in if they lived in a solar-powered mobile home. The lesson focuses on alternative energy sources and demonstrates how easily teachers can create their own activities using data of their choice. The goal is to move students away from using simplistic data. In the lessons offered through My NASA Data, students have the opportunity to graph and analyze the same data that NASA scientists might use. Data analysis and graphing skills are taken beyond the typical science classroom. Instead, students begin to use authentic NASA data—making science more relevant and exciting.

The lesson plan presented in this chapter provides valuable skills in graphic differentiation and can be used in both middle and high school classes during units on solar energy, electricity, and electromagnetic spectrum. It can also be used in science research classes as a data-analysis lesson, or in units on light, energy, alternative energy sources, technology, and Earth science systems.

ALTERNATIVE ENERGY SOURCES LESSON

Background

Alternative energy sources are increasing in popularity and becoming easier for consumers to use. Solar-powered walkway lighting, attic fans, and pond pumps are readily available at home improvement stores. Unlike fossil fuels that will one day be depleted, solar-powered homes and vehicles use renewable energy. Students see this technology as a positive step toward energy solutions in this country; however, some do not fully understand how such devices benefit consumers.

Photovoltaic (PV) cells—which are often found in roof-mounted solar panels, solar cell-phone chargers, and solar walkway lighting—gather energy from the Sun and convert it into electricity. Their effectiveness varies, depending on factors such as latitude and cloud coverage. Knowing how much solar energy is available for these cells can assist those who use them in their homes and businesses.

Solar energy data are gathered from various polar orbiting and geostationary weather satellites, including the Geostationary Operational Environmental Satellite (GOES) and the Polar Operational Environmental Satellite (POES). The "Monthly Surface All-Sky Shortwave Downward Flux" data are

compiled by the Surface Radiation Budget (SRB) project and offered through the My NASA Data Live Access Server (LAS; see "On the web"). The SRB project combines satellite data and models to provide a consistent data set for understanding Earth's surface radiation. For these data, the satellite measures the rate of transfer of solar energy (measured in watts [W] per unit area [m^2]) at the surface of the Earth—this determines how much of the Sun's energy gets to Earth's surface at a given location each month. More solar energy reaching the surface results in more electrical energy from the solar cell.

Solar panels' effectiveness changes throughout the year. By comparing monthly averages of surface downward radiation in various locations around the United States, students can analyze areas that would be more or less suited to solar panel use. This lesson introduces students to overlay and difference plots and how they can be used to analyze information. An overlay plot shows two or more sets of data on the same graph, but does not necessarily correlate one set of data to the other. A difference plot, on the other hand, shows the difference between two sets of data (zero means that the data are identical, while a high number indicates they are far apart). Comparing the same data on different types of graphs allows students to better understand what the data mean, and helps them develop the ability to analyze graphical data.

Lesson scenario

In this alternative energy lesson adapted from My NASA Data, students explore LAS data on a journey around the United States. Students begin with the following scenario:

> *Congratulations! You are the rock star you always wanted to be. However, the life of a star is different than most—it requires months on the road in a motor home. To decrease your carbon footprint, your motor home's electricity is powered by solar panels. As long as you have enough sunshine, you will be fine. Therefore, you need to determine the following:*
>
> 1. *Where will you live each month of the year (it must provide an average of 100 W/m^2 of solar energy)?*
> 2. *How does the monthly average solar energy compare to your hometown?*
> 3. *Find one place you might like to live during a particular season, but cannot because of insufficient solar energy levels.*

To answer these questions, students follow two procedures in which they compare the LAS data in their hometowns to the data at both a higher latitude location and a lower latitude location. In doing so, students explore parts

of the United States in which they might want to live, using solar energy to power their home. In the process, they create both an overlay graph (Figure 1) and a difference graph (Figure 2, p. 254) to analyze their data. Through graphic analysis, students identify locations that will receive enough solar radiation to meet their living requirements.

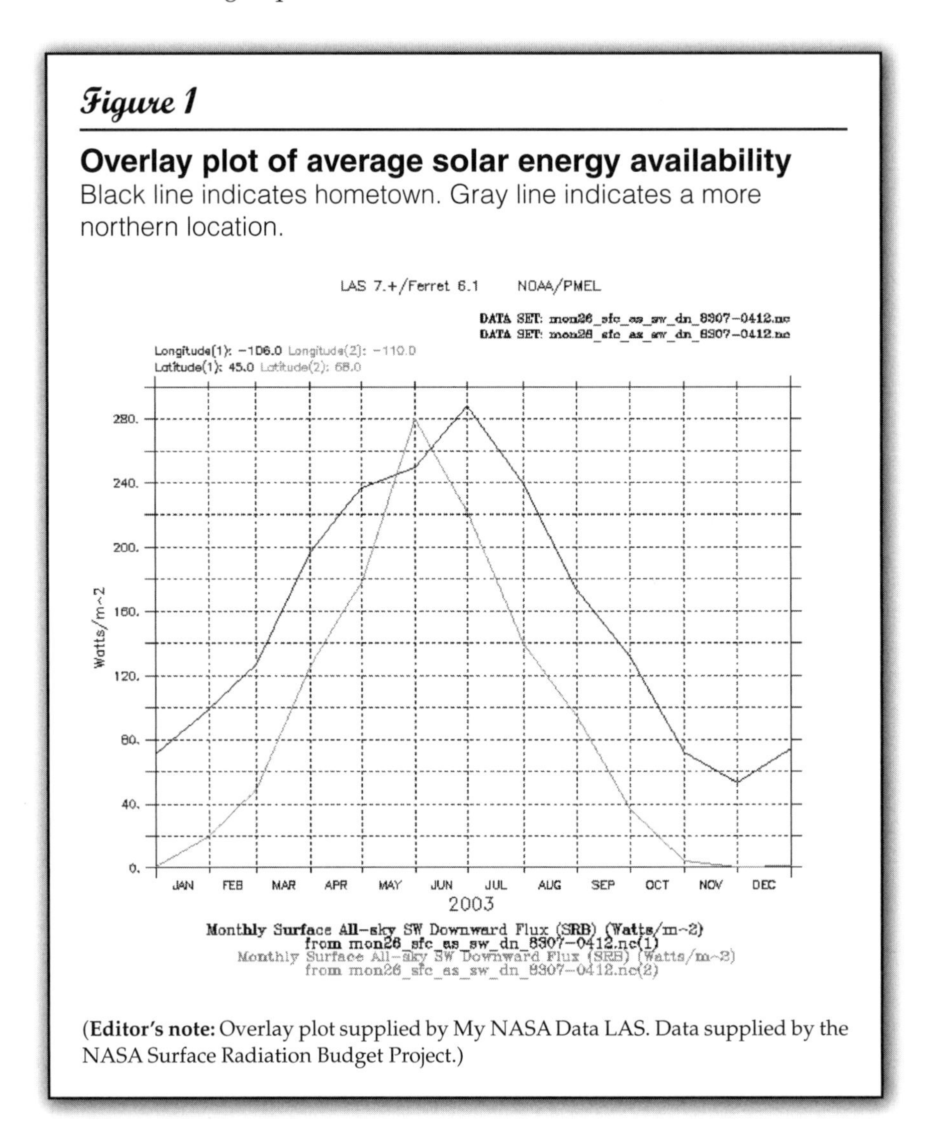

Figure 1

Overlay plot of average solar energy availability

Black line indicates hometown. Gray line indicates a more northern location.

(**Editor's note:** Overlay plot supplied by My NASA Data LAS. Data supplied by the NASA Surface Radiation Budget Project.)

Student objectives

In the alternative energy sources lesson, students should achieve the following objectives:

- Determine geographical locations of potential PV arrays.

- Differentiate between an overlay plot and a difference plot.
- Create a plan for solar use based on solar energy availability.
- Determine one place you would like to live, but cannot due to low solar energy availability.
- List factors that contribute to an area's higher solar radiation level.

Procedure 1: Higher latitude

Students analyze data for a location with a higher latitude than their hometown.

1. On the My NASA Data LAS website (see "On the web"), click on the data sets in the following order: "Atmosphere," "Atmospheric Radiation," and "Surface." Check the box for "Monthly Surface All-Sky SW Downward Flux (SRB)," and click "Next." This parameter gives an average of how much of the Sun's energy reaches a given location on Earth's surface each month.
2. Click on the "Compare Two" tab on the left-hand side to compare two different areas of the country.
3. Set the following parameters:
    a. Select view: "Time series (t)"
    b. Select output: "Overlay plot"
    c. Select region: "North America"
    d. Click on "Var 1" (variable 1) and input coordinates of your hometown. A white crosshair will identify this location on the map.
    e. Click on "Var 2" (variable 2) and pick somewhere north of your hometown (i.e., somewhere that has a higher latitude) at the same longitude. You may find it easier to click on the map rather than inputting coordinates. A yellow crosshair will appear.
    f. Select time range: Pick any 10-year period (e.g., January 1994 to January 2004). Click "Next." This will produce an overlay plot of the data set.
4. Reflect on the trends you see in the plotted graph (i.e., in the summer months the solar energy usually increases, and it decreases in the winter months).

5. Go back to the time range and this time, select one year (Figure 1). You should identify that the red line has a higher latitude, and the black line is your home location. Record your observations in the data table (Table 1).
6. To truly see the relationship between the two locations, go back and select output as a "Difference plot," and click "Next" (Figure 2). Depending on your browser, you may also have to respecify the latitude and longitude locations and the time range. Any part of the graph that reads above zero means there is more solar energy at that location. Parts of the graph located below zero mean that home has less solar energy than the secondary location.
7. Record your observations in a data table (Table 2).

Figure 2

Difference plot comparing hometown solar energy availability to a more northern location

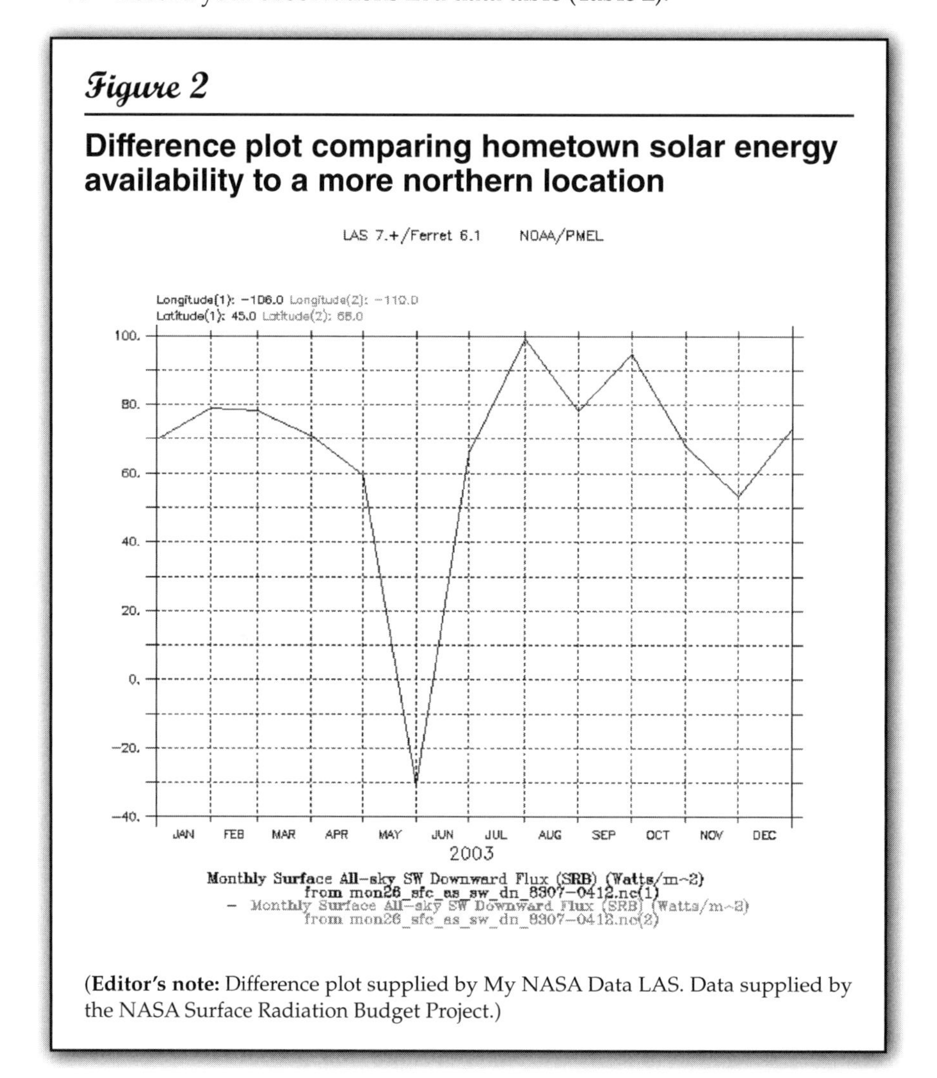

(**Editor's note:** Difference plot supplied by My NASA Data LAS. Data supplied by the NASA Surface Radiation Budget Project.)

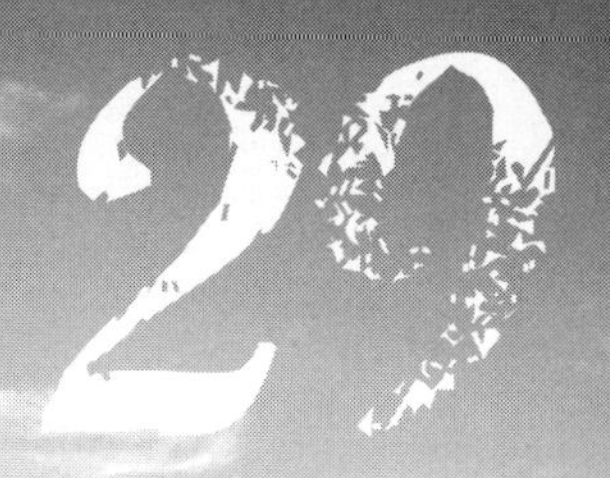

Procedure 2: Lower latitude

Students analyze data for a location with a lower latitude than their hometown.

1. Repeat "Procedure 1" with a more southern (lower) latitude location than your hometown. Note that latitude is one variable that plays a part in solar energy availability.
2. Record your observations in the data tables (Tables 1 and 2).

Table 1

Procedures 1 and 2 observations

Place a check mark in each box where there is sufficient solar energy to provide power for the motor home.

Location	Months with sufficient energy											
	Jan	Feb	Mar	Apr	May	Jun	Jul	Aug	Sep	Oct	Nov	Dec
1. *Hometown* Latitude: Longitude:												
2. *More northern* Latitude: Longitude:												
3. *More southern* Latitude: Longitude:												
4. *Other U.S. area* Latitude: Longitude:												

Table 2

Procedures 1 and 2 observations

Place a plus sign where the hometown gets more solar energy and a minus sign where the second location gets more solar energy.

Difference between two locations	Months of comparison											
	Jan	Feb	Mar	Apr	May	Jun	Jul	Aug	Sep	Oct	Nov	Dec
1 vs. 2												
1 vs. 3												
1 vs. 4												

Data exploration

Students locate multiple places they would like to live in addition to their hometowns and determine during which months they could live at each location.

1. Using the same method from "Procedure 1," investigate data and try to explain why certain locations might receive more solar radiation than others because of differences in average cloud coverage, topographic features, or other variables.
2. Finally, determine one location where you would like to live in a solar home, but cannot because of insufficient solar energy.

DISCUSSION

After completing the activity, discussion topics could include

- the differences in graphical representations of data,
- factors that contribute to high solar energy levels in certain areas of the country,
- pros and cons of geographical locations for various alternative energies, and
- other microsets of data available for student or group projects.

ASSESSMENTS AND OUTCOMES

Student assessment for this lesson plan is based on results. A minimum expectation is to complete "Procedure 1" and "Procedure 2." Data tables are then reviewed to determine if students correctly analyzed the difference graphs for their data. Successful students may share their results with peers before all students are provided an additional opportunity to continue the lesson.

Students who have not analyzed their difference graphs correctly should choose another location to analyze; those who have succeeded in this task are given an enrichment opportunity to explore locations around the world—rather than just in the United States. Some students may choose to explore the LAS data set and create their own data comparison for extra credit.

Ultimately, questions from this activity are further assessed on the content unit test.

CONCLUSION

Science goals for the 21st century include thinking critically and logically about the relationship between evidence and explanations. Using pure science data from My NASA Data gives students the opportunity to do just that.

Students are inquisitive and are excited to learn using NASA information. They are challenged with the amount and types of data available and become actively involved in their own learning.

The lesson presented in this chapter was easily created with the user-friendly data sets from My NASA Data. The website contains over 140 Earth system science parameters that can be used for student inquiry projects and about a dozen starter ideas for science projects. From here, you can select data sets to create a lesson visualizing one or more aspects of our dynamic planet. Lessons created by other teachers that are featured on My NASA Data include Evidence of the Change Near the Arctic Circle, Ocean Currents and Sea-Surface Temperature, Carbon Monoxide and Population Density, and Validation of Stratospheric Ozone. These resources demonstrate the many uses of the parameters available on My NASA Data. With the inclusion of a grade-level appropriate science glossary, relevant National Science Education Standards (NRC 1996), and other documentation, this website constitutes a rich resource for increasing the level of scientific inquiry and authentic experience in your classroom, with a broad selection of entry points for getting started.

Acknowledgment

The My NASA Data project was developed under NASA funding through the Research Education and Applications Solutions Network (REASoN) program.

On the web

My NASA data: *http://mynasadata.larc.nasa.gov*

My NASA data LAS: *http://mynasadata.larc.nasa.gov/las/servlets/dataset*

References

Chambers, L. H., E. J. Alston, D. D. Diones, S. W. Moore, P. C. Oots, C. S. Phelps, and F. M. Mims, III. 2008. The My NASA Data project. *Bulletin of the American Meteorological Society* 89 (4): 437–442.

National Research Council (NRC). 1996. *National science education standards.* Washington, DC: National Academies Press.

Chapter 30

SOLAR PANELS AND ALTERNATIVE ENERGY IN THE EIGHTH-GRADE CLASSROOM

By Laura Bruck

Before we started this unit, I barely knew anything about this topic, but now I feel like an expert!

—Student comment

For me, these kinds of comments represent the Holy Grail of science teaching. The challenge, however, lies in developing creative lessons and activities that bring about such positive results. How do I get students interested? Why should they care about the topics we discuss? What can I do to make science class fun and engaging? I decided that peer collaboration in a student-driven environment offered many solutions to my questions.

30

SOLAR PANELS AND ALTERNATIVE ENERGY IN THE EIGHTH-GRADE CLASSROOM

During our unit on energy and electricity production, I designed a problem-based learning project in which students worked collaboratively in peer groups. To incorporate authentic inquiry into their experiences, groups were free to investigate a topic of their choosing under the umbrella of solar energy, giving them freedom to work in an inquiry-based environment, while reinforcing important concepts about energy, energy production, and the social aspects of science. Upon completion of the project, I anticipated students would be able to

- explain/demonstrate proper assembly of series and parallel circuits,
- compare/contrast the current in solar-powered and battery-powered circuits,
- determine the best source of light for powering solar panels,
- appreciate the electricity used in everyday life and understand the role of alternative energy, and
- use experimental data and outside research to defend their conclusions.

As will be discussed later, student feedback generated from SALG (Student Assessment of their Learning Gains) surveys (Wisconsin Center for Educational Research) and pre- and posttest scores indicated that my students made advancements in these areas as a result of participating in this project.

ASSESSING STUDENTS' BACKGROUND KNOWLEDGE

As we know, students come from a variety of backgrounds and experiences. To assess their prior knowledge of circuitry and electronics, students completed a baseline SALG survey. The SALG is a free, online survey that teachers can use to gather information and feedback from students anonymously (*www.salgsite.org*). The SALG works by providing a statement, which students then rank on a Likert-like scale to indicate their mastery of the topic. For instance, if the statement reads, "I know how to assemble circuits," students would choose from the following rankings: "Not at all," "A little," "Somewhat," "A lot," or "A great deal." Teachers can use existing statements provided on the SALG or write their own (which I did, and recommend that others do as well to make it more relevant to your situation).

A baseline SALG is intended for use before a unit or course, and a "regular" SALG survey is used after the completion of the unit or course. The SALG is not a test worth points, but a way of seeing how students rate

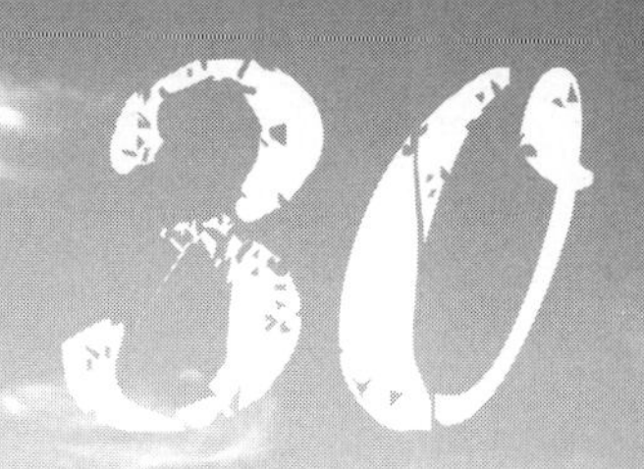

Table 1

Selected baseline SALG results; students rated their current understandings prior to the project and preproject activities.

Currently, I ...	N/A	Not at all	A little	Somewhat	A lot	A great deal
understand how solar panels work.	4%	9%	20%	30%	31%	6%
am able to make circuits work with a solar panel as a power source.	11%	50%	22%	13%	4%	0%
am able to fix problems with circuits.	13%	43%	26%	9%	4%	6%
am confident that I can understand this subject.	4%	4%	28%	35%	24%	6%
am comfortable working with complex ideas.	0%	6%	15%	30%	33%	17%
can find scientific information in books, online, or in journals.	0%	0%	9%	26%	30%	35%
can interpret data I collect in lab.	0%	0%	7%	33%	43%	17%
can make a sound argument and use data to support it.	0%	7%	9%	28%	39%	17%
can work effectively with others.	0%	2%	2%	7%	31%	57%

their understandings. Using the SALG baseline survey to gauge students' background knowledge was very beneficial. Sample questions and student responses to the baseline SALG are provided in Table 1.

Of 58 students, 54 (93%) completed the baseline SALG survey. The results of the teacher-edited baseline SALG survey indicated that my students had little or no prior knowledge of circuitry or solar energy. I was encouraged by the percentage of students who rated highly their confidence in trying new things, working in a group, finding information, and making sound arguments. In addition, students took a 17-question, teacher-designed pretest (multiple choice and free response). The pretest assessed

vocabulary, troubleshooting, and general knowledge. Sample questions are provided below:

1. An element in a circuit that reduces the current is called a/an
   a. Ohm
   b. Inhibitor
   c. Protector
   d. Resistor

2. The arrangement of the circuit shown below is
   a. Linear
   b. Straight
   c. Parallel
   d. Diagonal
   e. A series

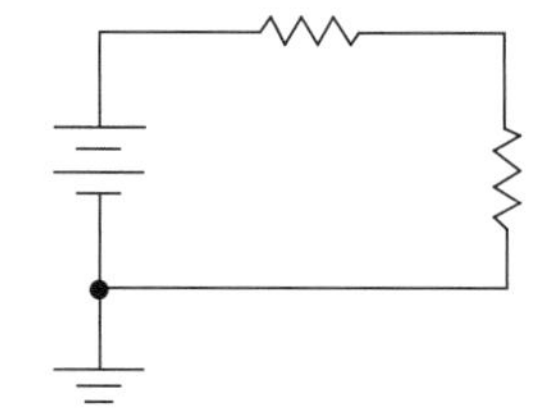

3. Explain the advantages and disadvantages of solar energy.

Of the 58 students, 56 (97%) completed the pretest, and the average score was a 38%. Although this was a low score, most students lost the majority of their points as a result of vocabulary, which I knew could be taught with few difficulties. From this, I concluded that this project and preproject activities were appropriate in difficulty and in design for the student population at our school.

PREPROJECT PREPARATION AND GROUP ASSIGNMENT

Students worked in groups of six or seven during the preproject activities to get used to working together. While this number may seem large, I chose to arrange students this way to ensure that all groups had equal access to the materials. In addition, I strategically assigned students to their groups so each group had a mix of personalities and skill levels.

I began with preproject background instruction. Students learned how coal-fired, hydroelectric, and nuclear power plants produce electricity, and compared this to alternative sources of energy. The class discussed the energy crisis our country faces and learned about the environmental effects of coal-fired power plants. We specifically discussed our own state, Indiana, one of the country's largest consumers of coal, and the consequential environmental impacts. These steps were taken so that the solar-energy project would be more relevant to their daily lives, and to help them understand the need for alternative energy.

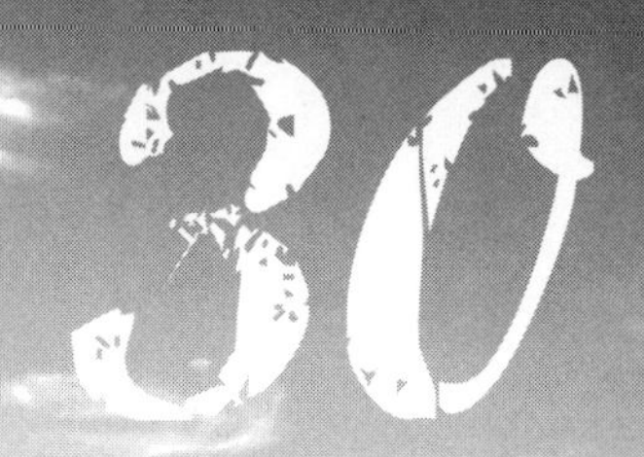

In addition, because solar-panel lab investigations require use of circuitry, our preproject activities included lab practice on how circuits are assembled and the safety considerations. Student groups completed structured lab investigations in which they built series and parallel circuits, added lightbulbs and buzzers to their circuits, and learned how to use the multimeter and integrate solar panels into circuits. Then, as the final part of the preproject activities, students completed more inquiry-based lab investigations of the same topics. These steps were taken to ensure students were comfortable assembling circuits and understood how to do so safely, and their progress informed me if students were ready for the fully open-ended project.

THE PROJECT—AN INTRODUCTION

To begin the project, students were given a planning guide (Figure 1, pp. 264-265) to assist with developing their driving questions and help groups organize their thoughts, delegate roles, and keep on track.

Students were assessed via a pretest (completion grade only), planning guides, homework assignments, a group lab report, a group presentation to the class, peer and self-evaluations of the group work as a whole, and a posttest (counted as a regular unit test). Students were given grading rubrics in advance for the group presentation and group lab report, and were given a topical study outline before the posttest. All grading information and expectations were discussed the day the project was introduced so that students would be aware of the grading criteria in advance.

GENERATING DRIVING QUESTIONS

For homework on the day the project was introduced, students individually devised three driving questions about solar energy that they could both find information about using resources as well as conduct experiments on using classroom materials. These were the only restrictions on the driving questions. The following day in their groups, students compared their individual questions and came to a group consensus on which question would be the group's driving question. This was exciting because, as it turned out, no two groups investigated the same question. Sample questions included "Are solar panels worth the cost?" and "How many solar panels does it take to power a toaster oven?"

I had to approve of all driving questions before further progress was made. To earn approval, teams explained what data they would collect during the experiment and how they would do so, and describe how they would conduct outside research (e.g., books, internet). The purpose of the outside research was to answer any questions that could not be determined by conducting an experiment, as well as to confirm their lab findings. Outside research was not

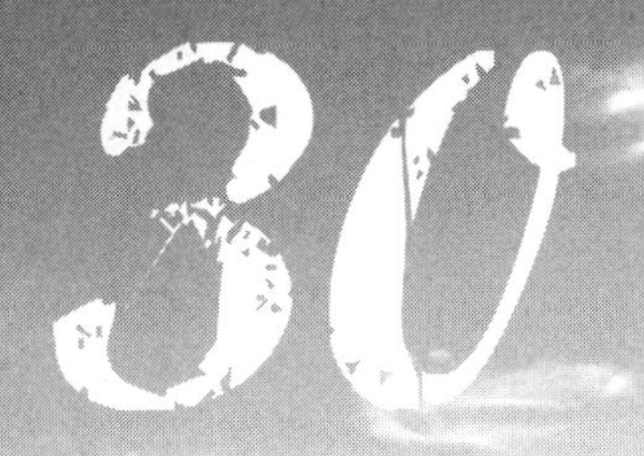

Figure 1

Solar-panel investigation planning guide

Name:
Team members:

Introduction
In this project, you will develop a researchable question and investigate it as a team. In addition, you will use the tools available in lab to collect your own data to help support or refute your driving question. When completed, you will make a presentation to the class on your findings and write a group lab report on your procedures, results, and suggestions for future work. Read the problem statement below, and then talk to your team members about possible driving questions.

Problem statement
Sustainable, clean energy is a huge problem in the United States. Every year, millions of tons of pollutants are added into our air and water sources as a result of electrical energy production. While this is unfortunate for our environment, electrical energy is a requirement for maintaining the lifestyles that Americans today enjoy. There are many sources for electrical energy (coal, nuclear, hydroelectric, wind, geothermal, and solar). In this project, you will be working with solar panels.

Driving question
On your own, devise three questions about solar panels and solar energy that you think are interesting and that you would like to investigate further using the internet, books, and other references, in addition to your own experiments in lab. List your three questions in the space provided.

1.
2.
3.

Compare your three questions to those of your team members, and pick one that you think would be the most researchable and give you the best results. This is the driving question for your solar panels project.

The driving question for this project is:

____

____

Background information
In the space provided, list the information you will need to research to answer your driving question. Also, list the experiments that you will need to conduct. This is a tentative list, and ideas can be added or subtracted from the list as your project progresses.

____

____

____

____

(continued)

Figure 1 (continued)

References
List the sources you plan to use as references for your driving question. You may add or remove references from this list as your project advances.

____

____

____

____

Implementing your project
As you carry out your research and conduct your experiments, use your science notebook to record all data, observations, and especially the experiments that failed. This information will be critical to the success of your project and your ability to answer your driving question.

to be the primary mode of collecting information, but rather, a supplement to the student-generated experiments conducted with their groups.

If students' questions were not appropriate, then I stepped in and helped them either narrow or broaden their question. For example, one group needed help refining what they would measure (such as current with a multimeter, the number of lightbulbs or types of devices that could be made to work, etc.). This planning information was recorded on the planning guide (Figure 1). Groups kept a copy for themselves, and submitted a copy to me.

ASSIGNING STUDENT ROLES

Many comment that effective collaborative learning occurs when each student in a group has a specific role to play (Lin 2006; Parr 2007; Steward and Swango 2004). In this project, I used a second planning guide to help students assign roles and keep them accountable. I provided a list of roles, but students had the freedom to add to the list. Roles were designated in relation to writing their final report so that all students would have a part to play. Like the first planning guide, students kept a copy for their records and submitted a copy for me, which was used to keep the teams accountable.

CREATING EXPERIMENTS

The planning guides were designed to make experiment creation simpler and more straightforward. The idea was that by going through the steps on the planning guides (Figures 1 and 2), students would discover for themselves the parameters that needed to be considered and the materials to be used. Of course, some student groups did need a little teacher assistance, especially with regard to ensuring that materials were not damaged. However, my involvement with designing their experiments was minimal. I made suggestions or modifications as needed but was not a major contributor to their creative process. For instance, in situations where more than one method would work, I did not tell students this but let them discover it for themselves. In other scenarios, students needed to be reminded that experiments must be repeatable, and therefore procedures should be executed at least twice.

From there, groups carried out their approved procedures. I used a very student-centered approach; the only questions I answered directly were ones pertaining to safety. While some groups found this frustrating initially, they quickly learned how to rely on each other's expertise and ideas to solve problems.

TIMELINE

Pretest (1 day)
Preproject teaching and activities (6 days)

- Electrical energy production methods (hydroelectric, coal fired, etc.)
- Circuitry vocabulary (series, parallel, voltage, current, circuit maps)
- Teacher-directed lab on assembling circuits
- Inquiry lab on assembling circuits
- Teacher-directed lab on using solar panels in circuits
- Inquiry lab on using solar panels in circuits

Completion of planning guide, writing procedures, teacher approval (3 days)
Experimentation (3 days)
Analysis and repeating experiment trials if needed (2 days)
Group lab report workday (1 day)
Presentation workday (1 day)
Group presentations (2 days)
Posttest and wrap-up (1 day)

IMPACT ON STUDENT LEARNING

At the conclusion of the project, students completed a "regular" (nonbaseline) SALG survey online and teacher-designed posttest.

Pre/posttests

The pre- and posttests were written to be similar, but not identical. The average score on the pretest was 38%, and the average score on the posttest was 93%. This difference was shown to be statistically significant ($t = -33.86$; $p < 0.001$).

Figure 2

Research paper outline/guide

Group members:

Use this guide to help construct an outline of the paper/lab report you are going to write as a group. Turn in one form per group. Keep one for your records. Everyone must have at least one role.

Title/driving question:

Introduction
Person/people responsible for writing the introduction:

___

1. Why is this question important or relevant?

___

___

2. Why did your team choose this question?

___

___

3. What background information do you already know about this topic? List what you already know.

___

___

4. What do you plan to learn/discover?

___

___

Methods
Person/people responsible for writing the methods:

___

1. How and where did you collect your information? Where did you get your information?

___

___

2. How will you decide if a source is reliable/credible?

___

___

3. Write a step-by-step procedure for the experiment(s) you plan to do in lab. The teacher must sign your procedure.

___

___

Analysis
Person/people responsible for writing the analysis:

___

1. How will you determine when you have completely answered the question? Can you answer it beyond a doubt?

___

___

___

2. What will you do if you don't feel like you have reached a firm conclusion? Explain.

___

___

___

(continued)

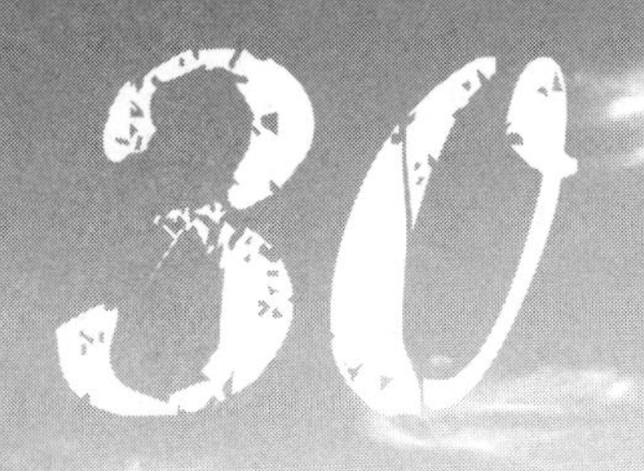

Figure 2 (continued)

3. What will you offer as evidence to prove that you have answered the question? List both the experimental results and the information collected from sources that you will use to demonstrate that your question has been answered.

____

____

____

Conclusions and future work
Person/people responsible for writing the conclusions:

____

1. How will you come to a final conclusion?

____

____

____

2. How will you decide what you would do differently in the future?

____

____

____

____

3. Explain how you think your research can be used by others.

____

____

____

____

____

____

____

____

SALG survey

The second SALG survey provided feedback on students' increases in knowledge and offered students an outlet for anonymous feedback. Of 58 students, 51 (88%) completed the second SALG survey (Table 2).

The posttest scores indicated a statistically significant gain in test scores, and the SALG survey results confirmed the expectation that student learning was significantly positively impacted by this project. These responses further illustrate how students' ideas changed as a result of the project.

CONCLUSION

In this project, students were challenged to develop a researchable question about solar energy and electronics and devise a means of answering it. Students worked cooperatively, with specific roles for each member, conducting research, conducting experiments, analyzing results, and writing the final research/lab report. Throughout the project, students were supervised by the classroom teacher, but were not provided direct answers to questions unless the safety of students or equipment was in question. Rather, they were

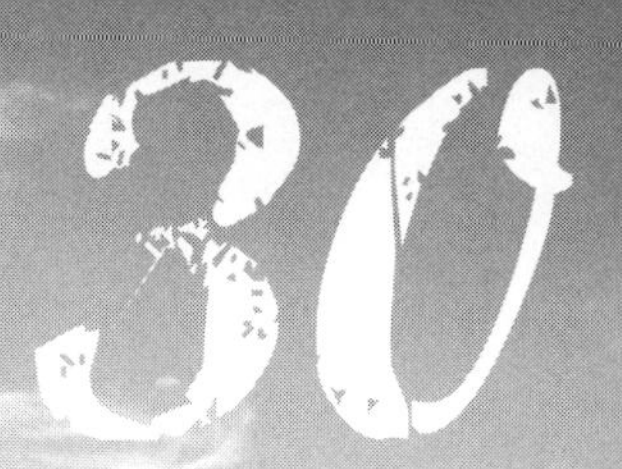

Table 2

Selected SALG posttest results; students rated their current understandings after completion of the preproject activities.

	No gains	Small gain	Moderate gains	Good gains	Great gains	Not applicable
How electricity works	0%	12%	14%	37%	37%	0%
How electronic devices work	0%	10%	20%	41%	29%	0%
How solar energy works	0%	0%	14%	35%	51%	0%
What a circuit is	0%	8%	8%	33%	51%	0%
How ideas from this class relate to the real world	2%	2%	12%	53%	31%	0%
How to assemble circuits with batteries as the power source	2%	4%	10%	37%	47%	0%
How to assemble circuits with solar panels as the power source	0%	4%	14%	31%	51%	0%
How to figure out what's wrong if a circuit is not working	4%	4%	12%	31%	49%	0%
Thinking about how I use electricity in my daily life	6%	16%	20%	33%	25%	0%
Thinking about how difficult complex circuits are	4%	12%	20%	33%	29%	0%

carefully guided to develop their own solutions to problems that emerged as they conducted experiments.

As a result of this project, students increased their understanding of circuits, electronics, and alternative energy, as indicated by pre- and posttest scores and feedback collected using the online SALG survey. In addition, this project stimulated high-achieving students, while providing an environment for lower-achieving students to thrive, as well.

This project can be adapted to the materials available in any classroom. For instance, if teachers do not have access to solar panels, the project could be revised to have students develop researchable questions for battery-powered circuits. I plan to reuse this strategy in other contexts, such as genetics and inheritance, by again having students develop a driving question and then conduct experiments and research to answer it. Based on my experience with this, I believe that when students are able to choose their own project ideas and implement them under supervision, the results are much more meaningful than when students are simply taught the material in a lecture format or if students follow a cookbook-style lab exercise.

MATERIALS

Note: All prices are as of 2009, and all items can be reused from year to year until they break or need to be replaced. This supply list is calculated for a classroom of 30 students, working in groups of five, with three solar panels per group, and sharing other items.

- 15 5 W STP 005S-12 solar panels ($60 each)
- 5 250 W clamp lights ($13 each)
- 1 pack of CFL white lightbulbs for clamp lights ($10 per pack)
- 1 300 W incandescent lightbulb ($4)
- 2 150 W incandescent lightbulbs ($3)
- red and blue lightbulbs ($3 each)
- 1 LED lightbulb ($11)
- 5 70 dB piezo buzzers (for preproject labs and testing circuits, $3.50 each)
- 5 microvibration motors (for preproject labs and testing circuits, $4 each)
- 20 2.4 V/360 mA incandescent bulbs (for preproject labs and testing circuits, $2 for two-pack)

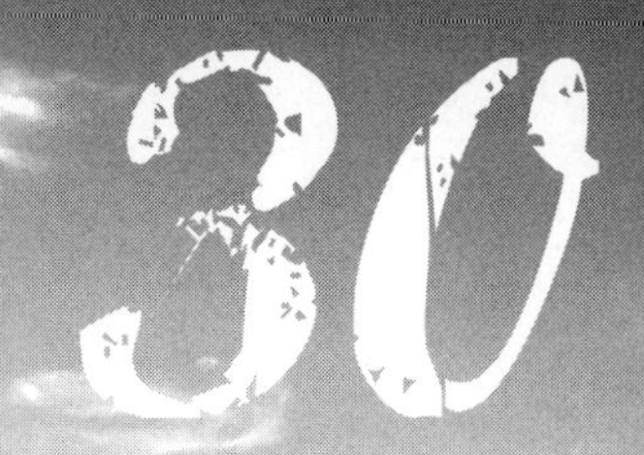

- cafeteria trays (borrowed) for easily transporting materials outside
- 1 pair of wire strippers ($13)
- 2 digital multimeters ($20 each)
- 3 rolls of electrical tape (for safely securing connections in circuits, $6.50 each)
- 8 six-outlet surge protectors (for safely plugging in light sources, $7 each)
- materials that students brought in for their solar circuit experiments (preapproved and safety inspected by the teacher prior to use) included toaster ovens, fans, alarm clocks, rechargeable flashlights, and remote control toy truck batteries (to recharge using the circuits)

Acknowledgment

This project was supported by a Toshiba America Foundation 7–12 classroom grant received in September 2009, This project would not have been possible without the support of Toshiba. For more information about Toshiba's grant program, please visit *www.toshiba.com/taf/612.jsp.*

References

Lin, E. 2006. Cooperative learning in the science classroom. *The Science Teacher* 73 (5): 34–39.

Parr, R. 2007. Improving science instruction through effective group interactions. *Science Scope* 31 (1): 19–21.

Steward, S., and J. Swango. 2004. The eight-step method to great group work. *Science Scope* 27 (7): 42–43.

Wisconsin Center for Educational Research. 2010. *Student Assessment of Their Learning Gains (SALG). www.salgsite.org.*

Chapter 31

WINDMILLS ARE GOING AROUND AGAIN

By Richard H. Moyer and Susan A. Everett

Do all things old really become new again? Depending on your age, your mental image of a windmill may be of the classic Dutch style, the ubiquitous American farm style of the 19th and early 20th centuries, or the giant-size wind turbines often seen today grouped in massive farms. While the design has changed over the centuries, the basic idea has remained—some type of blade captures the energy of the wind in order to turn a shaft that does some kind of work, such as turning a millstone or turning a coil of wire in a magnetic field to generate electricity. Wind is reemerging as a clean and reliable source of energy—primarily for the production of electricity.

In this 5E Instructional Model lesson, students will construct a simple pinwheel-type windmill to test the power generated by different designs. Students will compare three- and four-bladed pinwheels made from manila folders or plastic report covers. This lesson addresses the International

Technology and Engineering Educators Association standard "energy is the ability to do work using many processes" (ITEA 2002, p. 162). In addition, the National Science Education Standards for the middle level include the standard "energy is transferred in many ways" (NRC 1996, p. 155). Energy from the Sun is transferred to wind energy, which becomes mechanical energy as the windmill turns, and then lifts a weight storing gravitational potential energy. Students will also observe a pinwheel connected to a small electric motor (cost: approximately $5), which generates electricity that will be used in turn to power a smaller motor (cost: approximately $2; motors are available through most science supply distributors) to turn another pinwheel. Here, the mechanical energy of the windmill is converted to electrical by the first motor acting as a generator, and then back to mechanical by the second motor, which finally turns another pinwheel producing wind again.

HISTORY

People have used the energy of the wind for sailing purposes since as early as 5000 BC. Windmills were first used in China and the Middle East for pumping water and grinding grain around 200 BC. These ideas were later brought to Europe, where the Dutch improved on the design. The technology made its way to America and in the late 19th century was popular on farms and in rural areas, primarily to pump water and generate electricity. By the 1930s, most rural areas had been wired for electricity, and the use of windmills declined until the price of fossil fuels began to increase in the 1970s. Today, "wind energy is the world's fastest-growing energy source and will power industry, business and homes with clean, renewable energy for many years to come" (U.S. DOE 2005, p. 1).

TEACHER BACKGROUND INFORMATION: INVESTIGATING WINDMILLS

Engage

Review with students safety guidelines on the use of sharp objects such as scissors and pushpins. Students should use caution when working with glue guns and wear safety goggles at all times during the investigation. Use low-temperature hot-glue guns for this activity.

To engage students, you may want to show them a variety of windmill photos, easily found online by searching Google images for "windmills." Initiate a discussion of students' ideas regarding the function and purpose of windmills. Depending on your location, students may or may not have significant experience with windmills. You may also wish to have students construct a KWL chart with their ideas and questions about windmills.

Have students construct a pinwheel of their design. They should try to determine how moving air causes the blades to turn. This will provide some

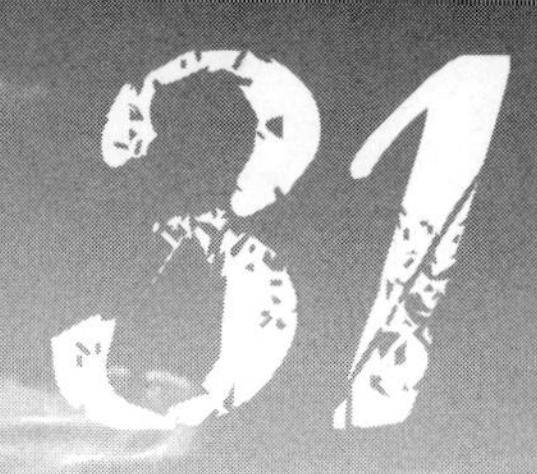

initial scaffolding experience in constructing a pinwheel before trying to make one in the Explore phase that is capable of lifting a weight.

Focus students' attention on the fact that windmills are of different design and have varying numbers of blades. Explain that in this lesson they will design an experiment to see whether a three-bladed or a four-bladed pinwheel-type windmill can lift a weight faster. If you would prefer to have students participate in a much more open inquiry and design process, use the ideas in the Evaluate section for the Explore component of the lesson.

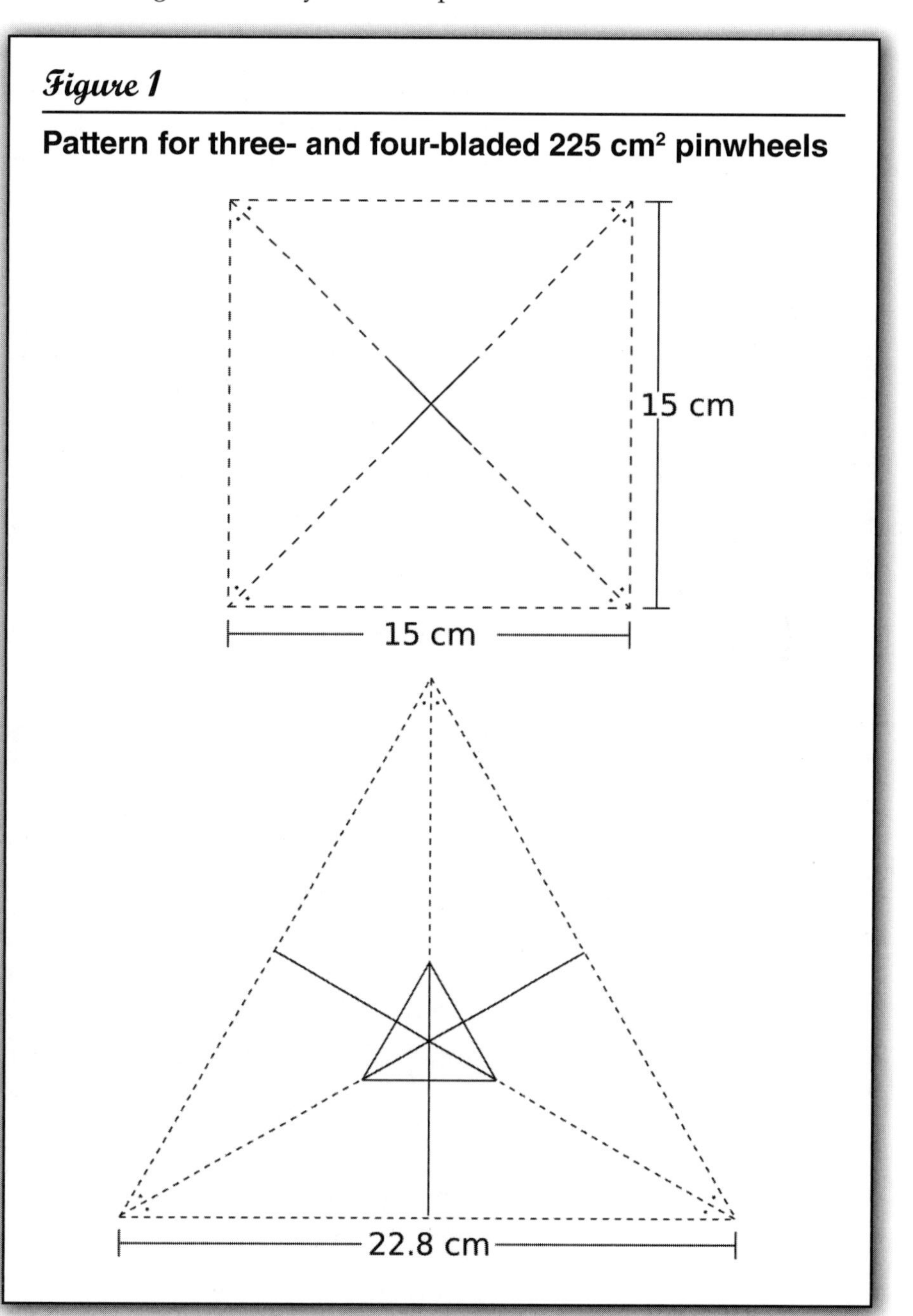

Figure 1

Pattern for three- and four-bladed 225 cm² pinwheels

Explore

Each group of two to four students will need

- 2 pencils with newer, flat-topped erasers
- 2 large milkshake straws (8 mm rather than the standard 5 mm diameter)
- 2 file folders or plastic report covers
- 2 pushpins, 100 cm of thread
- 2 paper clips
- 4 washers or other weights to hang on the paper clips
- ruler

- scissors
- tape
- stopwatch
- template for the three- and four bladed pinwheels (Figure 1), and
- access to a low-temperature hot-glue gun.

If you want each group to find the mass of the weights and the paper clips, then they will need a balance, as well. One large box fan per group is ideal, although groups can share them. Students should be reminded to keep fingers and clothing away from the fan blades.

It will be helpful if you show students completed pinwheels (Figures 2 and 3). You may choose to give students the option to design their own pinwheels or use the procedure described in the Activity Worksheet, see p. 282. Students may encounter some difficulty making the pinwheel turn a shaft to lift the weight. It is relatively easy to construct a pinwheel in which the blades turn, however, it takes much more tinkering and design work to construct a pinwheel-type windmill where the blades freely turn and cause the

Figure 2

Pinwheel details—the two longer straws are not glued to the pencil and act as bearings

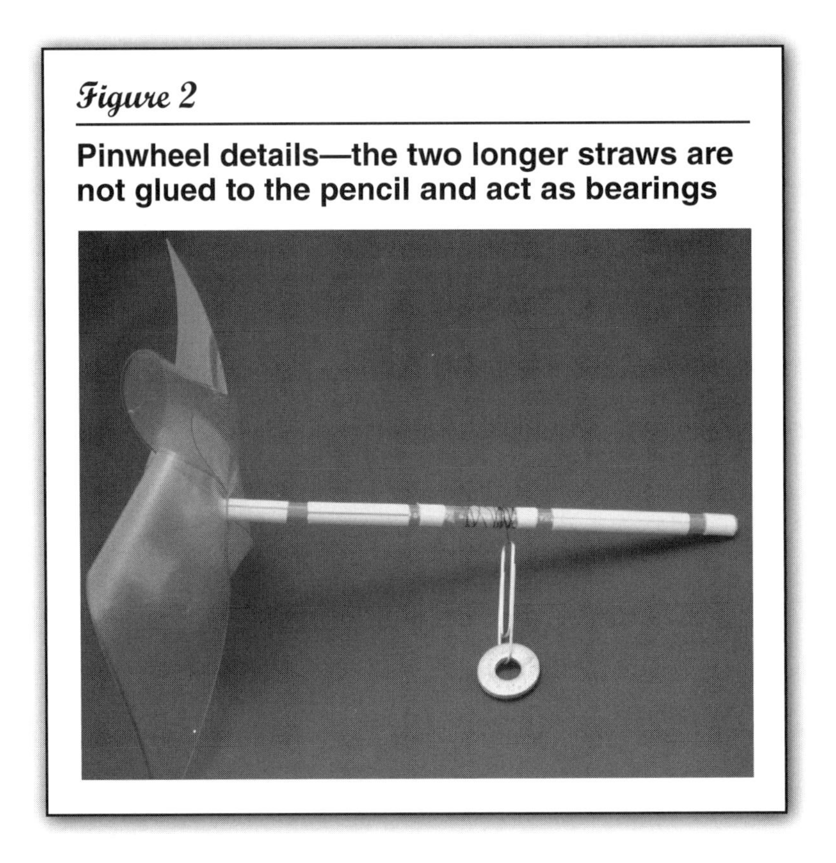

Figure 3

Completed pinwheel with detail of pushpin

shaft to turn, as well. Students must take care at all steps of the windmill construction. Wobbly shafts, bearings with too much friction, and unbalanced blades will all greatly interfere with the performance of the windmills. Your students will discover this whether you let them design their own windmills or follow the instructions provided in the Activity Worksheet. Depending on your schedule, you may choose to have students construct and troubleshoot windmills during one class period and then test them the next. Small pieces can be stored in a zipper plastic bag.

You may wish to have students design their own procedures for testing the three- and four-bladed pinwheels, or you can follow the sample procedure described here. Students should practice holding the pinwheel in front of a fan so that it spins freely and does not rub against their hand (Figure 4, p. 278). Remind students of fan safety and to continue to wear their goggles. Students should hold the pencil level by the two straw bearings and not squeeze too tightly for maximum performance. For all trials, the pinwheel should be held the same distance from the fan and in the same relative position with respect to the fan. It is also advisable for students to practice using the stopwatch to

Figure 4

Spinning pinwheel lifts the weights

Table 1

Sample data table for pinwheels

Type of pinwheel	Time (in seconds) to lift weight					
	Trial 1 (s)	Trial 2 (s)	Trial 3 (s)	Trial 4 (s)	Trial 5 (s)	Average (s)
Three blades	7.2	5.6	4.4	5	6.2	5.68
Four blades	1.8	2	1.8	3	2	2.12

time how long it takes to lift the weights. Students should conduct multiple trials and record their data into a table as shown in Table 1.

Explain

While students' data will likely vary, they should all determine that the four-bladed pinwheels lift the weights faster. For a pinwheel of this size, the curved area facing the wind is greater for the one with four blades than three, since the three-bladed pinwheel has a greater surface that is flat. It is this curved area that produces the torque that causes the blades to turn.

Help students understand the energy transfer they have just experienced. The motor on the box fan has changed electrical energy into mechanical energy powering the fan, which moved the air and created wind. The wind was transferred to mechanical energy of the spinning pinwheel, which turned the pencil and lifted the washers. The washers now have stored gravitational potential energy. If the fan is turned off, gravity will pull down on the washers, and this potential energy is transferred to the kinetic mechanical energy of the turning pencil and pinwheel, and this mechanical energy is finally transferred by the pinwheel to the air again.

If your students are not familiar with the scientific definition of work, you may need to review. Have students continue to work with their groups for this portion of the lesson to facilitate learning for those who may have difficulty with the calculations. Work is the product of a force acting through a distance and is commonly measured in units of newton meters, also known as joules. It is likely that your balance will measure mass in grams rather than weight in newtons. Either you or your students will have to convert grams to newtons. Since $w = mg$, a force in newtons can be found by multiplying the mass in kilograms by the gravitational constant, g, which is 9.8 m/s^2. In our example, the mass of the paper clip and two washers was 10.4 g (0.0104 kg); multiplying this by g results in a force of 0.102 N. We lifted the weights 40 cm, so the amount of work done is

$$\text{work} = \text{force} \times \text{distance} = 0.102\ \text{N} \times 0.4\ \text{m} = 0.041\ \text{N·m}$$

Since students lifted all of the weights the same distance (and the weights were identical), both of the pinwheels did the same amount of work. However, the four-bladed pinwheel was able to do the work substantially more quickly (nearly three times in our data) and therefore is more powerful. Power is a measure of the rate at which work is done. For the three-bladed pinwheel, students will find

$$\text{power} = \text{work/time} = 0.041\ \text{N·m} / 5.68\ \text{s} = 0.007\ \text{watts (W)}$$

For the four-bladed pinwheel, the power students will find is

$$\text{power} = 0.041\ \text{N·m} / 2.12\ \text{s} = 0.019\ \text{W}$$

Extend

In this section, you will need to construct two pinwheels to attach to the two small motors using a short piece of a coffee stirrer and a dab of hot glue. The

Figure 5

Pinwheel with generator turning motor and another pinwheel

two motors are connected to each other with wires and alligator clips. When the pinwheel attached to the larger motor is placed in front of the fan, it will cause the second pinwheel to turn (Figure 5). To help students understand the transfer of energy, connect one motor to a 1.5-volt cell so that the pinwheel spins. The battery transfers stored chemical energy into electrical energy that the motor transfers into mechanical energy that turns the pinwheel. An electric generator is essentially a motor working backward—if you turn the shaft of a motor, this mechanical energy can be turned into electrical energy. This is what causes the second pinwheel to spin. To review the energy transfer here, the wind causes the first pinwheel to turn, which turns the shaft of the larger motor, which produces an electric current that causes the second motor to turn as this electrical energy is transferred to mechanical, which turns the second pinwheel and causes the air to move, creating wind. Students may note that the second pinwheel is turning much more slowly due to inefficiencies of the transfers, with most of the "lost" energy going into heat from friction.

This lesson may be extended in many ways. Questions raised by students may deal with some of the current engineering, environmental, and societal issues related to the expanded use of windmills. Some engineering issues students could research are location of windmills relative to electrical needs and grid connections, initial cost compared to value of electricity generated, and long-term maintenance issues. Environmental questions might focus on potential harm to birds and bats and how the carbon footprint of the manufacture, transportation, installation, and maintenance of the windmill compares to traditional electrical power generation. Societal issues include what to some may be perceived as visual pollution and noise pollution, and their impacts on people and commerce.

Evaluate

There are many ways this investigation can be expanded, since there are many variables that can be tested by students: blade design (length, shape, and material), vertical versus horizontal shaft windmills, or improving the soda-straw bearings to reduce friction loss. You may wish to have students search the internet and the library for information on windmill design (see KidWind Project 2010).

CONCLUSION

Wind energy is a reemerging source of renewable energy that has served humans for much of our history. Students may not realize that many engineers work on environmental projects such as trying to maximize the amount of energy that can be transferred from the wind into electrical energy for use by society without the detrimental effects of burning fossil fuels.

Activity Worksheet: Investigating windmills

Note: Wear safety glasses during this activity. Use caution when working with the low-temperature hot-glue gun, cutting, using sharp objects, and using the fan for testing.

Engage

1. Have you ever seen a windmill? How do you think a windmill works? For what purpose do you think people use windmills? What questions do you have about windmills? Record your thinking. Discuss your ideas with your classmates.
2. Your teacher will provide you with materials to use to design a pinwheel. Construct your pinwheel and try to figure out how moving air causes the blades to turn.
3. Now that you have some experience in designing and constructing a pinwheel, you will design a test to determine if the number of blades on a pinwheel-type windmill will affect its power. The question you will explore is "Will the number of blades on a pinwheel affect how fast it can lift a weight?"

Explore

1. Using the provided materials, construct the two pinwheels, one with three blades and the other with four. (**Note to teachers:** Alternatively, you could provide students with the following procedure and the pattern provided in Figure 1 with the same surface area of 225 cm^2 for each. Procedure: Cut out each pinwheel along the dotted lines. Note that each is cut in from the corners about two-thirds of the way to the center. Use the pushpin to make the holes at the dots in the corners and in the center. Insert the pin through the corners and then the center to secure, and set aside. Cut six pieces of straw as shown in Figure 2. Slide them onto the pencil and glue all but the two longer pieces, which serve as bearings in which the pencil can freely spin. Use glue sparingly, as excessive glue will greatly impede the spinning of the pinwheel. Attach a 40 cm length of thread with a small piece of tape to the center of the pencil as shown. A paper clip and two washers are attached to the other end of the thread. Finally, push the pin and pinwheel into the eraser, securing with a small dab of glue.)
2. Which pinwheel do you think will lift the weight faster, the three- or four-bladed one, if you hold it in front of a large box fan?
3. Plan a way to test the two pinwheels. In your plan be sure to include multiple trials to find an average time. Think also about which variables you must keep constant for a fair test. Share your plan with your teacher before you begin. Keep fingers and clothing away from the moving fan blades and continue to wear your goggles.
4. Time how long it takes each pinwheel to lift the weight and then organize your data into a table.

Explain

1. Share your results and conclusion with the class. How do your results compare with other groups'? Did one pinwheel perform better than the other? What is your evidence for this? Look carefully at the two pinwheels and see if you can deduce why one lifted the weight faster.
2. How was energy transferred in the pinwheels you made? Where is the energy when the weight has been lifted all the way up to the pencil shaft? (**Hint:** If you turn off the fan, what happens to the weight and your pinwheel if you hold it gently by the straws?)
3. Work is the product of a force acting over a distance. If the force is measured in newtons

(continued)

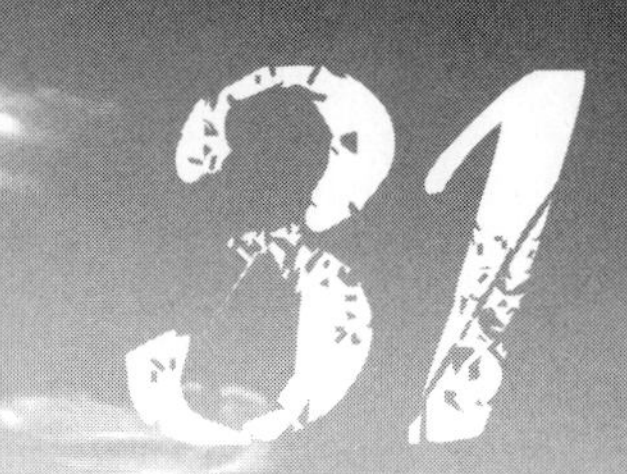

Activity Worksheet *(continued)*

and distance in meters, then the amount of work is in newton meters (also called joules). The formula for work is $W = Fd$. How much work did each pinwheel do? Your teacher will either give you the force in newtons or show you how to calculate it. The distance is the height your pinwheel lifted the weight.

4. If one of your pinwheels is able to do this work faster than the other, then it is said to be more powerful. Power (in watts) can be calculated by dividing the amount of work done, in joules, by the amount of time it took to do the work, in seconds: power = work/time. Calculate the amount of power in watts (joules per second) for both of your pinwheels.
5. Is one of your pinwheels more powerful than the other? How do you know? Compare your results with those of other groups in your class.

Extend

1. Observe what happens when the small motor attached to the pinwheel is connected to a battery. Explain the transfer of energy.
2. Predict what will happen if the motor is connected instead to a second, larger motor that is spinning in the wind from the box fan.
3. In the second case the wind caused the larger motor to turn. How did the system of the two motors connected together transfer this wind energy?

Evaluate

In this investigation, you found out what happened when you compared three- and four-bladed pinwheels. There are many other possible variables and windmill designs. Select a factor and design a test. Compare your results of the new design to the pinwheel model you investigated here. Determine the amount of power your design produces and share your work with your classmates.

Acknowledgments

The authors wish to acknowledge teachers Leah Walkuski and Cindy Pentland, Robert Simpson III for his photography, and physicist Paul Zitzewitz for his consultation.

References

International Technology Education Association (ITEA). 2002. *Standards for technological literacy: Content for the study of technology.* 2nd ed. Reston, VA: ITEA.

Kidwind Project. 2010. Wind turbine blade design. *http://learn.kidwind.org/learn/wind_turbine_variables_bladedesign.*

National Research Council (NRC). 1996. *National science education standards.* Washington, DC: National Academies Press.

U.S. Department of Energy (DOE). 2005. History of wind energy. *www1.eere.energy.gov/windandhydro/wind_history.html.*

Resources

Moyer, R. H., J. K. Hackett, and S. A. Everett. 2007. *Teaching science as investigations: Modeling inquiry through learning cycle lessons.* Upper Saddle River, NJ: Pearson/Merrill/Prentice Hall.

U.S. Department of Energy. 2005. Wind. *www.eere.energy.gov/topics/wind.html.*

Chapter 32

A FIRST ENERGY GRANT

PINWHEEL ELECTRICAL GENERATION

By John Schaefers

This is an interdisciplinary activity—with art, science, and math classes involved—where students design their own pinwheels, and then attach their design to a DC generator (motor).

I introduce the project by illustrating the physical layout and displaying an example of a completed pinwheel. Students research pinwheels, turbines, and their applications to other forms of energy (see "On the web").

Students decide on a pinwheel design and construct the pinwheel in art class from a square piece of construction paper. The model is traced and cut along the lines which, when bent and pinned to the center, will form the wings of the pinwheel. Then, a square of cardboard is glued to the back of the pinwheel and allowed to dry. The generator is hot glued to the center of the cardboard. Students are given safety guidelines for using the glue gun and scissors.

After completion of the pinwheels in art class, students bring them to math class where they calculate the diameter and surface area. Finally, students bring the pinwheels to science class. Groups of four students are formed, and students choose one of their pinwheel designs to test in a wind tunnel (fan) where they measure the voltage, current, and wind speed. Prior to testing their designs, students are introduced to basic circuitry, voltmeter usage, heat measurement, wind speeds, data collecting, recording, and analysis.

Allow two periods to complete this activity since it takes a few minutes for students to become adept at using the volt/amp meter and to get their

Figure 1

Student data table

Questions

With the data from six groups, answer the questions to the right and graph the data.

	Surface area of pinwheel	Diameter of pinwheel
Pinwheel		

	Fan speed slow	Fan speed medium	Fan speed fast
Voltage	mV	mV	mV
Current	mA	mA	mA
IR temp. pinwheel	°C	°C	°C
IR temp. fan	°C	°C	°C
Wind speed	mph	mph	mph
Revolutions	revs	revs	revs

1. How does the voltage/current compare to the fan speed (graph)?
2. How does the pinwheel weight compare to the voltage/current (graph)?
3. Is current and voltage linear (straight line increase)?
4. What happened to the IR temperature value of the pinwheel as the rotation rate increased?
5. What happened to the IR temperature value of the electric motor as the rotation rate increased?
6. Was the wind speed linear?
7. What happened to weight of the pinwheel compared to the voltage?
8. How can we determine the number of revolutions?
9. What happened to the IR temperature value of the electric motor as the rotation rate increased?
10. Was the wind speed linear?
11. What happened to weight of the pinwheel compared to the voltage?
12. How can we determine the number of revolutions?

Figure 2

Pinwheel model and related materials

Figure 3

Pinwheel setup and three-speed fan

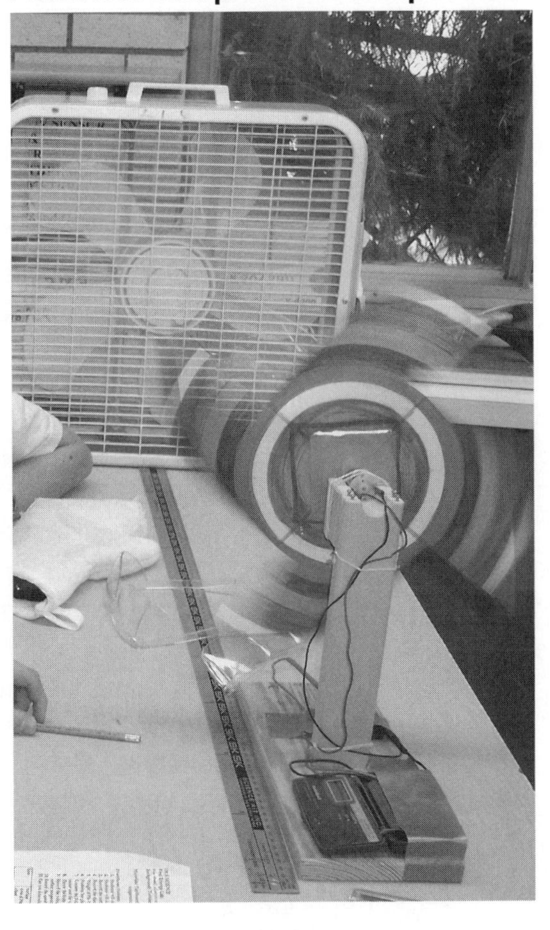

pinwheel to rotate properly. To test their designs, students connect the DC motor to the test stand, and connect the circuit alligator clip leads to the motor, voltmeter, and load (light) circuit. They then throw the knife switch and engage the fan at slow speed. Students are required to record the voltage generated at the three fan speeds after 15 seconds rotation and the surface temperature of the electric motor after the 15-second intervals. While testing their designs in the wind tunnel, students record data in a table and complete a set of questions (Figure 1).

A SUCCESSFUL PROJECT

Our students enjoy this project. The design of the pinwheels is a big success, as well as the determination of the pinwheels' area. Figures 2–4 show typical materials used and the setup for the activity. Problems that arose concerned the linkage to the volt meter; I applied tape to help hold the probes to the contacts. Rubber bands held the motor to the stand, which had to be modified to reach out farther so that the blades of the pinwheel didn't collide with the stand. I also placed a clipboard

Figure 4

Digital voltmeter

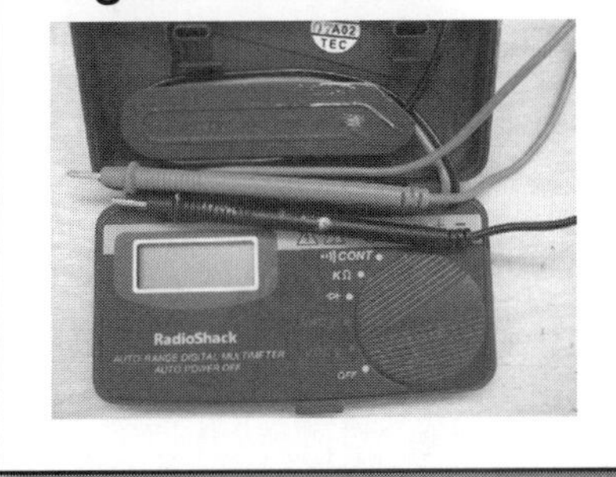

Figure 5

Materials and equipment for one setup and total cost

- 1 roll 18–22 gauge wire $4.00
- 2 electric knife switches $4.00
- 1 pack circuit alligator connectors $4.00
- 1 small DC motor (generator, 1.5–3.0) low torque $15.00
- 1 female and male connectors for the small motors $1.69
- 1 Auto read voltmeter, digital $20.00
- 1 IR thermo detectors (heat measurement) $41.00
- 1 package art supply (cardboard) $20.00
- 2 fans $40.00

Total = $149.69

Optional: Hand generators. I also use a large capacitor (1 farad) for the storage of the energy generated; this could be used to run the hand generators. $45.00

at the table so that students could record their results on it and use the data for their graph constructions.

As a follow-up to this activity, students study wind energy and learn about the Garett, Pennsylvania, wind farm and its wind-generated energy.

On the web

American Wind Energy Association: *www.awea.org*

Community Energy: *www.newwindenergy.com*

Energy Justice Network: *http://www.energyjustice.net/*

NASA, Ohms law: *www.grc.nasa.gov/WWW/K-12/Sample_Projects/Ohms_Law/ohmslaw.html*

Scied.net: *www.scied.net*

Voltage circuit simulator: *http://jersey.uoregon.edu/vlab/Voltage*

Wind and hydropower technology energy: *www1.eere.energy.gov/windandhydro*

Part 2

Student Projects and Case Studies

Chapter 33

DOING SCIENCE WITH PBS

By Steve Metz

Student projects have long been part of the teacher's toolbox, going back at least as far as the early 1900s, when John Dewey reported on the value of experiential learning and student-directed classrooms (1916, 1938). Most teachers understand the value of projects as a way to provide students with a means for taking responsibility for their own learning and actually "doing science."

Project-based science (PBS) is finding a place in more and more secondary school science programs as teachers discover its power to engage students and develop critical-thinking skills. PBS is firmly rooted in constructivism—the idea that individuals construct knowledge individually, through active and meaningful interactions with their environment, rather than by passively receiving transmitted information. PBS is also strongly indebted to the somewhat more obscure "constructionism," which, inspired by constructivism, suggests that learning is an active process and students learn most effectively when they are constructing a meaningful product.

Notes about PBS

Project-based learning in science is a method in need of an accepted title and acronym. The obvious and often-used choice is *PBL*, but this creates confusion with *problem*-based learning. The rather awkward *PjBL* is sometimes used to avoid this confusion. Possible alternatives include *PBI (project-based instruction)* and *PBSI* or *PBSL (project-based science instruction or learning)*. No single term seems yet to have gained universal acceptance. For this chapter, I have chosen *PBS—project-based science*—which seems to be the best tentative solution, despite its evocation of "Public Broadcasting System."

Project-based science (PBS) is sometimes confused with *problem*-based learning (PBL). In fact, the two methods share many commonalities in their emphasis on rich, open-ended, student-centered classrooms; authentic inquiry; collaborative learning; interdisciplinary approaches; and performance-based assessment. In a nutshell, PBL grew out of the case-based teaching used in educating health care, legal, and business professionals and uses ill-structured problems as the curricular focus. The essential elements of PBS involve a complex driving question and the use of tools and technology to create a tangible product.

Although there is no single accepted definition of PBS (see "Notes about PBS"), project-based classes do share several essential elements. In a well-designed project, students engage in extended inquiry by addressing complex, authentic questions and creating a meaningful product or artifact. In the best PBS scenarios, learning is driven by challenging, open-ended driving questions that connect with students' interests and lives. Students work in collaborative groups for research, design, and production; teachers become facilitators.

When students work together to create authentic, meaningful products, they appreciate the priority of using evidence to make decisions as they critically evaluate possible solutions. Like the authentic research problems encountered by practicing scientists—and the issues students will face in their own lives—good projects present an ever-changing variety of approaches and solutions. Dewey again proves prescient in his suggestion that students should "attack their topics directly, experimenting with methods that seem promising," instead of being lead to assume "that there is one fixed method to be followed. The teacher who does not permit and encourage diversity of operation in dealing with questions is imposing intellectual blinders upon pupils—restricting their vision to the one path the teacher's mind happens to approve" (Dewey 1916).

Class management issues are sometimes cited as a potential difficulty with open-ended projects—it can be challenging to manage a large class in which students have a high degree of autonomy. In my own experience, when students are truly and actively engaged in a project, classroom management issues often decrease. It can also be a challenge to balance time constraints and align projects with the requirements of district curriculums, science-content standards, and high-stakes testing. There is no question that, to be successful, projects must be central to the curriculum, not simply a peripheral, add-on activity.

The ability to successfully complete complex projects is an important life skill. Projects play an important role in business, industry, art, science, and everyday personal affairs. Each of us has our own special projects—so why not include PBS in your classes?

References

Dewey, J. 1916. *Democracy and education: An introduction to the philosophy of education.* New York: The Free Press.

Dewey, J. 1938. *Experience and education.* New York: Collier Books.

Chapter 34

STUDENTS FOR SUSTAINABLE ENERGY

INSPIRING STUDENTS TO TACKLE ENERGY PROJECTS IN THEIR SCHOOL AND COMMUNITY

By Regina Toolin and Anne Watson

Sustainable energy is one of the most critical issues facing our planet today. As the world struggles with fluctuating oil prices and rising green energy initiatives, students need to know that they have the power to effect change. At Montpelier High School (MHS) in Vermont, students are accustomed to making such changes in their school and community. Over the last few years, MHS students have participated in the Annual Winooski River Cleanup Project, the construction of a solar-powered greenhouse that provides produce for the school's cafeteria, and a thriving composting program used to fertilize the produce and plants grown inside the greenhouse.

This chapter describes the sustainable energy projects that MHS physics students designed during the spring semester of 2008. An overview of the project is followed by a description of the planning and implementation processes, examples of specific student energy projects, and a discussion of outcomes and lessons learned. We hope that our experiences will inspire other teachers and students to undertake similar projects and create positive changes in their own communities.

PROJECT PLANNING AND IMPLEMENTATION

After completing units on motion, force, and momentum, the Classical Physics course at MHS culminates with a unit on energy. During the spring semester, energy systems and energy transformation are the big underlying ideas (Wiggins and McTighe 2003), with a focus on applying energy concepts, equations, and theories to local sustainability initiatives. Students quickly learn that the ability to calculate the theoretical energy produced by a small wind turbine is just as important as the ability to build one. They come to appreciate the importance of calculating energy equations and applying this new knowledge in meaningful and relevant contexts.

Preparation, planning, and problem-posing

Student-driven sustainable energy projects were first introduced in the Classical Physics course in 2004. Given the projects' history and reputation, students, teachers, and administrators have become accustomed to the excitement it generates each spring. In fact, the projects are often the reason students enroll in the course in the first place.

Parents and community members are central to project development. Beyond being informed of project goals and objectives—through a letter and the school website—they are invited to participate in project development and to attend final presentations.

In 2008, there were 43 students—12 females and 31 males—enrolled in two sections of the Classical Physics course. Most were seniors, heterogeneously mixed in terms of their background and ability. Given the gender ratio in both sections of the course, we encouraged equitable participation in group discussions and project work. At the beginning of the energy unit, we built students' background knowledge and addressed their prior misconceptions. The first three weeks were dedicated to assessing and teaching the fundamental concepts of energy (i.e., kinetic and potential energy and conservation of energy) through inquiry, cooperative learning, and direct instruction.

Next, student projects were initiated through a brainstorming activity that examined the question "How can we reduce the need for energy or switch to alternative forms of energy consumption in the Northeast?" This activity served to spark student interest about the problems and limitations of energy resources, which led to questions about local efforts to reduce energy consumption. For example, after discussing hydropower as an alternative energy source, one student asked, "Don't we have a dam in Montpelier? Do you think it could be used to generate electricity? How much energy? Would the cost be worth it?" Similarly, after a discussion on the kilowatt usage of electric appliances, one student announced, "I bet the school's walk-in refrigerator uses a huge amount of energy. We could save a lot of energy and money by

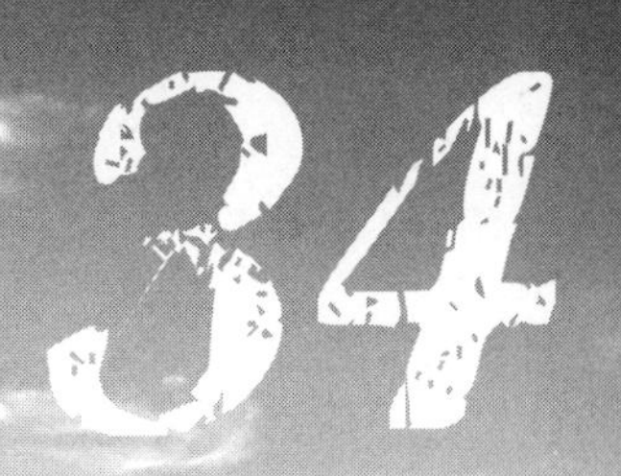

replacing it." From comments and questions such as these, the class developed a menu of possible projects to pursue (Table 1).

Table 1

Sustainable energy projects

Project title	Goal	Artifact or product
Solar hot water	Construct and estimate payback period of a solar hot water system.	Prototype
Hybrid cars	Compare hybrid car energy use with conventional car energy use.	Efficiency and feasibility report
Photovoltaic systems	Educate public about photovoltaic systems.	Efficiency and feasibility report
Wind turbine	Construct power-producing wind turbine. Estimate the payback period.	Prototype
Undershot water wheel	Construct power-producing undershot water wheel. Estimate payback period in the Montpelier section of the Winooski River.	Prototype
Connecting two segments of Montpelier bike path	Estimate carbon and financial payback of bike path through Montpelier.	Efficiency and feasibility report
Pyromex waste gasification	Evaluate benefits and hazards of a proposed power generation system for Montpelier.	Presentation to city planner
Refurbishing a derelict electric car	Refurbish an electric car used by the school.	Restored electrical system
Montpelier's hydroelectric dam	Evaluate potential power-generating capacity and financial payback of a local dam.	Efficiency and feasibility report
Creating an additional bus route in Montpelier	Collect data on student interest in a new bus route in Montpelier.	Efficiency and feasibility report
Creating an additional bus route from Montpelier to Burlington	Collect data on student interest in and potential energy savings of a new weekend bus route from Montpelier to Burlington, Vermont.	Efficiency and feasibility report
Montpelier High School's walk-in refrigerator	Evaluate efficiency of the school's walk-in refrigerator.	Efficiency and feasibility report
Refurbishing a bicycle-powered lightbulb system	Refurbish a bicycle generator. Calculate the power-generating capacity and payback period.	Efficiency and feasibility report
Pellet heating systems	Compare and educate public about pellet heating systems.	Efficiency and feasibility report
Electric efficiency for homeowners	Educate public about appliance efficiency. Calculate energy savings.	Efficiency and feasibility report

Project teams were then formed according to common interests that emerged during the brainstorming activity. Driving questions (Weizman, Schwartz, and Fortus 2008) were generated by each team and served to focus each project. As students engaged in problem-posing (e.g., "How much energy does this lightbulb use?" and "How can I make this fan generate electricity?"), they realized that learning new science concepts would be necessary to move forward in their projects. Whenever more than one group had a question regarding the same concept, we introduced all students to the concept through a brief lesson. In this vein, project-based science (PBS) not only provided motivation for learning new material but also for revisiting old material.

In preparing for these projects, standard safety practices were implemented. We "identif[ied] risks that [could] put students into harm's way" (Roy 2009, p. 12) and required that all students wear personal protective equipment (PPE) when working with tools and equipment. The safety precautions taken for each project were varied. For example, using power tools to construct wind turbines and solar panel boxes required the use of goggles or safety glasses and gloves (actual construction may also require hard hats), modeling of appropriate equipment use, and special instruction and supervision by the industrial arts teacher. Refurbishing the school's electric car also required supervision by a certified electrician and teacher to properly install potentially dangerous 12 V batteries. For these and other projects, appropriate experts were consulted to ascertain and implement the proper safety precautions. However, because it is ultimately the classroom teacher's responsibility to ensure safe practices, all procedures—even those suggested by experts—were checked in accordance with board of education safety policies, legal safety standards, and best safety practices and approved by the teacher. (**Safety note:** If students are working with acid batteries, an eyewash station within 10-second access is required, along with appropriate PPE.)

Project implementation

Actual project work began in the fourth week of the unit and was defined by the following introduction:

> As American society approaches its limits involving natural resources, energy has become one of the most important topics of our time. If we are to achieve sustainability, we will need to re-vision the future and change our energy policies and practices. This project is an opportunity for you to be a part of that change. The goal is to research a question of interest regarding energy sustainability, draw some conclusions about your research—based at least in part on your knowledge of physics concepts and calculations—and present your findings to the class and community.

We provided students with a set of flexible project criteria that addressed essential science standards, concepts, and skills—this was key to the success of the student-initiated projects. During a class brainstorming session, students helped develop a project rubric. The rubric included a detailed description of the project, consultation with a community expert, and a discussion of the underlying physics concepts (see "On the web.") Students were also evaluated on the projects' professionalism and a discussion of its feasibility and potential obstacles.

Community experts from local businesses played a valuable role in contributing to the success of student projects. Over the last few years, we have developed an expanding list of contacts that includes parents, representatives from local businesses, and the city planning commission. Beyond the list of contacts we provide, students have found their own experts through car dealerships, out-of-state companies, additional parents, and school employees. Students also used the internet, when appropriate, to gather information.

Most experts interacted with students via e-mail or phone. Others, such as the certified electrician, actually came to the school to help with student projects. Prior to contacting any organization, students were instructed on proper etiquette and procedures for contacting and interacting with professionals. They were also directed to report any issues encountered to the classroom teacher; issues were few and were mostly limited to lack of phone call or e-mail response. Students did not give out personal information or meet anyone in person without teacher approval.

Project example

All of the student-designed projects were engaging and informative. However, one project deserves a closer look—it was conducted by a team of three students that studied the efficiency and feasibility of restoring the hydro-turbine at the Lane Shops Dam in Montpelier. The team's goal was to research the dam's operation and calculate the energy efficiency, environmental impact, and payback period needed to restore the hydro-turbine; the "payback period" refers to the time necessary to recoup the initial investment. Consultation with the town planner revealed the hydro-turbine's basic components and operation as well as essential data to calculate energy efficiency, impact, and payback. The team obtained the yearly average flow rate and height of the dam to calculate the number of kilowatt-hours (kWh) that would be generated by the dam per year (Figure 1, p. 298).

Through information obtained from the team's expert (i.e., the town planner), students found that Montpelier would require approximately 476 million kWh per year for its electricity needs. Students calculated that the restored dam could produce 893,000 kWh per year—enough electricity

Figure 1

Energy calculations for the hydro-turbine at the Lane Shops Dam in Montpelier

Yearly average flow rate	3.88 m^3/sec = 3,880 kg/s = mass
Gravity	9.80 m/s^2
Height	3.35 m
Efficiency coefficient (a)	0.800 (80%)
Formula	Power= Mgha/t = J/s = watts
Calculation	102,000 W = 893,000 kWh per year

The Lane Shops Dam in Montpelier

A student measures the height of the Lane Shops Dam

for 218 working couple households, 290 single person households, or 165 households with two children. Finally, if the estimated cost to restore the dam was $500,000 with a kWh cost of $0.13, students determined that about $116,000 per year would be saved by the restored hydro-turbine with a payback period of 4.3 years.

PROJECT ASSESSMENTS

Performance and peer-assessments and self-assessments are integral to evaluating student understanding and achievement in the project-based classroom. In the early phases of the project, students were engaged in a brainstorming session to determine the essential criteria for project performance. The following questions summarize the outcome of this session and became the rubric criteria (see "On the web") by which students were evaluated for the sustainable energy project:

- What are your driving questions and goals?
- What is your plan?
- What are your results?
- What are your recommendations?
- What is the science behind it all?
- Who are your references and consultants?
- How well do you convey your ideas?

There were varying degrees of frustration along the path to success for students involved in the energy projects. For those who built prototypes, for example, trial and error was often the best teacher. In analyzing class evaluation data for the project, 14 out of 16 project teams met or exceeded all standards for the project.

Another essential feature of PBS is the opportunity for students to participate in an open-ended survey to assess overall project goals and objectives. In our classes, this included the following questions:

- What did you find most valuable about your project work?
- What recommendations would you offer to improve the project?
- What makes this project different from others you have done in this or other classes?

In the survey, students reported being motivated by their ability to build a hands-on model and to discover the relevance of their projects to global and local energy issues. A student engaged in the walk-in refrigerator project (Table 1, p. 295) wrote, "It was a real-world project. We had the ability to improve the efficiency of an aspect of our school." A student involved in the home electrical efficiency project (Table 1) was excited by the fact that "little changes could really add up and save a lot of money." Still another student who took part in the pellet stove project (Table 1) commented that her project would help to "make the world a better place on a small scale."

Students had various responses to the question "What makes this project different from others you have done in this or other classes?" Many appreciated that the projects had to do with the "real world" and that "it connected to a real-life situation." One student summed up her experience with the following quote: "It meant something to me. I was passionate about the outcomes. And, the outcome was not just a grade."

OUTCOMES AND LESSONS LEARNED

Engaging in student-initiated projects often requires teachers to relinquish the role of expert and let others step in. In the project-based classroom, the teacher becomes a manager or coach who learns alongside students and directs them toward worthy and credible resources. Often, the most successful projects involve a community expert who is as invested as students are in the project. Teachers should be prepared to help students generate a sizeable list of experts and resources. When new project topics emerge, students should be encouraged to find the appropriate source on the internet or in the local community. With guidance, students learn to critically evaluate sources of information.

To effectively manage the implementation of a vast number of student projects over time, teachers must be on board with every project and commit to guiding students through each phase. The establishment of milestones throughout a project is essential. Milestones help students readily complete project tasks and help teachers provide frequent formative assessment and feedback. Teachers considering such projects for the first time might have multiple student groups work on the same question. This allows groups to compare project plans and results and thus learn from one another.

Projects of this nature can be conducted on a limited budget. Our entire budget for 16 projects was approximately $150. Feasibility studies—such as those described in Table 1—can be conducted with little to no expense. Materials needed to build prototypes can be donated or acquired at a local hardware store for a small cost.

Most of the projects undertaken in 2008 entailed residential-scale changes. By inviting parents and the community to listen to project presentations, students had the opportunity to speak directly to an audience for whom their work was most meaningful. Ultimately, it is the parents and adults who make

decisions about whether to buy a pellet stove or solar hot water heater, or conduct a weatherization audit. Through these presentations, families in the Montpelier community became more educated about local issues and thus were more likely to make informed decisions about home-based energy consumption.

Students who addressed current civic concerns, such as restoring the local dam or the waste gasification project, were invited to share their results with related decision-making committees. This helped the community's decision makers to be more informed; it also communicated to students that their education is valuable and that adults do care about what they have to say. In addition to learning important physics concepts, students came to understand the applications of these ideas to important community, national, and global issues.

Conducting long-term, standards-based projects that are interesting and relevant is essential to science teaching and learning today (Krajcik, Czerniak, and Berger 2002). The projects described in this article are a means to engage students in true scientific inquiry. By posing and answering questions that are relevant to their own lives and communities, students ultimately produce tangible products that can have meaning far beyond the walls of the science classroom (Colley 2008).

Elements of PBS

Good PBS experiences include

- a rich, complex driving question that is relevant to students' lives,
- production of artifacts,
- student-centered learning,
- collaboration,
- accountability,
- use of technology,
- appropriate safety considerations,
- interdisciplinary and cross-disciplinary inquiry,
- extended time frame, and
- reliable performance-based assessment.

This list is adapted from Colley 2008

Acknowledgment

The authors wish to thank Montpelier High School (MHS) teachers Tom Sabo and Karen Smereka and the MHS Earth Group for their leadership and participation in the Annual Winooski River Cleanup and the MHS composting and solar-powered greenhouse projects.

On the web

Sustainable energy project rubric: *http://www.nsta.org/highschool/connections/201004SustainableRubric.pdf.*

References

Colley, K. 2008. Project-based science instruction: A primer. *The Science Teacher* 75 (8): 23–28.

Krajcik, J., C. Czerniak, and C. Berger. 2002. *Teaching children science: A project-based approach.* Boston: McGraw Hill College.

Roy, K. 2009. Taking responsibility for safety. *The Science Teacher* 76 (9): 12–13.

Weizman, A., Y. Schwartz, and D. Fortus. 2008. The driving question board. *The Science Teacher* 75 (8): 33–37.

Wiggins, G., and T. McTighe. 2003. *Understanding by design.* Alexandria, VA: Association for Supervision and Curriculum Development.

Chapter 35

THE STATE HIGH BIODIESEL PROJECT

REDUCING LOCAL WASTE WHILE LEARNING ABOUT ALTERNATIVE ENERGY

By Paul L. Heasley and William G. Van Der Sluys

Through a collaborative project in Pennsylvania, high school students have developed a method for converting batches of their cafeteria's waste fryer oil into biodiesel using a 190 L (50 gal) reactor. While the biodiesel is used to supplement the school district's heating and transportation energy needs, the byproduct—glycerol—is used to make hand soap to sell in the school store. Proceeds from both of these products are used to support the continuation of the project, in which students learn the science behind biodiesel and its relation to our environment.

Students have accomplished this work as part of the State High Biodiesel Project. The project allows students to conduct a series of small-scale experiments to synthesize biodiesel using the transesterification process, in which an alcohol reacts with an ester to produce a different alcohol and ester (Figure 1a). Students attending State College Area High School (State High) learn to characterize biodiesel through a variety of quality tests (e.g., infrared [IR] spectroscopy, measurement of refractive index and specific gravity); they also determine the heat of combustion of biodiesel and compare it to other fuels.

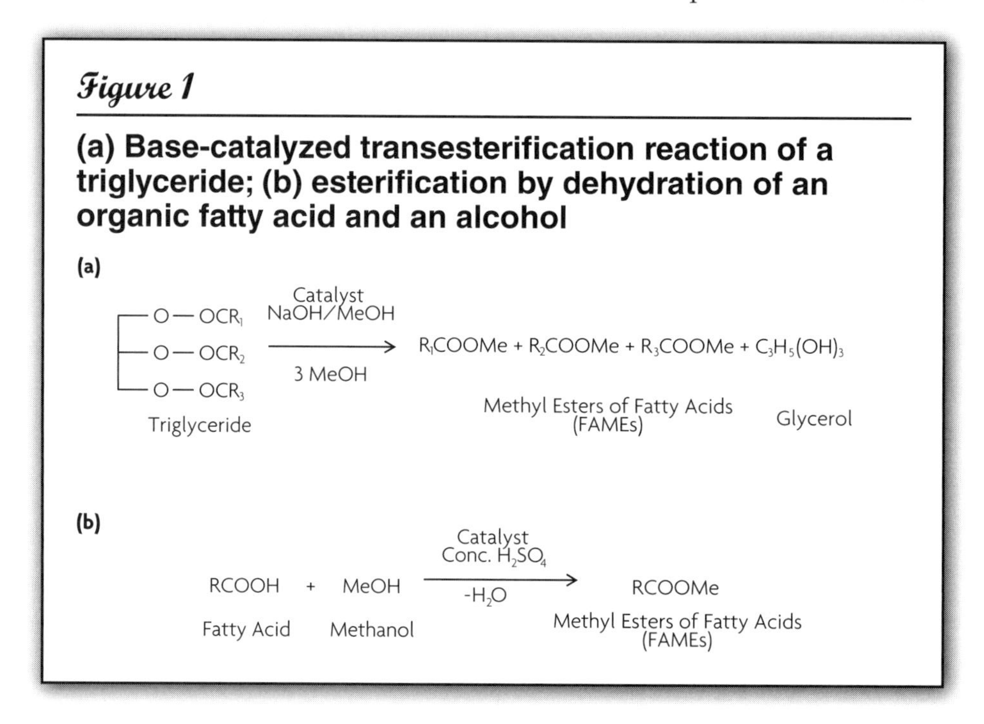

Figure 1

(a) Base-catalyzed transesterification reaction of a triglyceride; (b) esterification by dehydration of an organic fatty acid and an alcohol

This chapter provides an overview of the State High Biodiesel Project with the hope of inspiring other school districts to develop their own programs. Complete details for the activities described in this article, along with background and safety information, are available on the Market Place for the Mind (MPM) website (see "On the web")—an educational resource created by the Pennsylvania Department of Agriculture (PDA) in cooperation with the Pennsylvania Department of Education.

AN ALTERNATIVE ENERGY OPPORTUNITY

Several years ago, we began developing a series of alternative energy laboratories to be used in our 10th- and 11th-grade Integrated Chemistry and Physics course (see MPM, "On the web"). One of these labs, for example, involved the synthesis and characterization of biodiesel from new vegetable oil, which was particularly significant at that time because of a recent spike in fuel costs and the associated interest in alternative renewable energy sources (Pahl 2005; Tickell 2003).

During this initial phase of the project, it became clear that there was a significant opportunity to develop a collaborative effort between our high school's Agricultural Sciences Program and Science Department to teach applied science to students who often avoid mainstream science courses. Refreshingly, these students often bring a different kind of practical hands-on knowledge and experience to the project, qualities that might not be exhibited by traditionally academically successful students. The activities involved in the project are consistent with the current initiative in our school district, state, and the nation to encourage more students to pursue science, technology, engineering, and mathematics (STEM) careers.

OVERVIEW OF LABS

Biodiesel can be prepared from most triglycerides (see "On the web"), which are commonly known as "fats" or "oils" and are essentially esters of glycerol and long-chain fatty acids. The physical properties of the fat or oil are significantly affected by the degree of unsaturation in the alkyl chain, with liquid oils being favored by the introduction of carbon-carbon double bonds that are characteristic of unsaturated alkyl chains. To understand and visualize the three-dimensional structures of these properties, students build molecular models of glycerol, fatty acids, methanol, and various triglycerides before making biodiesel (Figure 2).

Figure 2

Students examine molecular models of compounds related to the project.

Biodiesel is easy to make using the base-catalyzed transesterification reaction between any commercially available fresh vegetable oil (e.g., soy or canola) and excess methanol: These reactants produce a complex *fatty acid methyl ester* mixture (FAME)—a biodiesel—and a glycerol byproduct. (See Figure 1 for the chemical reaction equation.) The homogeneous mixture of glycerol and excess methanol is denser than the FAMEs and easily separates, providing a convenient method of purification in which the biodiesel is decanted or placed in a separatory funnel. The resulting biodiesel is washed with tap, distilled, or deionized water to remove impurities, such as soap and methanol.

Students measure a number of the biodiesel's physical properties, including the density, and compare it to water and vegetable oil. They also measure biodiesel's heat of combustion using the equipment shown in Figure 3. This is most conveniently done using fuel-containing oil lamps to heat known amounts of water in stainless steel containers. As a control, other fuels, such as petroleum diesel, are measured and compared to the biodiesel. While the experimentally obtained heats of combustion are considerably lower than literature values (Table 1), students discuss their relation and significance. The low values are presumably due to the excessive loss of heat to the surroundings. This error could be corrected by determining the heat capacity of

Figure 3

Experimental setup used to measure heats of combustion of biodiesel and other fuels (See "Safety concerns.")

Table 1

Student-measured heats of combustion for biodiesel and petroleum diesel

Compound	Heat of combustion (H, kJ/g)	Literature value (kJ/g)	%E
Biodiesel	15.5 +/– 1.0	40.2 (Schumacher et al. 1995)	-61%
Petroleum diesel	12.1 +/– 0.5	45.8 (Schumacher et al. 1995)	-74%

the calorimeter using a fuel with a known heat of combustion (MPM, see "On the web"). In addition, students note that the biodiesel produces much less soot than petroleum diesel (Figure 4), resulting in a more complete combustion reaction under our conditions, presumably due to the presence of oxygen in the molecular structure. Further characterization can be done using our Science Department's IR spectrophotometer to detect hydroxyl molecular vibrations (v O–H, 3000–4000 cm^{-1}) of impurities such as water, glycerol, fatty acids, and mono- and diglycerides that might affect the combustion process.

Figure 4

A comparison of the combustion of biodiesel, methanol, and petroleum diesel using oil lamps

RECYCLING WASTE OIL

At our high school, a natural extension of the laboratories is to prepare biodiesel from waste fryer oil obtained from our cafeteria. The use of waste oil presents some challenges, since it is typically contaminated with other food products—such as water, various oil-decomposition materials, fatty acids, and rancid fat—that can neutralize the catalyst. Therefore, additional effort is often required to avoid excessive soap formation, including filtration, drying, and a titration to determine how much sodium hydroxide to use.

Safety note

Only the teacher, using appropriate personal protective equipment, should handle the concentrated sulfuric acid. (For other safety considerations, see "Safety concerns.")

In an effort to avoid some of these pitfalls (Kac 2001; Pahl 2005), we recently modified our procedure to create a two-step process, in which the contaminant fatty acids are initially converted to FAME by esterification (Figure 1b). This involves dehydration synthesis using concentrated sulfuric acid, followed by neutralization and transesterification using a methanolic solution of sodium hydroxide. The sodium sulfate is removed along with the glycerol in subsequent water washing steps.

We have been able to produce good quality biodiesel using this method, the determination of which is based on IR spectroscopy and various other characterization tests, such as gas and liquid chromatography; the latter two are performed by the Pennsylvania State University's (PSU) Chemical Engineering and Agricultural Sciences Departments. We have not been willing to incur the significant cost of the American Society for Testing and Materials (ASTM) testing (see "On the web"): We do not plan to sell our product to the general public and therefore are not compelled to meet these standards.

Our initial attempts to make 19 L (5 gal) batches of biodiesel made use of an old recreational vehicle (RV) water heater, standard hose fittings, and a water pump, all purchased locally. The PDA funded our request for a Rural Youth Grant of $2,000 to build a 190 L (50 gal) biodiesel reactor (Alovert 2003). We estimate that a similar reactor could be built for considerably less expense by using a discarded water heater, pipe fittings and valves available from a local hardware store, a conventional oil pump, and a recycled 208 L (55 gal) drum for washing. Students in our high school's Agricultural Engineering class assembled our reactor (Figure 5) and periodically prepare 151–189 L (40–50 gal) batches of biodiesel from the cafeteria's used fryer oil. We estimate that we will be able to make between 1,514–1,893 L (400–500 gal) of biodiesel a year.

Figure 5

The State College Area School District biodiesel reactor

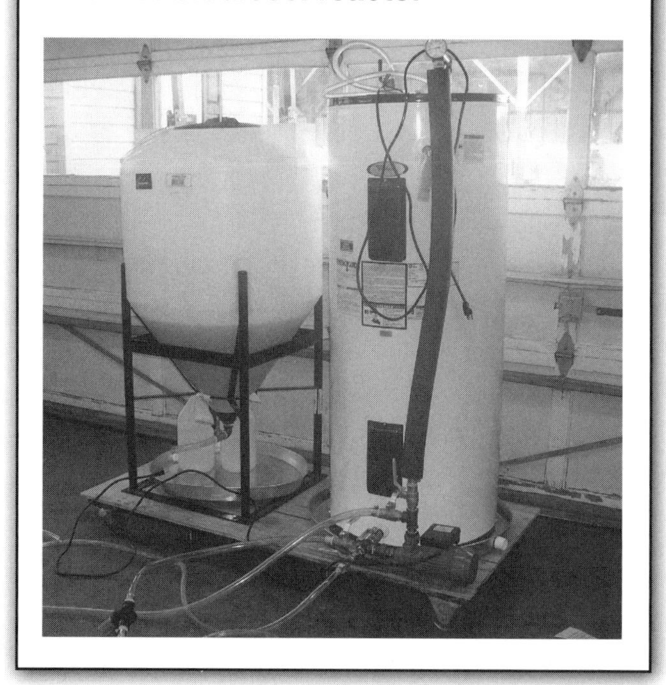

Biodiesel can be used in any conventional diesel engine and can be used as a heating fuel, which supplements our district's energy budget. Previously our school district incurred a substantial expense of $500 per year to dispose of waste fryer oil. The money the district provides to fund our project is based on splitting the difference between our production costs per gallon and the market price for biodiesel. The profits are used to purchase additional equipment and consumable supplies such as sodium hydroxide and methanol, which we currently purchase at a 208 L (55 gal) bulk rate of $6.25/4 L ($5.95/gal) of methanol. The methanol is stored in a secure and well-ventilated outside storage shed. We estimate that our current costs for producing biodiesel are approximately $1.00/3.8 L ($1.00/gal).

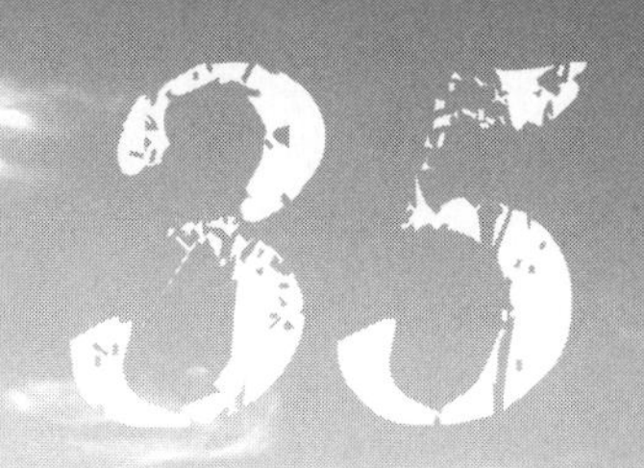

Perhaps the most unusual aspect of this project has been the preparation of "butter biodiesel" from the 2008 Pennsylvania Farm Show's butter sculpture (Figure 6; Pauling 2008). This sculpture depicted two rural school children and their cow waiting for a school bus. This project was extremely challenging due to the complicated nature of butter fat (Helz and Bosworth 1936), and we do not recommend that butter routinely be used to make biodiesel. Our students prepared about 150 L (40 gal) of butter biodiesel, and we produced 236 ml (8 oz.) jar samples with labels that included a picture of the sculpture, a short description of our program, the chemical process, and IR spectrum. These samples, along with biodiesel prepared from our waste fryer oil, are given to visiting dignitaries and other people interested in our project. (**Safety note:** These are just for display purposes and should be labeled "not for human consumption.")

Figure 6

Pennsylvania Farm Show's 2008 butter sculpture

GLYCERIN SOAP PRODUCTION

In the process of creating biodiesel, by-products include a combination of excess methanol and glycerol, which are separated through distillation. The excess methanol is then stored over molecular sieves to remove any water and reused to make subsequent batches of biodiesel.

The resulting glycerol contains some biodiesel, as well as mono- and diglycerides, and can be made into soap by adding aqueous solutions of sodium or potassium hydroxide. Adding sodium hydroxide results in a light brown, hard bar soap, while adding potassium hydroxide produces a clear, amber gelatinous soap. Both types of soap have extremely good lathering, moisturizing, and grease-cutting properties and can be used for a variety of purposes.

Safety concerns

There are a number of safety issues related to this project that should be noted. The major concerns associated with the preparation of biodiesel involve the use of methanol and sodium hydroxide. Students should be made aware of the flammability and toxic nature of wood alcohol (methanol) if ingested and the caustic nature of lye (sodium hydroxide). Prior to doing laboratory work, teachers should review and share with students *material safety data sheet* (MSDS) information related to safe handling of these hazardous chemicals. The sodium methoxide catalyst used in the transesterification process is itself a very strong base and should be handled with extreme caution. (Sodium methoxide is highly flammable, reacts violently with water, can cause severe eye damage, is harmful if inhaled or swallowed, and is extremely destructive of mucous membranes.) Therefore, we recommend that students always wear appropriate chemical-splash goggles, aprons, and nitrile gloves while preparing biodiesel. Appropriate laboratory ventilation is required, all sources of ignition should be removed from the lab, and all experiments should be carried out in a fire-resistant fume hood.

While there have been reports of accidents when performing similar experiments using volatile organic fuels such as alcohols, the vapor pressures of biodiesel and petroleum diesel are quite low, which significantly reduce the hazards and risks associated with this experiment. Specifically, in order for a fuel to burn, it must be in the vapor phase and mixed with an appropriate amount of air. In fact, diesel fuels are rather difficult to ignite under ambient conditions, making them relatively safe as compared with other more volatile fuels, such as gasoline or kerosene.

The teacher, and not students, should perform the comparison of the three fuels—biodiesel, methanol, and petroleum diesel—shown in Figure 5 as a demonstration. The demonstration should be done in a fume hood, and the teacher should take extreme care not to knock over or break the methanol lamp, which would increase the surface area of the liquid and the evaporation rate, possibly resulting in an uncontrollable fire or explosion. Fire suppression equipment (e.g., sprinkler system, fire extinguisher, fire blanket) are required. Students should be at a safe distance—at least about 5–6 m from the demonstration—and the teacher should use a safety shield. All students and the teacher must have on appropriate personal protective equipment (e.g., chemical-splash goggles, aprons).

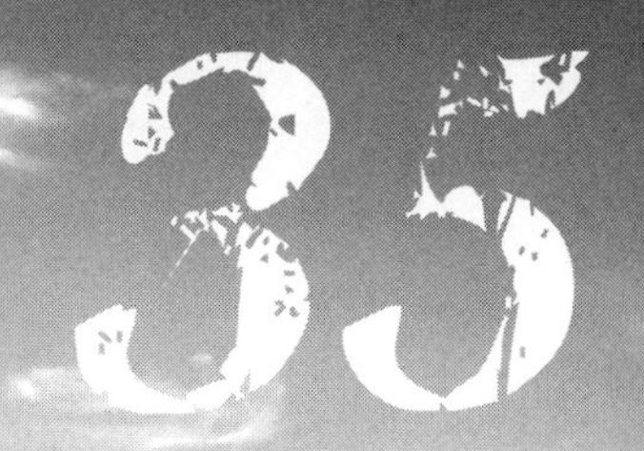

We prepare these soaps in our chemistry classes several weeks before the winter break; students occasionally use the bars of soap as holiday gifts. We also sell the soap in our school store, which further helps reduce costs associated with the project. Prior to use, the soap should be allowed to age to ensure complete reaction of the lye. If excess unreacted lye is present, it will eventually react with carbon dioxide in the air to form a white crust of sodium bicarbonate.

ENVIRONMENTALLY RELATED ISSUES

This project provides a forum for discussion of a variety of issues, ranging from the use of foreign oil and its effect on our economy to global warming and the greenhouse effect (due to anthropogenic production of carbon dioxide from combustion of fossil fuels). Biodiesel is not a substantial contributor to global warming: It is produced from a renewable plant resource and therefore is a nearly carbon-neutral energy source. In addition, biodiesel has extremely low amounts of sulfur, which helps reduce smog and air pollution. Biodiesel also undergoes a more complete combustion reaction than petroleum diesel, making it a potentially superior transportation fuel.

However, there are some disadvantages to using biodiesel as a fuel. In particular, the gel point for most FAMEs is well above that of petroleum diesel—in cold northern climates, using pure 100% biodiesel may not be sensible. However, it is possible to mix biodiesel with petroleum diesel or to install fuel tank heaters to ensure that the fuel remains liquid. For instance, we can store our vehicles in a warm garage and use a 20% biodiesel/80% petroleum diesel mixture during the winter.

CONCLUSION

The State High Biodiesel Project has been a relatively high profile endeavor, resulting in state and national press coverage. Our students have been given numerous opportunities to display their newfound knowledge to the public—they have presented material to the press and local, state, and national government officials. This has included poster presentations at our state capitol in Harrisburg and a poster presentation at the U.S. Capitol in Washington, D.C. We (the authors) have also presented several related continuing education workshops for educators, sponsored by PSU and the PDA.

We are particularly proud of the State High Biodiesel Project's ability to reduce our school district's waste while simultaneously educating students about a number of alternative energy issues. We have now taken a waste product—used fryer oil—and turned it into two useful and value-added products. We hope our students will take these lessons to heart and apply them appropriately in their own lives. We also hope that other school districts will consider similar initiatives.

Acknowledgments

The authors would like to thank the State of Pennsylvania Departments of Agriculture and Education for funding this project, the State College Area School District School Board and administrators for their support, and the members of the PSU biodiesel team for their helpful discussions and analytical work.

On the web

ASTM: *www.astm.org/STATQA/index.html*

Make your own biodiesel: *http://journeytoforever.org/biodiesel_make.html*

MPM: *http://www.marketplaceforthemind.com/LessonPlanBlobStream.aspx?IdLessonPlan=39*

References

Alovert, M. 2003. Collaborated biodiesel tutorial: Appleseed processor. *www.biodieselcommunity.org/appleseedprocessor*

Helz, G. E., and A. W. Bosworth. 1936. The higher saturated fatty acids of butter fat. *Journal of Biological Chemistry* 1 (16): 203–208.

Kac, A. 2001. The foolproof way to make biodiesel. *http://journeytoforever.org/biodiesel_aleksnew.html*

Pahl, G. 2005. *Biodiesel: Growing a new energy economy.* White River Junction, VT: Chelsea Green.

Pauling, D. 2008. Butter sculpture to meet end in State College. *Centre Daily Times.* January 4.

Schumacher, L., A. Chellappa, W. Wetherell, and M. Russel. 1995. The physical and chemical characterization of biodiesel and low sulfur diesel fuel blends. Report to National Biodiesel Board. University of Missouri.

Tickell, J. 2003. *From the fryer to the fuel tank: The complete guide to using vegetable oil as an alternative fuel.* 3rd ed. New Orleans: Joshua Tickell Media Productions.

Chapter 36

THE SIDEWALK PROJECT

STUDENTS WORK WITH THEIR COMMUNITY TO INVENT A HEATED SIDEWALK POWERED BY AN ALTERNATIVE ENERGY SOURCE

By William Church

If you have the opportunity to visit New Hampshire in winter, be sure to stop by Littleton High School and drop a snowball on the sidewalk. Despite the freezing temperatures outside, the snowball will melt once it hits the sidewalk. This phenomenon is the result of what is now called "the sidewalk project," a collaboration of more than 40 high school physics students, 10 local mentors, and a few regional and national organizations who worked together to invent a way to heat a sidewalk with an alternative energy source.

HOW IT BEGAN

The sidewalk project began in the spring of 2002, when I learned about Lemelson-MIT InvenTeams (*www.inventeams.org*). This program provides grants of up to $10,000 to high schools to invent something that addresses a town or school need. After learning more about the grants, I coordinated several meetings with interested people from the school and town, including 10th- and 11th-grade students who were signed up for my physics classes.

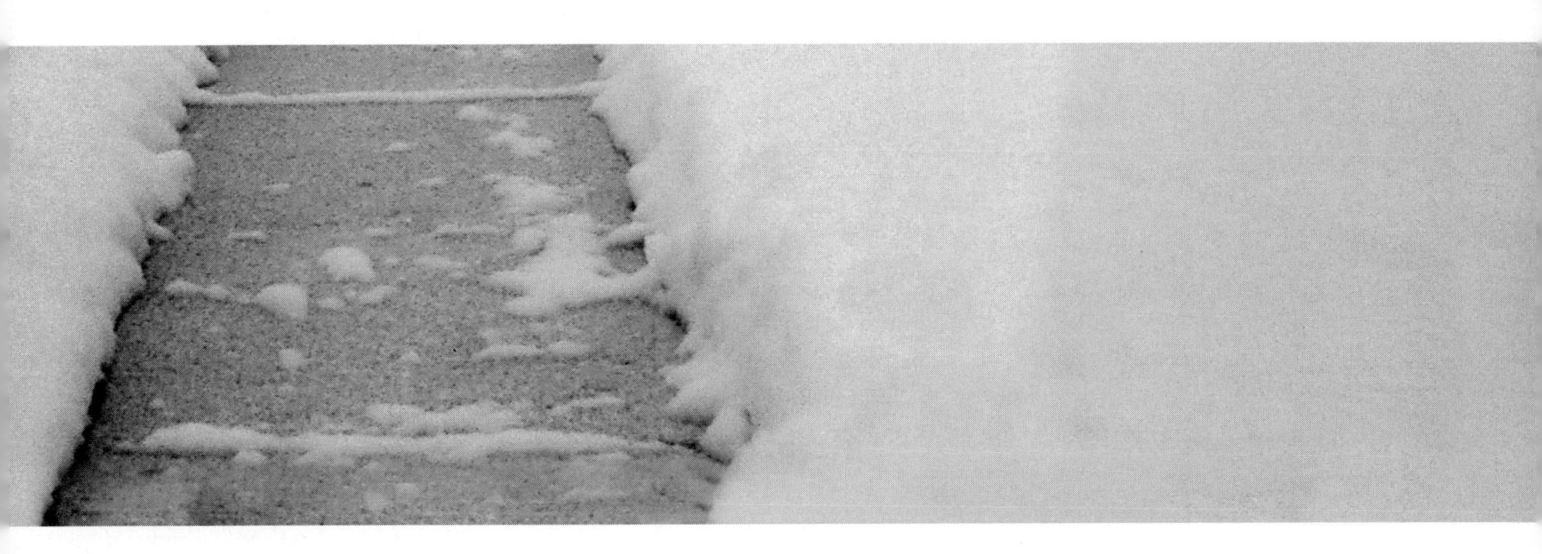

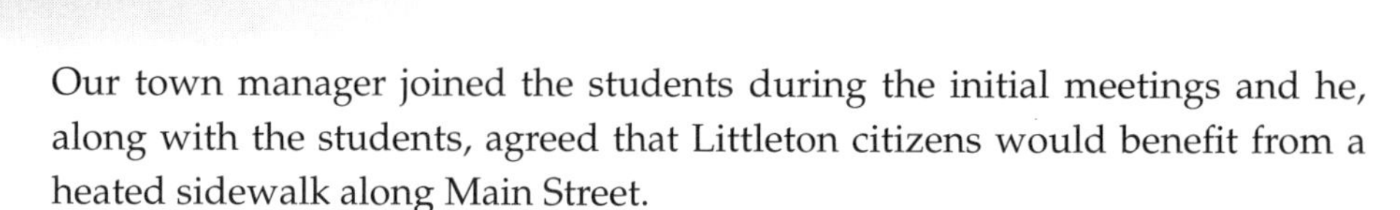

Our town manager joined the students during the initial meetings and he, along with the students, agreed that Littleton citizens would benefit from a heated sidewalk along Main Street.

Students decided to power the sidewalk with small-scale alternative energy sources. The town manager ended the first meeting with one key point: If the project was feasible and worked, there was a good chance that the heated sidewalks could be installed in Littleton's downtown during upcoming construction. This opportunity was a great motivator for students.

MODIFYING THE CURRICULUM

Students felt they had identified a real need in our town and brainstormed ways to power the heated sidewalk. Once our grant proposal was selected for InvenTeams funding, I set out to teach students the concepts and skills necessary to complete their invention. To accomplish this, I incorporated the sidewalk project into my physics curriculum as part of the course lab component. The project provided a vehicle for teaching applied physics concepts such as energy and its many forms, thermodynamics, and electromagnetism. The project also offered authentic ways for students to engage in meaningful experimental and engineering design and to practice their scientific writing skills.

To emphasize curriculum that centered on specific physics concepts, I assigned lab investigations on physics topics related to the research areas identified by students. For example, when students investigated possible heating sources to use, they studied concepts such as the nature of heat, temperature, convection, radiation, conduction, insulation, and the first and second laws of thermodynamics. When students investigated an energy transport system, they calculated area, volume, flow rate, and heat transfer. And when students explored powering pumps with solar energy, they learned about electricity, electric current, magnetism, AC and DC circuits, solar cells, energy, resistance, and power.

HELP FROM MENTORS

As students' investigations began to require larger experimental setups and specialized skills, we relied on community mentors for help. Mentors shared their own expertise or connected students with other experts so they could safely build and test their ideas on a larger scale. For example, an engineer from a local manufacturing plant helped students move their "waste heat" experiments from beakers and plastic tubing to a barrel stove and copper tubing. Our town's highway superintendent helped students move from conducting experiments with cardboard to investigations on a 400 kg slab of concrete.

Over the course of many (difficult) months, a full-scale prototype began to emerge. The stress of trying to do something challenging with very real deadlines and with the whole town watching was significant. One student who summarized the experience well said, "This project is like Godzilla and I'm Tokyo!" But, with the help of mentors, flexibility on my part, and positive local media coverage, students managed the stress and kept working. The payoff for all their hard work came so quickly one day in January that another student said, "We spent so many months working and working and not seeing anything big get done until bam everything came together in two days!" Seeing the pieces come together really motivated students to work on the next challenge, because they needed to make sure their invention actually worked!

With the help of mentors, students assembled a complex prototype outside of the high school (Figure 1). The alternative energy source students decided to use was the waste heat from the chimney of a wood-fired barrel stove. Based on results of their previous classroom experiments, students created a helix out of copper tubing and placed it into the chimney. With the help of a local plumbing contractor, students created a piping system that ran from the chimney to a large concrete slab that was equal in size and composition to a section of sidewalk. Students pumped fluid through the tubing in the chimney, where the hot gases from the chimney heated the fluid. The fluid was then pumped to insulated piping that was pressed firmly against the bottom of the concrete slab.

Figure 1

Students developed a working prototype of their sidewalk heating system.

DEVELOPING PROTOTYPES

Learning from techniques used in radiant heating systems, students knew that if the fluid was at or greater than a specific temperature, it would transfer enough energy to the concrete to melt snow as it fell on its surface. At a separate experiment site, students explored the use of a 100 W photovoltaic system to generate enough energy to power the pumps they were using.

To assess the performance of the system, students used data loggers. These small computers allowed students to automatically collect and store air and concrete slab temperatures during a range of conditions and through many experiments. Students analyzed these data and implemented solutions to the problems they identified. For example, students observed large fluctuations in the system's fluid temperatures because of the heating and cooling cycles of the woodstove. To address this, students installed a mixing valve

that kept the system temperatures constant. Students also observed that the first arrangement of the tubes beneath the concrete caused the concrete to be heated unevenly. Students addressed this with a new tubing arrangement that distributed the heat over the whole surface of the concrete. Students celebrated success when the air temperature measured between 5 and 10° below freezing and snow melted on the slab.

Our working prototype inspired the town and school to work together to install a second prototype in Littleton High School's front sidewalk. This system was installed last year and is an improved site on which to test the invention. Current physics students now have a laboratory in which they continue to collect data to answer the question—will it work on Main Street? This powerful question continues to unite many members of our town and school in a very rich learning experience.

References

National Research Council (NRC). 1996. *National science education standards.* Washington, DC: National Academies Press.

New Hampshire Department of Education. 1996. K–12 science curriculum frameworks. *http://www.education.nh.gov/instruction/curriculum/science/documents/framework.pdf*

Chapter 37

ASKING AUTHENTIC QUESTIONS WITH TANGIBLE CONSEQUENCES

By Anne Watson

As a physics teacher, it seems irresponsible to teach energy without asking students hard, relevant questions such as, "What will we do when oil becomes prohibitively expensive?" Therefore, in the fall of 2005, I asked my physics students to identify some energy-related problems in our community that we could solve. During brainstorming sessions, students in my senior-level physics course generated a huge list of suggestions. Even though students came up with excellent ideas, the suggestions were a bit too complicated and expensive. Eventually, with help from other faculty, we came up with a suitable project: Students would research options for a renewable onsite power source for the new water pump we would soon buy for our school's greenhouse. My students were immediately excited about this project and we went right to work.

Student-generated image of solar panels on the school's greenhouse

EXPLORING DIFFERENT OPTIONS

The greenhouse at our high school was completed in 2004 and biology students began using it in the fall of 2005 to produce salad greens for the school lunch program. Compost is collected from the cafeteria and trucked to Vermont Compost Company, which in turn provides a cheaper rate for soil to grow more plants in the greenhouse.

In keeping with the sustainable ideals of our school and community, we were committed to providing the electricity for the pump from a renewable, localized system. The first step for our project would involve conducting a feasibility study regarding our onsite electric generation options. Students were perfectly capable of doing the research and they proposed many potential sources—wind, solar, hydro, and even bicycle generators.

The feasibility study became the culminating project for a unit on energy and power. Students, working alone or in pairs, chose one possible power source to research. The final activity of the project was to present the research to the class with a recommendation to the school to pursue the option or not. The presentation had to include energy and power calculations (linking this project to traditional science standards), as well as documented evidence that students had contacted an expert in the community regarding their research topic. After all, if the school was really going to invest in one of these options we needed expert advice. The students squirmed at the thought of calling an adult they didn't know, but I assured them that it was a skill they would use throughout their lives.

As students began to contact experts, we found ourselves flooded with support from the community. For example, the president from Solar Works Inc. (a local photovoltaic company) came to our classroom to discuss the benefits and challenges of solar panels. Other groups that helped inform our students' presentations included Windstream Power LLC, Northern Power Systems Inc., and Central Vermont Solar and Wind.

After four weeks, students gave their presentations, all of which were top quality. The authenticity of the question provided a refreshing motivation to the students. Over the course of two class periods, students presented their research to the school's principal, head custodian, a handful of science teachers, and the service learning coordinator. After an extensive discussion, the students recommended:

- Acquiring as many solar panels as we could fund to be connected to our local power grid, using an inverter and fail-safe system;
- Building bicycle generators; and
- Continuing to explore hydropower in the coming year.

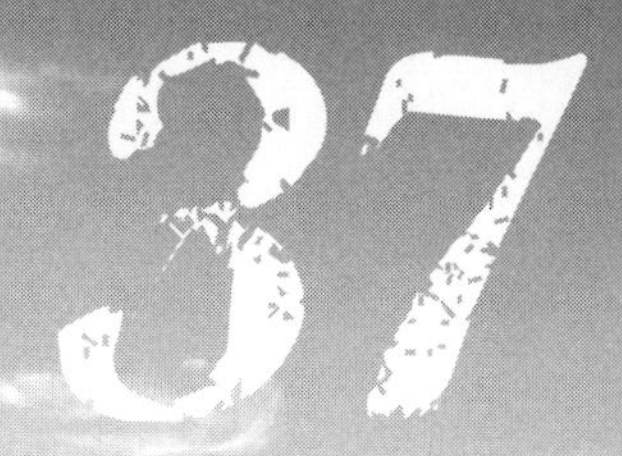

SECURING FUNDING

Solar Works Inc., provided a cost estimate for a 2 kW photovoltaic system, and agreed to work alongside students during the installation process once funding had been secured. During summer 2006, we held a two-week summer school session that was entirely dedicated to building bicycle generators. I co-taught this course with a shop instructor who taught students how to weld, and we had a professional electrician from Common Ground Audio help us with the electrical connections.

A student powers a stereo system by pedaling!

By the end of the summer session we had five operational bicycle generators. A photograph in the local newspaper announced a public reception to show off the generators, which was attended by many members of the community, as well as parents and school staff.

Through one of the contacts made by my students, a member of the Montpelier Energy Team (MET) heard about the energy project. The MET is a group of local experts that advise the city government about energy decisions and climate change. He invited a group of students to present their results at one of their meetings. The students were delighted and surprised that someone outside of our school system would be interested in their work. After the presentation, the MET promised to help us realize our goals and has since helped us find funding.

Knowles Science Teaching Foundation also asked that my students and I present our project at their summer fellows' meeting. We also received a grant to purchase the solar panels from the Sustainable Future Fund with the Vermont Community Foundation for $2,000.

The whole experience taught us that there are people in the community that are more than happy to be involved in classroom efforts. More importantly, it helped me understand that when I started asking my students real questions, with unknown but significant answers, I needed the expertise of the community to support the learning in my physics class.

Chapter 38

THE QUIET SKIES PROJECT

STUDENTS COLLECT, ANALYZE, AND MONITOR DATA ON RADIO FREQUENCY INTERFERENCE

By Steve Rapp

The radio signals received from astronomical objects are extremely weak. Because of this, radio sources are easily shrouded by interference from devices such as satellites and cell phone towers. Radio astronomy is very susceptible to this radio frequency interference (RFI). Possibly even worse than complete veiling, weaker interfering signals can contaminate the data collected by radio telescopes, possibly leading astronomers to mistaken interpretations.

To help promote student awareness of the connection between radio astronomy and RFI, an inquiry-based science curriculum was developed to allow high school students to determine RFI levels in their communities. The Quiet Skies Project—the result of a collaboration between the National Aeronautics and Space Administration (NASA), the National Science Foundation (NSF), and the National Radio Astronomy Observatory (NRAO)—encourages

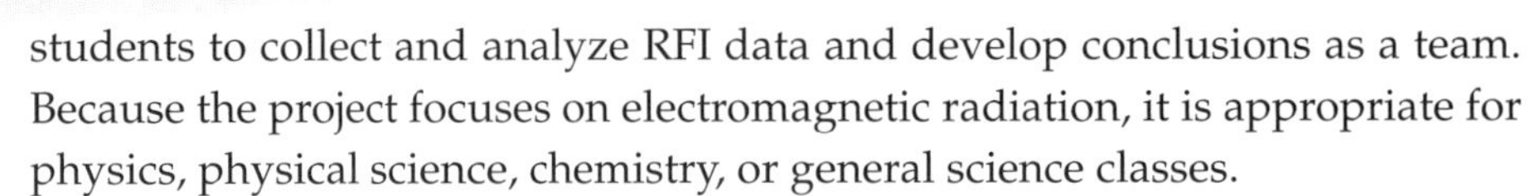

students to collect and analyze RFI data and develop conclusions as a team. Because the project focuses on electromagnetic radiation, it is appropriate for physics, physical science, chemistry, or general science classes.

My students—about 50 from 15 southwest Virginia high schools—participated in the Quiet Skies Project and were pioneers in the use of the beta version of the Quiet Skies Detector (QSD), which is used to detect RFI. In this chapter I explain more about RFI, the Quiet Skies Project, and my students' experiences with the project.

WHAT IS RFI?

RFI is any electromagnetic signal not of cosmic origin that interferes with another use of the spectrum. This is analogous to light pollution; nighttime lighting is helpful to all of us but detrimental to those trying to do optical astronomy. A cell phone is one example of a device that produces RFI. According to an e-mail correspondence with Ron Maddalena, an astronomer at NRAO, "a cell phone (1 W transmitter) with a 10 kHz bandwidth at 1 km is a 10^{16} Jansky (Jy) (1 Jy = 10^{-26} W/m^2/Hz) source. (**Note:** Named after the pioneering radio astronomer Karl Jansky and often used in radio astronomy, a Jansky is a non-SI unit of electromagnetic flux density equal to 10^{-26} watts per square meter per hertz [W/m^2/Hz].) This is 5×10^{12} times stronger than the brightest astronomical radio source at the same frequency (the supernova remnant Cassiopeia A, ~6000 Jy), and 10^{19} times stronger than the typical source observed with today's radio telescopes."

Figure 1

The Robert C. Byrd Green Bank Telescope

Radio astronomers occasionally find their investigations disturbed by RFI.

In addition to cell phones and satellites, RFI also can be created by electric fences, computers, cordless phones, defective television cable boxes, faulty insulators on power poles, broken lightning arrestors, and burned-out power transformers. Anything that produces an electric arc produces broadband radio frequencies. And, as previously mentioned, radio astronomy is very susceptible to RFI. For example, one night I was at the control center for The Robert C. Byrd Green Bank Telescope (Figure 1) observing an astronomer collecting data on pulsars. All of a sudden, the look of the data changed rapidly and small sharp spikes appeared on the computer screen (Figure 2). The data had become contaminated by RFI.

Figure 2

The spikes on the power density versus frequency graph indicate RFI

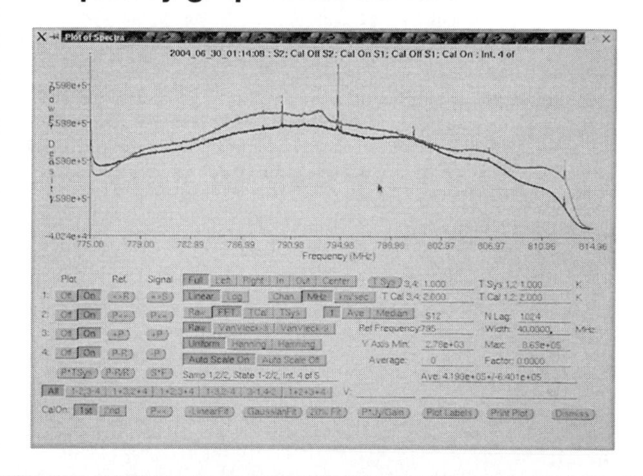

BETA TESTING

With the Quiet Skies Project, students learn about electromagnetic radiation, RFI, and radio astronomy by determining RFI levels in their communities. One of the best parts about this project is that it costs nothing. Teachers can contact NRAO (see "On the web") and apply for a loaner RFI kit to be shipped to them. The kit contains two QSDs, batteries and battery chargers, two Global Positioning System (GPS) units, two tripods to mount the detectors on, a portable transmitter for students to detect RFI, spectrum charts, and an explanation of how everything works. NRAO will even pay return shipping on the kit.

My students pioneered the use of the QSD's beta version. They field-tested the detector during the 2005–2006 school year. A prototype was created and tested at four frequencies: 800, 900, 1420, and 1665 MHz. Students used the beta version to explore the occurrence of RFI at their schools and in their communities.

With input from my students, another version of the QSD was completed in late 2006, which was much more portable and had a much longer battery life (Figure 3). Students reported it was a breeze to use and very easy to tote around. Students then collected data during the 2006–2008 school years (Table 1, p. 324).

During our 2007 spring field trip, students recorded RFI data at the NRAO in Green Bank, West Virginia (Figure 4). Students were to take

Figure 3

The version of the RFI detector used by students to collect data

Students turn on the detector using the knob on the left and also select narrow band or broadband frequency ranges; they tune the radio frequency using the dial on the right and read power on the small screen.

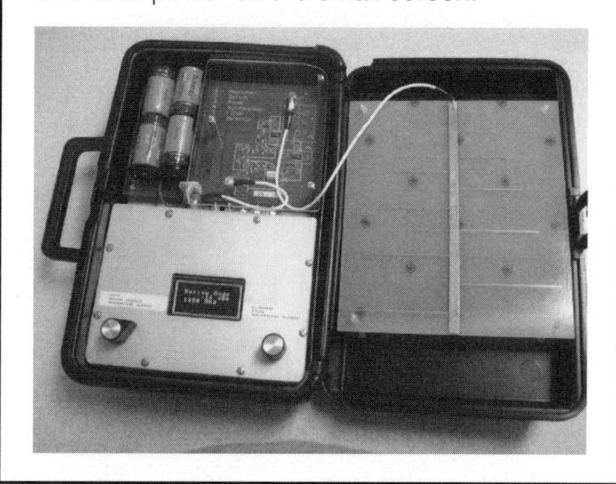

Table 1

Sample RFI data at 800, 900, 1420, 1665 MHz, and hotspots with high RFI (0 to +7 dBm)
(**Note:** dBm units measure the power ratio in decibels referenced to 1 mW. For a complete table, and a rubric for this investigation, visit *www.nsta.org/highschool/connections/200810Figure4Extension.pdf.*)

High school/ Community (**Note:** When tuning, students tune the detector from 800–1665 MHz looking for departure in "normal" readings.)	**Radio frequency** (MHz)	**Average power ratio** (dBm) (0 dBm = 1 mW)	**Elevation** (m)	**Relative RFI**	**Latitude** (d, m, s)	**Longitude** (d, m, s)
2006–2007 data						
NRAO Green Bank, WV	800	-42	844	LOW	38.26.02N	79.49.05W
Radio Quiet Zone	900	-43	844	LOW		
100 ft. from 40 ft. scope	1420	-40	844	LOW		
Lauren, Ben, Adam	1665	-40	844	LOW		
2007–2008 data						
Grundy High	800	-35	736	LOW	37.16.36N	82.50.42W
Kristina	900	-37	736	LOW		
	1420	-39	736	LOW		
	1665	-41	736	LOW		
Abingdon High (soccer field)	1420	-17	653	LOW	36.43.25N	81.57.20W
Sarah K. and Sarah G. (7 a.m.)	1665	-23	653	LOW		
Tuning	882	5	653	HIGH		
Tuning (6 p.m.)	882	4	653	HIGH		
J.J. Kelly High	800	-28	797	LOW	36.59.04N	82.32.04W
Zhanna and Rachel	900	-28	797	LOW		
	1420	-19	797	LOW		
	1665	-23	797	LOW		
Tuning	880	0	797	HIGH		
	956	0	797	HIGH		
	960	5	797	HIGH		
	1050	6	797	HIGH		
	1054	3	797	HIGH		
	1198	4	797	HIGH		
	1234	-8	797	LOW		

(continued)

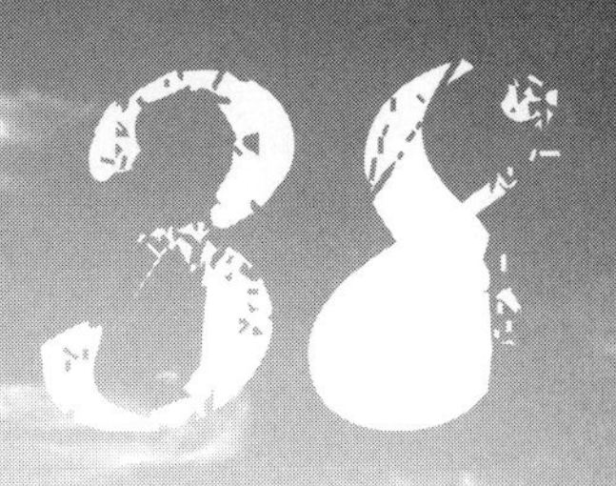

Table 1 *(continued)*

High school/ Community (**Note:** When tuning, students tune the detector from 800–1665 MHz looking for departure in "normal" readings.)	**Radio frequency** (MHz)	**Average power ratio** (dBm) (0 dBm = 1 mW)	**Elevation** (m)	**Relative RFI**	**Latitude** (d, m, s)	**Longitude** (d, m, s)
	1262	0	797	HIGH		
Council High (baseball field)	800	-39	500	LOW	37.04.85N	82.04.42W
Selena and Dax (morning)	900	0	732	HIGH		
	1420	-35	732	LOW		
	1665	-36	732	LOW		
Linwood Holton Gov. Sch.	880	0	626	HIGH	36.41.47N	82.00.01W
Steve	1018	2	626	HIGH		
Tuning	1060	0	626	HIGH		
	1062	0	626	HIGH		
	1066	0	626	HIGH		
	1080	7	626	HIGH		
	1082	0	626	HIGH		
	1086	0	626	HIGH		
	1250	4	626	HIGH		
	1260	0	626	HIGH		

12 points of data at 30° intervals in a horizontal circle. Zero degrees was aligned with cardinal north. Students again collected data for four frequencies: 800, 900, 1420, and 1665 MHz. An average was obtained for the 12 data points; sample data is shown in Table 1. These frequencies were chosen because they are common frequencies at which radio astronomers observe. Pulsars radiate in the range of 800–900 MHz while atomic hydrogen gives off radio waves at 1420 MHz. The hydroxyl molecular radical emanates radio energy at 1665 MHz. Students were also asked to tune through the range of 800–1665 MHz to see if they could find any "hotspots" where the signal suddenly changed to or approached a positive dBm reading, which is an abbreviation for the power

Students collecting data at NRAO during the 2007 field trip

ratio in decibels (dB) of the measured power referenced to one milliwatt (mW).

Students also determined the latitude, longitude, and elevation with a GPS unit. The elevations above sea level were recorded in meters and the average power of the signal was measured in dBm.

ANALYZING THE DATA

Once we had the data, I challenged students to take an inquiry approach to see if they could detect any relationships among the data. This was somewhat more difficult in my situation because my high school students (mostly juniors and seniors) are located in different schools throughout southwest Virginia. Students log into my physics, astronomy, robotics, or engineering class via the internet every day. For example, physics students log in every morning at 7:20 a.m., while astronomy students log in at 8:20 a.m., and engineering students log in at 11:40 a.m. They receive college credit for the courses.

I set up online discussion boards for students so they could chat outside of class and discuss the data. Students noticed some interesting trends in the data. The following are comments from a few of my students:

- *Of all of the RFI readings, only 16 frequencies were positive, and only 14 frequencies were at 0 dBm. (Positive numbers indicate there is great likelihood of interference since this is a higher power level than those readings with negative numbers.) The highest reading of RFI frequency was 19 dBm at 1420 MHz, with elevation of 940 m at 36.39.35 N, 81.32.41 W. Most of the positive readings also occurred at Linwood Holton Governor's School and J.J. Kelly High School.*
- *The first trend I noticed was that elevation makes little or no difference in the RFI levels. Second, I found that the time of day has little or no effect on the readings.*

- *When I graphed the morning readings of 800, 900, 1420, and 1665 MHz and the evening readings of the same MHz, I noticed that they all looked about the same at the same locations. I do not think that people use radio frequencies any more or less in the mornings as compared to the evenings. From my observations I feel that the lower frequencies, such as the 800 MHz and 900 MHz, experience more interference than the higher ones.*
- *One interesting thing I noticed while looking at the RFI data is that 1420 MHz and 1665 MHz are virtually tied with the greatest amount of lowest (negative) dBm readings. (This means these areas are not likely to experience RFI at those frequencies.) In Abingdon, Powell Valley, St. Paul, Mount Rogers, and J.J. Kelly, the lowest reading was usually at 1420 MHz or tied at 1420 and 1665 MHz. At Rocky Gap and Pound, the lowest readings were the same at 1420 MHz and 1665 MHz. At Patrick Henry, Pound, Twin Valley, Tazewell, and Council the lowest readings were usually found at 1665 MHz and sometimes even 900 MHz.*
- *The largest factor of high and low dBm readings seems to be variables such as radio quiet zones and the presence of cell phones. Data that are taken when cell phones are present have much higher dBm readings than most other data.*
- *Most of the significant changes in dBm readings occurred at about 800 MHz and about 1000 MHz. Even though the data was taken from many different locations, the dBm readings remained about the same in their respective radio frequencies. Most of the dBm readings are somewhere between –20 and –40 dBm.*

COMING TO A CONSENSUS

After many online class discussions and forums, students came to a consensus about what they had learned. The frequencies of 1420 MHz and 1665 MHz seemed to have the lowest RFI. This was to be expected because this is the allocated frequency range for radio communication services such as the Aeronautical Radionavigation Service, Broadcasting Service (a radiocommunication service in which the transmissions [e.g., sound and television] are intended for direct reception by the general public), Meteorological Aids Service, and Meteorological Satellite Service. (**Note:** These services are global agencies that are allocated certain radio frequencies.)

The frequencies of 800–900 MHz had the most RFI. This is understandable because cell phone towers are abundant in southwest Virginia, and cell phones use frequencies in the 824–894 MHz range. (**Note:** A second cell phone band uses frequencies from 1850–1990 MHz, above the frequencies tested by the students.) As a matter of fact, some schools have cell phone towers within one mile of the school, as illustrated in the data from Abingdon High School and Linwood Holton Governor's School. Students researched cell phone tower locations to discover proximities to their schools and communities.

The area of lowest RFI was at NRAO. This was no surprise because NRAO is in a unique National Radio Quiet Zone (NRQZ). Basically this means that anyone wishing to install a transmitter within the approximately 13,000 square miles "zone" must apply to NRAO for permission. The NRQZ was established by the Federal Communications Commission and by the Interdepartment Radio Advisory Committee in 1958 to minimize possible harmful interference to the NRAO and the radio receiving facilities for the United States Navy in Sugar Grove, West Virginia (see "On the web" for more information on NRQZ).

There seems to be no relationship between elevation and RFI. Students thought that this trend needed more investigation. They pointed out that some data seems to support a correlation but some data does not. Perhaps a study can be done next year with additional data being collected.

RFI of high magnitude was found when tuning from 800 to 1665 MHz. When a frequency was found that had a dBm reading of 0 or positive value, it was considered high magnitude. The frequency range where RFI was detected was 840–1260 MHz. Frequencies with the highest positive dBm readings were in the 1050–1250 MHz range. Students consulted radio spectrum charts (see "On the web") to determine the cause of this RFI. They found that the region from 960–1215 MHz is allocated to air traffic control radar and aeronautical radionavigation. Radiolocation and GPS use 1215–1300 MHz of the radio spectrum. Amateur radio has the range of 1240–1300 MHz dedicated to its use.

Students seemed to be very confident that there was no relationship between cardinal direction and RFI. Not one student found any evidence that RFI readings changed significantly as they turned the detector through 360°. Some students were very sure that were was no relationship between the time of detection and RFI. Others pointed out a few data sets did show some different readings when the data was collected in the morning versus the evening. Students thought that this was something that needed to be further investigated.

Acknowledgments

Special thanks to NASA, NSF, and NRAO in Green Bank for making this project possible. The Quiet Skies Project was funded by a grant from the NASA IDEAS Program. The author is deeply appreciative of the help and guidance from the following: Sue Ann Heatherly, Ron Maddelena, the entire NRAO staff, and Tom Gergely and Andy Cleg at NSF. The author would also like to thank all of his students who persevered though the data-collection process. Without them, there would be no project!

On the web

Allocation of radio spectrum in the United States: *www.jneuhaus.com/fccindex/spectrum.html*

Directions for the Quiet Skies Project: *http://steverapp.pageout.net/user/www/s/t/steverapp/Quiet_Skies_Project_Directions.htm*

Directions for using the QSD: *http://steverapp.pageout.net/user/www/s/t/steverapp/Quiet_Skies_Detector_Directions.doc*

NRAO Quiet Skies Project: *www.gb.nrao.edu/php/quietskies/interestPage.html*

NRQZ: *www.gb.nrao.edu/nrqz*

The Committee on Radio Astronomy Frequencies: *www.craf.eu/service.htm*

Part 3

Issues in Depth

Chapter 39

IN THE HOT SEAT

ANALYZING YOUR HEATING OPTIONS

By Janna Palliser

When winter rolls around, keeping yourself and your home warm is of the utmost importance. Heating your home seems like a simple subject to tackle, but there are many heating systems available, requiring differing fuels, installations, and costs. The various fuel types and their environmental footprints is the focus of this chapter.

TYPES OF HOME HEATING SYSTEMS

There are many different types of home heating systems. The five principal systems are highlighted below:

1. Forced air system (a furnace heats air using fuels such as natural gas, propane, oil, or electricity)
2. Radiant heating system (uses a variety of forms: in-floor systems use hot water heated by a boiler, which is fueled by natural gas, propane, oil, or electricity; heating stoves may use wood or coal)
3. Hot water baseboard system (uses hot water heated by a boiler to heat a space by a combination of radiation and convection; boiler is fueled by natural gas, propane, oil, or electricity)

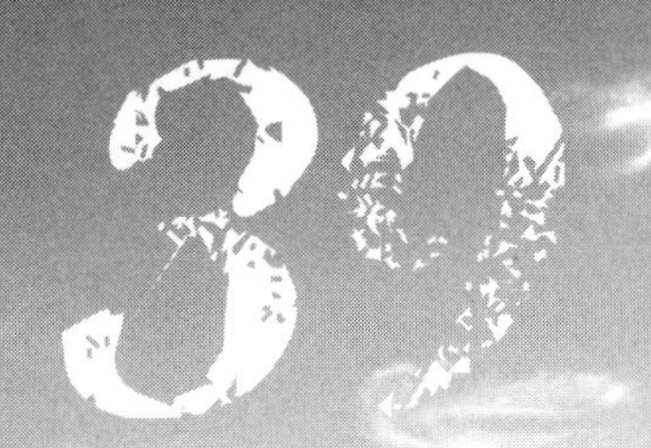

4. Steam radiant heating system (cast-iron upright radiators radiating heat with steam; steam boiler is fueled by natural gas, propane, oil, or electricity)

5. Geothermal heat pump (with ground loop geothermal systems, heat is taken from or deposited to the Earth by use of a ground loop pipe) (Formisano 2007)

FUEL TYPES

There are several types of fuel used in American households: oil, electricity, natural gas, propane, wood, pellet fuels, kerosene, coal, and solar (U.S. DOE 2010a). See Table 1 for average British thermal units (BTUs) per fuel type. One BTU is the amount of energy it takes to raise the temperature of one pound of water 1°F when water is at 39°F (U.S. DOE 2010a). See Table 2 for average fuel conversion efficiency of common heating appliances and Table 3 for fuel-cost comparisons.

Table 1

Average BTU content of fuels (U.S. DOE 2010a)

Fuel type	Number of BTUs/unit
Fuel oil	140,000/gallon
Electricity	3,412/kilowatt hour
Natural gas	1,025,000/thousand cubic feet
Propane	91,330/gallon
Wood (air dried)*	20,000,000/cord or 8,000/pound
Pellets (for pellet stoves; premium)	16,500,000/ton
Kerosene	135,000/gallon
Coal	28,000,000/ton

* Wood heating values vary significantly.

Oil

Heating oil, a petroleum product, is used by many Americans to heat their homes. Of 111 million American households, 8.5 million use heating oil as their main source of heat (U.S. EIA 2008a). Residential space heating is the primary use of heating oil, and the Northeast is the most reliant on heating

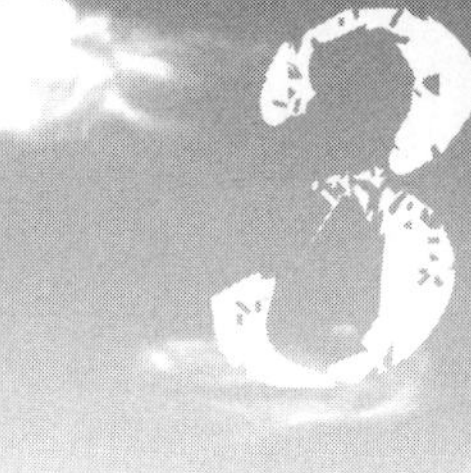

Table 2

Estimated average fuel conversion efficiency of common heating appliances (U.S. DOE 2010a)

Fuel type: Heating equipment	Efficiency (%)
Coal (bituminous)	
Central heating, hand-fired	45
Central heating, stoker-fired	60
Water heating, pot stove (50 gallon)	14.5
Oil	
High-efficiency central heating	89
Typical central heating	80
Water heater (50 gallon)	59.5
Gas	
High-efficiency central furnace	97
Typical central boiler	85
Minimum efficiency central furnace	78
Room heater, unvented	99
Room heater, vented	65
Water heater (50 gallon)	62
Electricity	
Baseboard, resistance	99
Central heating, forced air	97
Central heating, heat pump	200+
Ground source heat pump	300+
Water heaters (50 gallon)	97
Wood and pellets	
Franklin stoves	30–40
Stoves with circulating fans	40–70
Catalytic stoves	65–75
Pellet stoves	85–90

Table 3

Comparison of fuel costs

Fuel	Cost	Appliance efficiency	Cost per million BTUs
Wood pellets	Cost per ton in dollars: $275	80%	$20.96
Fuel oil	Cost per gallon in dollars: $2.73	78%	$25.36
Electricity	Cost per kilowatt hour in cents: 12¢	100%	$35.17
Natural gas	Cost per therm in dollars: $1.48	78%	$18.51
Propane	Cost per gallon in dollars: $2.19	78%	$30.74
Hardwood (air dried)	Cost per cord in dollars: $200	60%	$16.67
Coal	Cost per ton in dollars: $250	75%	$10.89

oil from October through March (U.S. EIA 2008a). Crude oil is a yellowish or black liquid found in underground reservoirs. To find oil, scientists and engineers test rock samples taken from the Earth. If oil is found, drilling begins and a well is established. About 53% of the oil products consumed in the United States are imported. Some of the top oil-producing countries are Russia, Saudi Arabia, Iran, China, and the United States. U.S. oil-producing states include Texas, California, Louisiana, Alaska, and North Dakota (U.S. EIA 2010).

Electric resistant heat

Electricity is the flow of electrical power or charge. Electricity is a secondary energy source; it is derived from the conversion of other sources of energy (coal, nuclear, solar). Most of the electricity in the United States is produced using steam turbines. A turbine converts the kinetic energy of a moving fluid (liquid or gas) to mechanical energy. In a steam turbine, steam is forced against a series of blades mounted on a shaft, which rotates the shaft, which is connected to a generator. The generator converts its mechanical energy to electrical energy based on the relationship between magnetism and electricity. In steam turbines powered by fossil fuels (coal, petroleum [oil], natural gas), the fuel is burned in a furnace to heat water in a boiler to produce steam. U.S. electricity is provided by the following:

- Coal = 48%
- Natural gas = 20%,
- Petroleum = 1%
- Nuclear power = 21%
- Hydropower = 6%
- Biomass (burned or converted to gas heat) = 1%
- Wind power = 1%
- Geothermal power = 1%

- Solar power = < 1%
  (U.S. EIA 2010)

Most electricity is produced from oil, gas, or coal generators that convert about 30% of the fuel's energy into electricity. Because of electricity generation and transmission losses, electric heat is often more expensive than heat produced using combustion appliances, such as natural gas, propane, and oil furnaces. Electric resistant heat is supplied by centralized forced-air electric furnaces or by heaters in each room. Room heaters include electric baseboard heaters, electric wall heaters, electric radiant heat, or electric space heaters (U.S. DOE 2010b).

Propane

Propane occurs naturally as a gas at atmospheric pressure, but can be liquefied if subjected to moderately increased pressure. It is stored and transported in its compressed liquid form, but by releasing propane from a pressurized container (with a valve), it is vaporized into a gas. Propane is nontoxic, odorless, and colorless, but an odor is added so that it can be detected by consumers (U.S. EIA 2008b). Propane is a by-product of natural-gas processing and petroleum refining. In natural-gas processing, propane is extracted to prevent operational problems in gas lines. In oil refineries, propane is produced as a by-product of heating oil and gasoline production. Demand is also met by imports (10%) and stored inventories. Of 111 million U.S. households, about 5 million use propane as their primary heating source (U.S. EIA 2008b). Most of the propane used in the United States is domestic, though some is imported from Canada via pipelines (U.S. EIA 2010).

Wood

Wood was the main source of heating fuel for the United States and the rest of the world until the mid-1800s and continues to be used in developing countries. About 16% of the wood consumed in the United States is used by residents to cook and for home heating (U.S. EIA 2010). A newer technology uses pellet-fuel appliances to burn small pellets made from compacted sawdust, woodchips, bark, agricultural crop waste, waste paper, and other organic materials. New pellet- and wood-burning appliances are cleaner and more efficient and powerful than traditional fireplaces or woodstoves (U.S. DOE 2010c).

Coal

Coal is a combustible, black or brownish sedimentary rock composed mostly of carbon and hydrocarbons. Coal is nonrenewable, as it took millions of years to create the coal that is mined today. Coal is extracted via surface or

underground mining. In surface mining, machines remove the topsoil and layers of rock ("overburden") to expose the coal seam. Once mining is completed, the soil and rocks are returned to the pit, the topsoil is replaced, and the area is replanted. Underground, or deep, mining is used when coal is far below the ground (sometimes thousands of feet deep). Machines below ground dig up the coal and it is transported to the surface via elevators and miners in the mine shaft. Coal is used to create almost half of all electricity generated in the United States (U.S. EIA 2010).

Natural gas

Natural gas consists mainly of methane, but propane and butane are also by-products. Natural gas occurs naturally underground or offshore and is accessed by drilling. Geologists and engineers study rock samples and take measurements at chosen areas to determine if natural gas is present. Once natural gas is found, drilling begins and a well is established. Gas flows up through a well to the surface and into large pipelines. Natural gas can also be stored and transported as a liquid (once it has been cooled to –260°F). Most of the natural gas used in the United States is produced in the United States, though some is imported. Natural gas is colorless, odorless, and tasteless; the chemical mercaptan is added to give it the rotten-egg smell so that natural gas can be detected. Slightly more than half of U.S. consumers use natural gas as their heating source (U.S. EIA 2010).

ENVIRONMENTAL CONSIDERATIONS

Heating your home probably generates more greenhouse gases than any other activity (U.S. DOE 2010a). Emissions that result from the combustion of fossil fuels include the following:

- Carbon dioxide (CO_2)
- Carbon monoxide (CO)
- Sulfur dioxide (SO_2)
- Nitrogen oxides (NO_X)
- Particulate matter
- Heavy metals (e.g., mercury)

Nearly all combustion by-products have negative impacts on the environment and human health:

- CO_2 is a greenhouse gas and a source of global warming.

- SO_2 causes acid rain, which is harmful to plants and to animals that live in water, and it worsens or causes respiratory illnesses and heart diseases, particularly in children and the elderly.
- NO_X contributes to ground-level ozone, which irritates and damages the lungs.
- Particulate matter results in hazy conditions in cities and scenic areas, and, along with ozone, contributes to asthma and chronic bronchitis, especially in children and the elderly. Very small particulate matter is thought to cause emphysema and lung cancer.
- Heavy metals (e.g., mercury) can be hazardous to human and animal health (U.S. EIA 2010).

Power plants that burn fossil fuels and some geothermal power plants contribute about 40% of the total U.S. carbon dioxide emissions (U.S. EIA 2010). The cleanest type of heating is solar energy, which produces no air pollution (U.S. DOE 2010a).

Oil

Fuel oil burned in home heating units produces small amounts of nitrous oxide, carbon monoxide, soot, sulfur dioxide, methane, and volatile organic compounds. Drilling for oil can disturb land and ocean habitats, and oil spills are extremely harmful to marine life (U.S. EIA 2010). Oil harms wildlife through physical contact, ingestion, inhalation, and physical contact. Floating oil can contaminate plankton, which includes algae, fish eggs, and various invertebrates. Fish feeding on these organisms can become contaminated, as well as the animals higher up in the food chain when the fish are consumed (U.S. FWS 2004). Oil spills can cause shifts in population structure, species abundance and diversity, and distribution (U.S. FWS 2004). The *Exxon Valdez* oil spill in 1989 released 11 million gallons of crude oil across 1,300 miles of coastline in Alaska, killing an estimated 250,000 seabirds, 2,800 sea otters, 300 harbor seals, 250 bald eagles, up to 22 killer whales, and billions of salmon and herring eggs (EVOSTC). The April 20, 2010, BP oil spill consistently spewed oil into the Gulf of Mexico for months, devastating wildlife, birds, fish, and people. Also, the improper disposal of large volumes of saline water produced with oil and gas, from accidental hydrocarbon and produced water releases, and from abandoned oil wells that were incorrectly sealed, can have major detrimental impacts to U.S. soils, groundwater and surface waters, and local ecosystems (Kharaka and Otton 2003).

Table 4

A year's worth of emissions from various heating systems (U.S. EIA 1994)

Factor	Indirect electric utility emissions from fossil fuels (lbs. per year)	Direct residential emissions (lbs. per year)	
	Heat pump	Fuel oil furnace*	Natural gas furnace*
Soot (particulates)	N/A	0.14	N/A
Sulfur dioxide (SO_2)	80	0.03–0.16	0.04
Nitrous oxide (N_2O)	42	8.11	5.69
Carbon monoxide (CO)	N/A	2.25	2.42
Organic compounds	N/A	N/A	0.67
Carbon	2,704	2,718	1,979

*The furnace and boiler emissions are for direct combustion and exclude the emissions from the generation of electricity required to power furnace fans and boiler pumps.

N/A = The data source does not provide this emission estimate.

Electric resistant heat

Electricity used in home heating produces no atmospheric emissions at the home. However, electricity generated by the burning of fossil fuels—the most common source of electricity—produces large quantities of pollutants at the generating plant sites (Table 4; U.S. EIA 1994). Figure 4 compares a year's worth of emissions from an oil furnace, a gas furnace, and an average heat pump. The values in Table 4 are calculated for heating in a moderate climate using 50 million BTUs per year (U.S. EIA 1994).

Electricity is also generated at nuclear power plants. Nuclear power does not generate carbon dioxide or air pollution, but mining and refining processes of uranium require large amounts of energy (U.S. EIA 2010). The main environmental concern with nuclear power is radioactive wastes such as uranium mill tailings, spent (used) reactor fuel, and other radioactive wastes. These materials can remain radioactive and dangerous to human health for thousands of years (U.S. EIA 2010). There is currently no disposal facility for high-level radioactive waste in the United States; high-level waste is stored at nuclear plants (U.S. EIA 2010). An uncontrolled nuclear reaction at a plant could contaminate air and water with radioactivity for hundreds of miles (U.S. EIA 2010).

Propane

Propane is a nonrenewable fossil fuel and produces fewer emissions than other fossil fuels. However, like all fossil fuels, when burned it emits carbon dioxide (U.S. EIA 2010).

Wood and pellet heating

Wood smoke contains hundreds of chemical compounds, including carbon monoxide, nitrous oxides, organic gases, and particulate matter, which pollute the air and may have adverse effects on human health. In some areas, wood-burning appliances have been banned due to the pollutants they generate. New wood-burning appliances are cleaner and more efficient and include a catalytic combustor that allows combustion gases to burn at lower temperatures, thereby cleaning the exhaust gas while generating more heat (U.S. DOE 2010c).

Coal

Ninety-three percent of all coal used in the United States is for producing electricity. Coal mining has a negative impact on ecosystems and water quality, and landscapes and scenic views are altered. Mountaintop removal (the tops of mountains are blown off and permanently removed) can create debris that chokes mountain streams. Acidic water can drain from abandoned underground mines and groundwater and surface waters can be tainted (U.S. EIA 2010). As with other fossil fuels, the combustion of coal produces sulfur dioxide, nitrogen oxides, particulates, carbon dioxide, and mercury. Mercury has been linked with both neurological and developmental damage in humans and other animals. Mercury concentrations in the air are usually low and of little direct concern. However, when mercury enters water—either directly or through deposition from the air—biological processes transform it into methylmercury, a highly toxic chemical that accumulates in fish and the animals (including humans) that eat fish (U.S. EIA 2010).

Natural gas

Natural gas is nonrenewable and burns cleaner than oil or coal (U.S. EIA 2010). Natural gas burned in home heating units produces small amounts of nitrous oxide and carbon monoxide and small amounts of soot, sulfur dioxide, methane, and volatile organic compounds. Natural gas is carbon based and produces large quantities of carbon dioxide when burned in a home heating system. However, natural gas produces nearly 30% less carbon dioxide than either heating oil or the average electric utility generation plant (U.S. EIA 1994). Natural gas consists mainly of methane (a greenhouse gas), which can leak into the atmosphere from wells, storage tanks, and pipelines (U.S. EIA 2010). When natural gas leaks, it can cause explosions, and land and marine

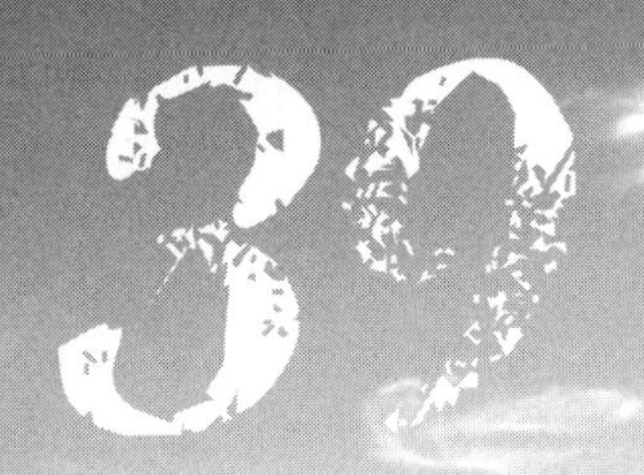

environments can be damaged during drilling for natural gas. Natural gas affects the environment when it is produced, transported, and stored (U.S. EIA 2010).

ALTERNATIVES

Thanks to environmental awareness, as well as financial concerns, there are ways to improve your current system or upgrade to a new and more efficient heating system that reduces environmental impact and costs. Also, the federal government currently offers tax credits for consumers who invest in renewable energy (e.g., solar panels) or increase the efficiency of their home (e.g., through energy-efficient windows, heating and cooling equipment, insulation).

Solar energy: Photovoltaic systems or solar cells can be installed to use the Sun's rays to heat water and provide heat. Initial installation can cost up to $25,000 and it can take up to a dozen years for homeowners to reap the financial rewards, but solar energy is the most environmentally friendly type of home heating available (*Boston Globe*). The U.S. Department of Energy's website has more information on the types of solar heating systems available (see "On the web").

Geothermal or geoexchange systems: These systems use a loop of underground water piping to transfer heat from the Earth and use 30%–60% less energy, run quietly, and require less maintenance than HVAC systems. The million plus homes using geothermal systems save nine billion pounds of carbon dioxide emissions (Planet Green 2009).

Fuel pellet stoves: These stoves produce little waste, use inexpensive fuel, and are 50% less expensive than propane and electricity and 35% cheaper than heating oil (*Boston Globe*).

Other ways to reduce environmental damage and lower costs include the following:

- upgrading your current heating system with HeatManager (a boiler-control system that improves the efficiency of gas-, oil-, and propane-fired residential boiler home heating systems by adjusting the burner run pattern to match the system's heat load) or with a hot water heating system fuel economizer (which adjusts the burner run pattern to match the system's heat load);
- using a fireplace insert (steel, iron, or glass) to increase efficiency and make the area airtight;
- installing radiant heat, which can reduce heating costs 20%–40%;
- using home heating oil that contains biofuel (vegetable oils);

- using space heaters in small areas;
- lowering your thermostat by 10–15° for eight hours (when you sleep or leave), which can reduce heating costs by 10%;
- weatherproofing your home by caulking and applying weather stripping in drafty areas;
- installing double-paned windows; and
- using ENERGY STAR appliances (thermostat, furnaces, etc.) (*Boston Globe*).

On the web

Active solar heating: *www.energysavers.gov/your_home/space_heating_cooling/index.cfm/mytopic=12490*

Energy Kids: Renewable solar: *www.eia.doe.gov/kids/energy.cfm?page=solar_home-basics*

Energy savers: Tax credits for energy efficiency: *www.energysavers.gov/financial/70010.html*

References

Boston Globe. Green home heating alternatives for this winter. *www.boston.com/lifestyle/green/gallery/home_heating_alternatives.*

Exxon Valdez Oil Spill Trustee Council (EVOSTC). Oil spill facts. *www.evostc.state.ak.us/facts.*

Formisano, B. 2007. Types of home heating systems. *http://homerepair.about.com/od/heatingcoolingrepair/ss/heating_types.htm.*

Kharaka, Y. K., and J. K. Otton, eds. 2003. *Environmental impacts of petroleum production: Initial results from the Osage-Skiatook petroleum environmental research sites, Osage County, Oklahoma.* Denver, CO: U.S. Geological Survey. *http://pubs.usgs.gov/wri/wri03-4260/pdf/WRIR03-4260.pdf.*

Planet Green. 2009. Home heating: Getting techie. *http://planetgreen.discovery.com/go-green/home-heating/home-heating-techie-definitions.html.*

U.S. Department of Energy (DOE). 2010a. Selecting heating fuel and system types. *www.energysavers.gov/your_home/space_heating_cooling/index.cfm/mytopic=12330.*

U.S. DOE. 2010b. Electric resistance heating. *www.energysavers.gov/your_home/space_heating_cooling/index.cfm/mytopic=12520.*

U.S. DOE. 2010c. Wood and pellet heating. *www.energysavers.gov/your_home/space_heating_cooling/index.cfm/mytopic=12570.*

U.S. Energy Information Administration (EIA). 1994. Reducing home heating and cooling costs. *http://tonto.eia.doe.gov/ftproot/service/emeu9401.pdf.*

U.S. EIA. 2008a. Residential heating oil prices: What consumers should know. *ftp://ftp.eia.doe.gov/brochures/heatingoil/index.html.*

U.S. EIA. 208b. Propane prices: What consumers should know. *ftp://ftp.eia.doe.gov/brochures/propane07/index.html.*

U.S. EIA. 2010. Energy sources. *www.eia.doe.gov/kids/energy.cfm?page=2.*

U.S. Fish and Wildlife Service (FWS). 2004. Effects of oil spills on wildlife and habitat: Alaska region. *http://alaska.fws.gov/media/unalaska/Oil Spill Fact Sheet.pdf.*

Chapter 40

CONNECT THE SPHERES WITH THE COAL CYCLE

By Renee Clary and James Wandersee

Coal fueled the Industrial Revolution and, as a result, changed the course of human history. However, the geologic history of coal is much, much longer than that which is recorded by humans. In your classroom, the coal cycle can be used to trace the formation of this important economic resource from its plant origins, through its lithification, or rock-forming changes, to its final recovery as a fossil fuel. The coal cycle also incorporates Earth-system science: Its event sequence integrates the various spheres of the Earth system, including the atmosphere, biosphere, hydrosphere, geosphere, and even a human-created "technosphere." Likewise, the coal cycle, as a subset of the carbon cycle, also integrates with the hydrologic cycle, and the rock cycle.

BACKGROUND

Coal is defined as a biochemical sedimentary rock. Unlike many rocks that are composed of one or more minerals, coal's components are not minerals because they originated as living, organic material. Coal forms when the remains of plants and animals accumulate in a low-oxygen, swampy environment. Over time, these organic layers are compacted and changed to produce a range of coal types, or ranks. After the original plant and animal material experiences some alteration and compression, peat can result. The peat can be further altered and compressed to form lignite, a low-rank coal composed of 25%–35% carbon. Further alteration can yield bituminous coal, then anthracite, a metamorphosed coal with the highest percentage of carbon (over 90%). Whereas lignite and bituminous coal are considered biochemical sedimentary rocks, anthracite can be classified as a metamorphic rock because of its higher degree of alteration (Figure 1). Anthracite represents only a small percentage of the coal found in the United States and is present in the Appalachian region.

Figure 1

How the compression of organic material yields coal

Once organic material is compressed and altered, peat can result. Peat can be further compressed and altered to produce lignite, a low-rank coal. A higher degree of alteration results in bituminous coal, while additional alteration yields anthracite coal, a metamorphic rock with the highest carbon content.

Formation of coal from organic material requires time—millions of years—and specific conditions. Fortunately, the conditions for deposition of the original organic layers existed at various times throughout Earth's history. During the Pennsylvanian period, approximately 300 to 320 million years ago, continental ice accumulation and the melting of glaciers resulted in fluctuating sea levels. When massive glaciers melted, the seas rose. These rising waters drowned the lush vegetation in swampy areas, depositing the organic layers necessary for future coal formation. When glaciers again

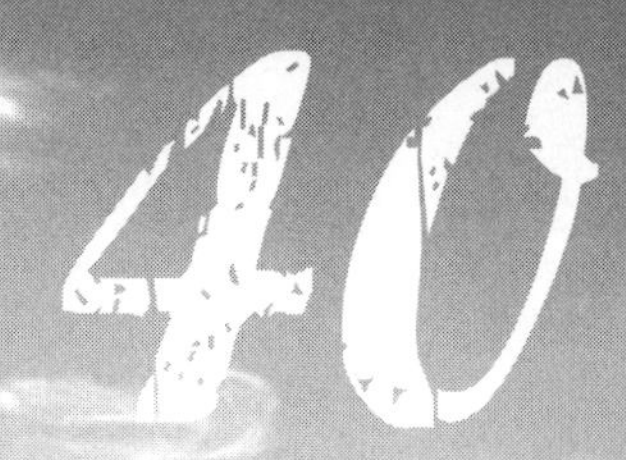

Figure 2

U.S. coal regions, 1996

The most altered coal, anthracite, is found in the Appalachian region; however, most Appalachian coal is bituminous.

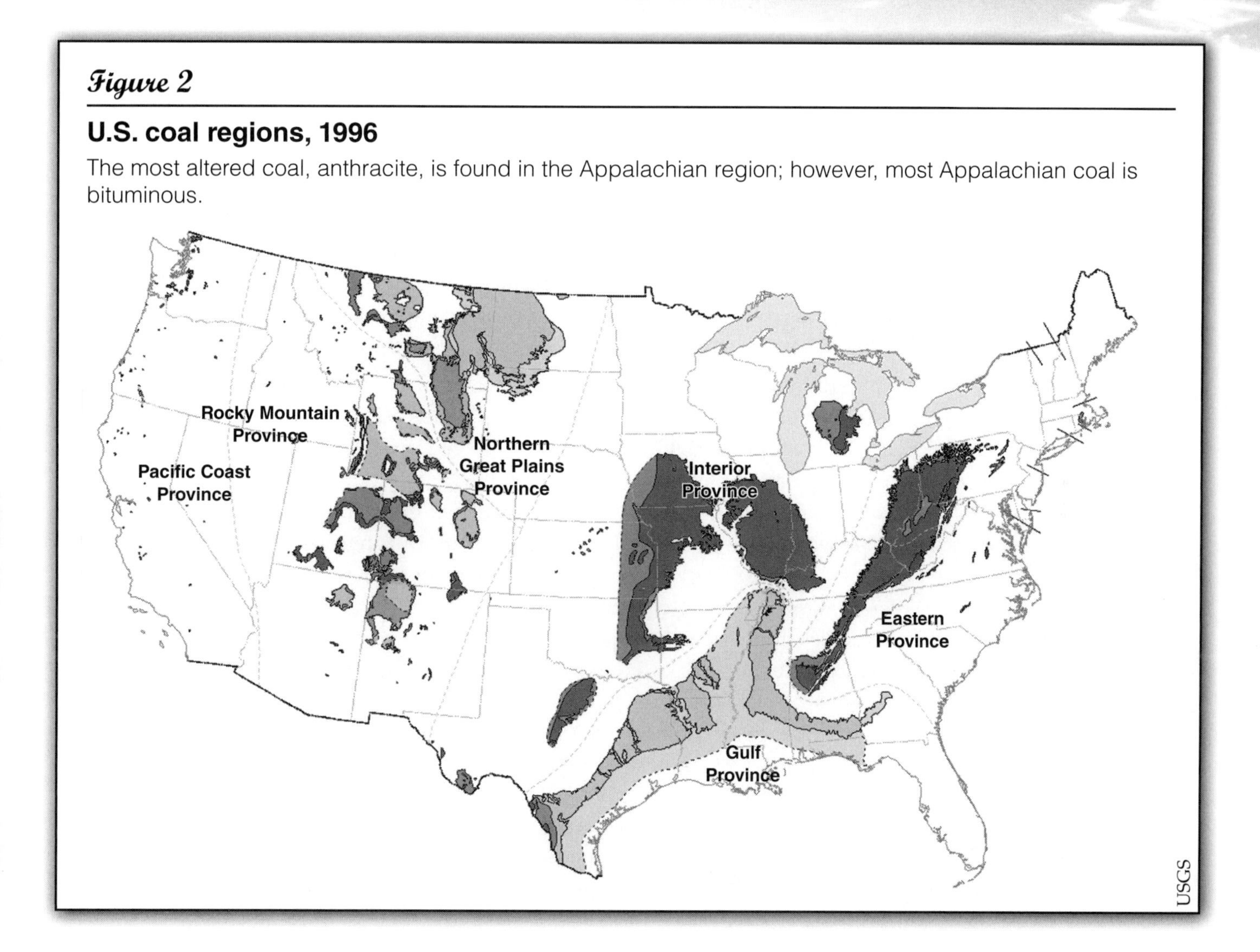

formed within polar regions, sea levels dropped, and vegetation could again grow in the once-flooded areas. The rising and falling sea level created *cyclothems* in the rock record, which are repeating marine and nonmarine rock units with coal interbedded within. In the United States, Pennsylvanian-age coals are found in the Appalachian region, as well as in the Midwest. Coals also formed more recently. Mesozoic-era coals (66–251 million years ago) are found in U.S. Rocky Mountain states and western North America. Lignite and sub-bituminous coals continued to be deposited in the early Paleogene period (23–66 million years ago) in the northern Great Plains states, and in the southern United States. Figure 2 maps the locations of the U.S. coal regions.

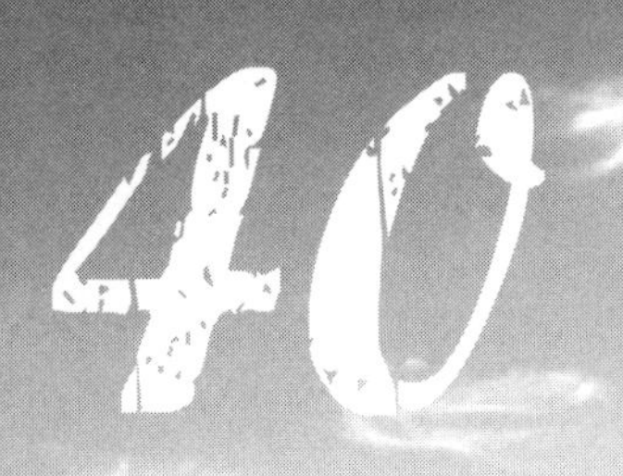

EXTENDING THE COAL CYCLE TO THE EARTH'S SPHERES

The coal cycle effectively focuses on one component of the carbon cycle (Figure 3). Furthermore, coal formation, extraction, and consumption illustrate how the coal cycle affects Earth's spheres, including the atmosphere, biosphere, hydrosphere, and geosphere. Photosynthesizing plants originally used carbon dioxide from the atmosphere to fix carbon for energy. These plants of the swamp forests are part of the biosphere, and they provided the original source material for the organic layers that eventually formed coal. Upon the plants' death, some of the carbon was returned to the atmosphere in the form of carbon dioxide as the plants decayed. However, the plant material that was buried effectively sequestered carbon—and as burial and alteration occurred, the plant material became part of the geosphere when it was included in the rock layers of our planet.

The hydrosphere is also involved in the early coal cycle. The original plant materials were part of a water-logged swamp, and further interacted with the hydrologic cycle through the transgression and regression of the seas during the formation of the coal-containing cyclothems.

The interaction of a coal cycle with Earth's spheres does not conclude with coal formation, either. Coal mining impacts the geosphere through the removal of large quantities of rock material as humans extract this economic resource. Unless the area mined for coal is reclaimed, the spent land left behind is unproductive, and sometimes polluted. During mining operations, coal dust enters the atmosphere, and can severely impact humans through diseases such as black lung. Water used in the extraction process can become contaminated, and metals and chemicals must be removed before the water is returned to the environment. Methane (CH_4) is released during mining operations, as well, and is one of the greenhouse gases implicated in climate change. Methane affects our atmosphere, which, in turn, affects our biosphere. The U.S. Environmental Protection Agency (1999) estimated in 1997 that 10% of anthropogenic methane emissions resulted from coal mines.

Finally, our consumption of coal affects Earth's spheres, too. Burning coal produces carbon dioxide, the major anthropogenic greenhouse gas implicated in climate change. The sulfur contained in coal is also released upon combustion, and this can result in the formation of sulfuric acid. Sulfuric acid affects the hydrosphere and can result in acid rain, which can damage crops and forests in the biosphere. Acid rain also affects the geosphere in the form of accelerated chemical weathering. In addition to carbon dioxide and sulfur compounds, an electricity-generating coal plant also releases mercury, arsenic, lead, and other toxic metals into the atmosphere (Union of Concerned Scientists 2009).

Figure 3

Carbon cycle

The carbon cycle follows carbon as it moves through the atmosphere, biosphere, hydrosphere, and geosphere. The coal cycle represents a specialized example of the carbon cycle and can be used to address all of Earth's spheres during the formation, extraction, and consumption of this fossil fuel. The arrows in this diagram show the flow direction of carbon, in billions of tons per year, as it is released in the atmosphere, sequestered by plants, and exchanged with the ocean.

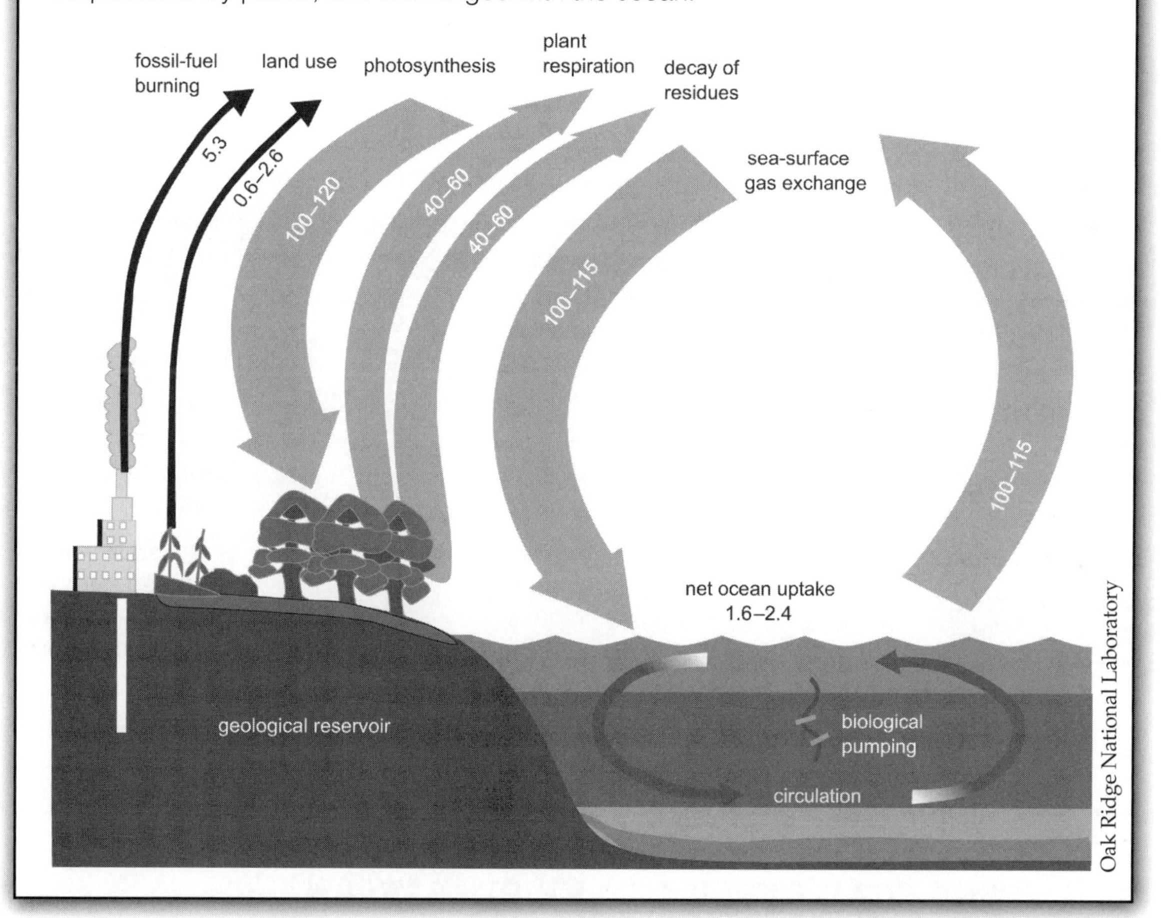

Oak Ridge National Laboratory

Some scientists identify a new "sphere" of Earth that is defined by the incredible impact that one species—*Homo sapiens*—has had on our planet: The "technosphere" includes all human-produced infrastructure and technology. Coal fueled both the Industrial Revolution and the technosphere as an inexpensive form of energy that spurred the development of machinery and electrical appliances, and powered an industrialized civilization (Figure 4).

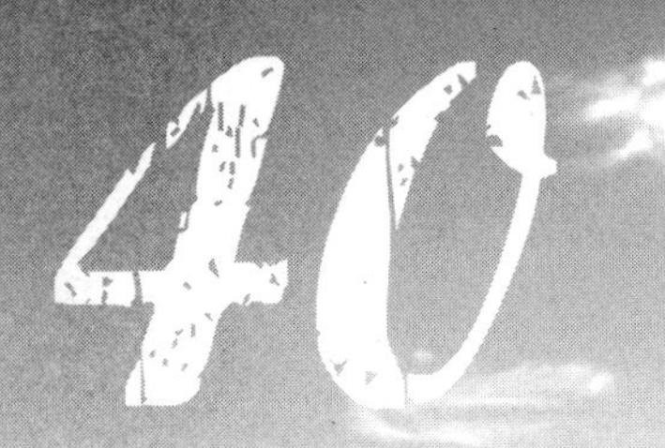

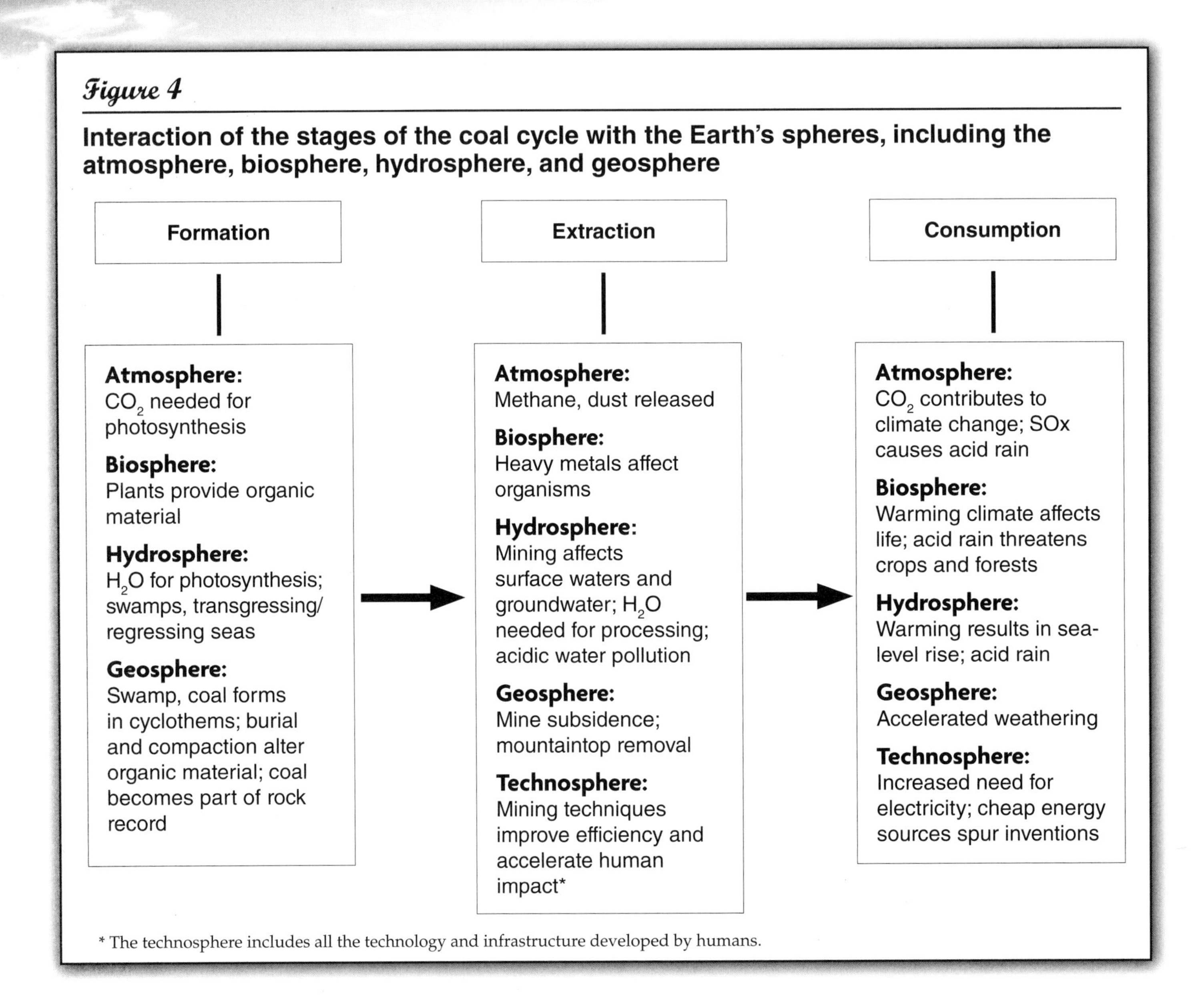

Figure 4

Interaction of the stages of the coal cycle with the Earth's spheres, including the atmosphere, biosphere, hydrosphere, and geosphere

* The technosphere includes all the technology and infrastructure developed by humans.

THE COAL CYCLE IN THE CLASSROOM

The coal cycle provides a specific portal through which to study the carbon cycle, the rock cycle, and the interactions of coal within the various spheres of the Earth—in its formation, extraction, and consumption. The coal cycle can be used to address important concepts in biology, ecology, geology, and chemistry, for an interdisciplinary science approach. Additionally, the coal cycle can be used to study the current issues of climate change, and environmental stewardship of our planet.

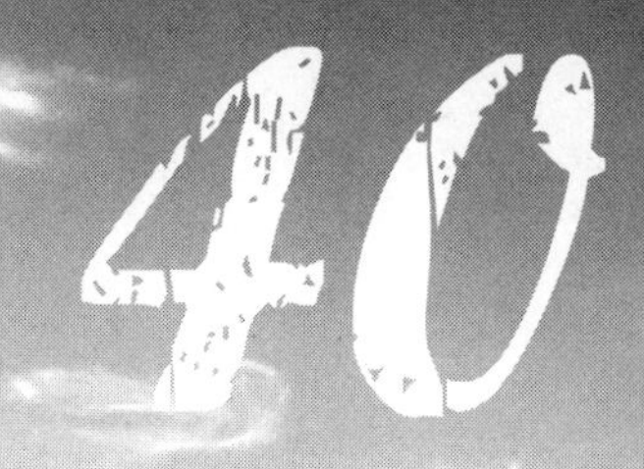

Introductory classroom activities

There are many ways in which the coal cycle can be introduced in classrooms. Table 1 (pp. 352–353) summarizes the various coal cycle instructional themes that are discussed below, with website resources and identification of the Earth spheres and cycles that each activity addresses. We suggest that teachers investigate the various resources presented, and choose the instructional pathway that best fits the curriculum.

Because the climate-change issue is at the forefront of science and politics, many resources for teachers exist, including lesson plans, multimedia sources, and interesting student activities. An overview of coal for students at the middle school level is available through the Need Project (*www.need.org/needpdf/infobook_activities/IntInfo/CoalI.pdf*). We suggest a classroom introduction to the coal cycle with an investigation into coal's formation and inclusion of the geologic time involved from the original plant origins, through coal formation, to the modern fossil-fuel resource. Students can also use coal to investigate various coal ranks, the rock cycle, and the carbon cycle.

Paleoenvironment reconstruction

Investigate the Pennsylvanian-age plants that constituted the majority of the coal swamps. Instead of oak and pine trees—which had not yet evolved—these Paleozoic plants included huge tree ferns and seed ferns! The Plant Fossils of West Virginia website (*www.geocraft.com/WVFossils/TableOfCont.html*) contains links to some of the more common plants that lived during the Pennsylvanian Period in the Appalachian area. For extraordinary photographs of the actual plant fossils from a 300-million-year-old Pennsylvania forest, visit the Illinois State Geological Survey's website (*www.isgs.illinois.edu/research/coal/fossil-forest/fossil-forest.shtml*).

Additional information can be found at the University of Wyoming Natural Science Program's website (*www.wsgs.uwyo.edu/coalweb/swamp),* which offers good discussions of the paleoenvironmental conditions necessary for original plant growth and deposition, and the resultant processes involved in coalification, or coal formation. Photographs of modern-day swamp analogues are included. We like the University of Kentucky–Kentucky Geological Survey's generalized diagram for coal formation (*www.uky.edu/KGS/coal/coalform.htm*). The Hooper Virtual Natural History Museum (*http://hoopermuseum.earthsci.carleton.ca/carbocoal/FRONTPG.HTM*) offers a virtual field trip that takes students through Paleozoic swamp conditions, burial and metamorphism, and mining processes. We suggest that teachers provide a context for the Paleozoic Era by demonstrating geologic time in the classroom (Clary and Wandersee 2009).

Table 1

Summary of instructional themes using the coal cycle, including resources and integration with Earth's spheres and cycles

Theme	Resource	Spheres and cycles
Introductory classroom activities		
Formation, mining history, mining techniques	*www.need.org/needpdf/infobook_activities/IntInfo/CoalI.pdf*	Biosphere, geosphere, hydrosphere, rock cycle
Paleoenvironment reconstruction		
Fossil plants	*www.geocraft.com/WVFossils/TableOfCont.html*	Biosphere, geologic time
Fossil forest website	*www.isgs.illinois.edu/research/coal/fossil-forest/fossil-forest.shtml*	Biosphere, geosphere
Swamp conditions, coalification	*www.wsgs.uwyo.edu/coalweb/swamp/*	Biosphere, geosphere, hydrosphere, rock cycle
Coal formation	*www.uky.edu/KGS/coal/coalform.htm*	Biosphere, geosphere, rock cycle
Coal formation, geology, coal classification, mining	*http://hoopermuseum.earthsci.carleton.ca/carbocoal/FRONTPG.HTM*	Biosphere, geosphere, geologic time, rock cycle
Geologic time demonstrations	*Clary and Wandersee (2009)*	Geologic time
Coal types and properties		
Coal samples to order (under $20)	*www.miningusa.com/store/coal/coal_samples_price_list.asp* *www.wardsci.com*	Geosphere
Microscopic views of coal	*http://geology.com/articles/coal-through-a-microscope.shtml*	Biosphere, geosphere, rock cycle
Coal formation animation	*www.classzone.com/books/earth_science/terc/content/visualizations/es0701/es0701page01.cfm?chapter_no=07*	Biosphere, geosphere, rock cycle
Coal rank identification, laboratory activity	*www.teachcoal.org/teacherstore/documents/CoalActivitySecondary.pdf*	Geosphere
Coal as a sedimentary rock		
Sedimentary rock construction, hands-on	*www.coaleducation.org/lessons/sme/elem/7.htm*	Biosphere, geosphere, rock cycle
Rock cycle animation	*www.geolsoc.org.uk/gsl/cache/offonce/education/resources/rockcycle*	Geosphere, rock cycle
The carbon cycle		
Carbon cycle, interactive web game	*www.windows2universe.org/earth/Water/co2_cycle.html&edu=elem*	Biosphere, geosphere, atmosphere, hydrosphere
Carbon cycle animation	*http://upload.wikimedia.org/wikipedia/commons/c/c8/Carbon_Cycle-animated_forest.gif*	Biosphere, atmosphere, hydrosphere

(continued)

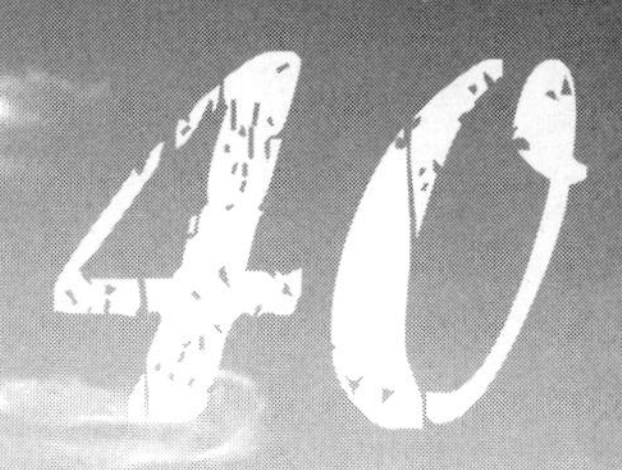

Table 1 (continued)

Theme	Resource	Spheres and cycles
Connecting coal to current issues		
Harmful effects of mining, burning, wastes	*http://wiseenergyforvirginia.org/the-coal-cycle*	Biosphere, geosphere, atmosphere, hydrosphere
Coal industry responses	*www.worldcoal.org/coal-the-environment*	Biosphere, geosphere, atmosphere, hydrosphere
Carbon capture and storage	*www.worldcoal.org/carbon-capture-storage*	Geosphere, atmosphere
The mining dilemma		
Muffin mining, hands-on	*http://www.darylscience.com/Demos/MuffinMining.html*	Geosphere
Kinesthetic mining activity	*http://edu.earthday.org/sites/default/files/Mountaintop_Removal_Coal_Mining_Lesson_Plan.pdf*	Geosphere
Mountaintop removal video	*www.greengorilla.com/video-post/turn-it-up-day*	Geosphere
Local student connection to mountaintop mining	*www.ilovemountains.org/myconnection*	Geosphere, hydrosphere
Earth footprint calculator	*http://files.earthday.net/footprint/flash.html*	Biosphere, geosphere, hydrosphere
Acid wastewater activity	*www.aep.com/citizenship/community/educationInit/gallons.aspx*	Biosphere, geosphere, hydrosphere
Overview of U.S. mining disasters	*www.nytimes.com/2010/04/06/us/06disasters.html*	Biosphere, geosphere
2010 West Virginia coal mine explosion	*www.nytimes.com/2010/04/06/us/06westvirginia.html*	Biosphere, geosphere
China mining disasters	*www.cbc.ca/news/background/china/mine_disaster.html*	Biosphere, geosphere
Coal and history		
U.S. timeline of coal	*http://www.teachcoal.org/lessonplans/pdf/coal_timeline.pdf*	Biosphere, geosphere
Coal timeline activity	*www.coaleducation.org/lessons/primary/general/impact_of_mining.htm*	Biosphere, geosphere
Coal and economics		
West Virginia coal video	*www.youtube.com/watch?v=JArYF8axBVY*	Biosphere, geosphere
Supply and demand, hands-on	*www.teachcoal.org/lessonplans/supply_demand.html*	Biosphere, geosphere
Coal in art and jewelry		
Jet formation and jewelry	*www.cst.cmich.edu/users/dietr1rv/jet.htm*	Biosphere, geosphere
Whitby jet, jet mining	*www.whitbyjet.co.uk/about-jet*	Biosphere, geosphere

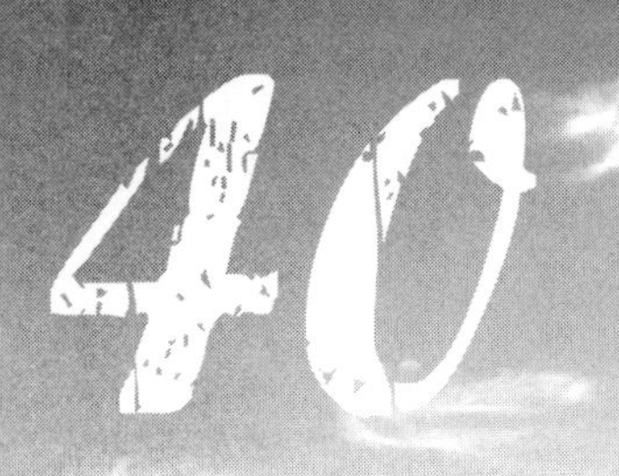

Coal types and properties

Students can compare samples of peat, lignite, and bituminous and anthracite coal, and describe the color, luster, hardness, streak, and density of each sample. (Mining Internet Services offers a teacher's kit with the different coal ranks at *www.miningusa.com/store/coal/coal_samples_price_list.asp;* coal samples can also be procured from Ward's Natural Science at *www.wardsci.com.*) To investigate microscopic views of coal, visit geology.com (*http://geology.com/articles/coal-through-a-microscope.shtml*) for photomicrographs of coal thin sections. An animation that shows the formation of the different coal types is available at Exploring Earth (*www.classzone.com/books/earth_science/terc/content/visualizations/es0701/es0701page01.cfm?chapter_no=07*).

An investigation into the speed of ignition, flame color, and speed of burning can be conducted with various coal ranks, provided the science classroom is properly equipped (*www.teachcoal.org/teacherstore/documents/CoalActivitySecondary.pdf*).

Coal as a sedimentary rock

To investigate different sedimentary rocks, students can make their own versions of sandstone, limestone, and coal. The Kentucky Coal Education website (*www.coaleducation.org/lessons/sme/elem/7.htm*) provides activity directions. (Students should wear protective eye equipment and gloves when handling cement.) To view where sedimentary rocks fit into the bigger rock cycle, visit the Rock Cycle website (*www.geolsoc.org.uk/gsl/cache/offonce/education/resources/rockcycle*), sponsored by the Geological Society of London. The site includes animations and information on various rock types.

The carbon cycle

Use the Windows to the Universe website to investigate the carbon cycle at *www.windows2universe.org/earth/Water/co2_cycle.html&edu=elem*. The website includes an interactive carbon-cycle game with various difficulty levels available. For an animated version of the carbon cycle in the U.S. National Park Service's Olympic Forest, visit *http://upload.wikimedia.org/wikipedia/commons/c/c8/Carbon_Cycle-animated_forest.gif.*

CONNECTING COAL TO CURRENT ISSUES

Climate change and environmental sustainability are key current issues, both politically and scientifically. The coal cycle can address both these issues in the classroom through a variety of methods. We try to present a balanced assortment of activities from both coal proponents as well as coal opponents in this discussion, but teachers should always make students aware of the source of the materials to reveal the potential for conflicting viewpoints.

Wise Energy for Virginia offers an interactive website (*http://wiseenergyforvirginia.org/the-coal-cycle*) with some of the potentially harmful effects of coal retrieval, consumption, and waste disposal. The World Coal Institute presents the global industry response to the concerns of climate change, pollution, and global warming at *www.worldcoal.org/coal-the-environment*. This website also includes a section on carbon capture and storage (*www.worldcoal.org/carbon-capture-storage*).

The mining dilemma

Strip-mining processes impact the geosphere through the removal of large amounts of rock and soil to retrieve coal. Underground mining can produce subsidence and collapse of the overlying land. Students can investigate the harmful effects of mining with the Muffin Mining activity, in which they attempt to retrieve blueberries from a muffin without impacting the muffin's shape (*http://www.darylscience.com/Demos/MuffinMining.html*). (**Note:** This activity should be done in an approved classroom or home economics laboratory, and not in a science laboratory where chemicals have been used.) A variation on the Muffin Mining Activity, the Coal Mining Calamity activity, can be found on the Earth Day Network (*http://edu.earthday.org/sites/default/files/Mountaintop Removal Coal Mining_Lesson Plan.pdf*). This activity is a kinesthetic one, as students search for different color bandanas within a pile of shoes (or books), without affecting the shape of the pile. A suggested follow-up activity is the "Turn It Up Day" Gorilla in the Greenhouse video (*www.greengorilla.com/video-post/turn-it-up-day*). Students can visit iLoveMountains.org to discover their local community connection to coal recovery via mountaintop removal (*www.ilovemountains.org/myconnection*—typing in the local school zip code reveals the community's connection to mountaintop coal mining). Students can calculate their Earth footprint (*http://files.earthday.net/footprint/flash.html*) and discuss ways to reduce their energy needs.

American Electric Power outlines a middle school activity that recaps an actual 1993 event that resulted from acidic mining waters. The website provides background materials, experiment suggestions for demonstrating coal water treatment, and assessment ideas (*www.aep.com/citizenship/community/educationInit/gallons.aspx*).

Mining does not come without work-related risks, as well. The *New York Times* outlined the deadliest mining disasters at *www.nytimes.com/2010/04/06/us/06disasters.html*, and the 2010 mining explosion in West Virginia is discussed at *www.nytimes.com/2010/04/06/us/06westvirginia.html*. Outside the United States, mining-related fatalities are especially high in developing countries. The worst coal-mining disaster, which claimed 1,549 lives in 1942, occurred in China. CBC News outlines the most tragic mining accidents in China at *www.cbc.ca/news/background/china/mine_disaster.html*. Students can investigate the mining disasters around the world in more detail, and pinpoint the sites on a world map.

EXTENDING COAL BEYOND THE SCIENCES

Coal and history

Human use of coal has a long history. The first coal-mining operations are attributed to the Chinese and the Greek scientist Theophrastus noted that blacksmiths in the Middle East used coal several hundred years before the Common Era. Romans used coal in England in the second and third centuries, and the Hopi tribe employed it for baking, cooking, and heating. In the United States, coal mining has been present since the late 1600s, and the first commercial mine opened in Virginia in 1748. The American Coal Foundation supplies a timeline of the history of coal in the United States at *http://www.teachcoal.org/lessonplans/pdf/coal_timeline.pdf*. Students can create a timeline using a short history of the Impact of Mining on American History (*www.coaleducation.org/lessons/primary/general/impact_of_mining.htm*).

Coal and economics

The story of coal in West Virginia has been re-created in a video, which teachers may want to embed within a PowerPoint presentation to limit student access to YouTube (Table 1). A coal supply and demand activity is available on the American Coal Foundation website, in which students search for "natural resources" in the form of beads, rice, and beans around the classroom, and then graph the results of their "recovery" (*www.teachcoal.org/lessonplans/supply_demand.html*).

Coal in art and jewelry

Coal can also be used in art. Jet, a form of lignite, is a semiprecious gemstone that is classified as a mineraloid because of its organic origins. With a hardness of only 2.5–4, it is easily carved into jewelry. Jet has a metallic luster, and was popularized by Queen Victoria as "mourning jewelry" after Prince Albert died (Figure 5). General information on jet (*www.cst.cmich.edu/users/dietr1rv/jet.htm*) and the famous UK Whitby Jet jewelry establishment (*www.whitbyjet.co.uk/about-jet*) are available on the internet for teachers and students.

Figure 5

Mourning jewelry

Jet, a form of lignite, became popular in jewelry because it was easy to carve, and was an acceptable adornment as "mourning jewelry" during the Victorian era.

Photograph courtesy of Detlef Thomas

DISCUSSION

The coal cycle provides a specific example through which the carbon cycle can be investigated in classrooms. Through coal, teachers can incorporate Earth-system science and reveal to students how the formation, extraction, and consumption of this fossil fuel involve the planet's atmosphere, biosphere, hydrosphere, and geosphere. Because coal is often mentioned in current climate-change discussions, a variety of resources are available, from both the proponents of coal-fueled power and the environmentalists who caution against its use. Many of these resources can be tailored for a good curriculum fit. The coal cycle provides a geobiological portal through which students not only study science but also become more environmentally conscious of human effects on our planet.

References

Clary, R. M., and J. H. Wandersee. 2009. How old? Tested and trouble-free ways to convey geologic time. *Science Scope* 33 (4): 62–66.

Union of Concerned Scientists. 2009. Environmental impacts of coal power: Air pollution. *www.ucsusa.org/clean_energy/coalvswind/c02c.html.*

U.S. Environmental Protection Agency. 1999. U.S. methane emissions 1990–2020: Inventories, projections, and opportunities for reductions. *http://epa.gov/methane/reports/04-coal.pdf.*

Chapter 41

PETROLEUM AND THE ENVIRONMENT

TEACHING ABOUT PETROLEUM AND THE FUTURE OF ENERGY RESOURCES

By Travis Hudson and Geoffrey Camphire

Your students live in a world that is powered by petroleum and other energy resources to an unsurpassed degree. But do they know where all the energy that they readily use on a daily basis comes from? Will they know where to find it tomorrow?

In 2005, the United States consumed more than 24% of all the energy used in the world—and about 60% of that energy was provided by petroleum (oil and natural gas). The availability of abundant, inexpensive energy is the main reason that our nation's standard of living leads the world. Americans can travel just about anywhere anytime, run all types of appliances and electronic gadgets, and remain comfortable regardless of the weather outside. It's a lifestyle shared by relatively few of the world's inhabitants.

But environmental issues and concerns accompany both the production and consumption of petroleum, as your students should understand.

If today's students are to have a role in decisions to meet the energy demands of tomorrow, they must understand petroleum's importance, its sources, how it is processed and

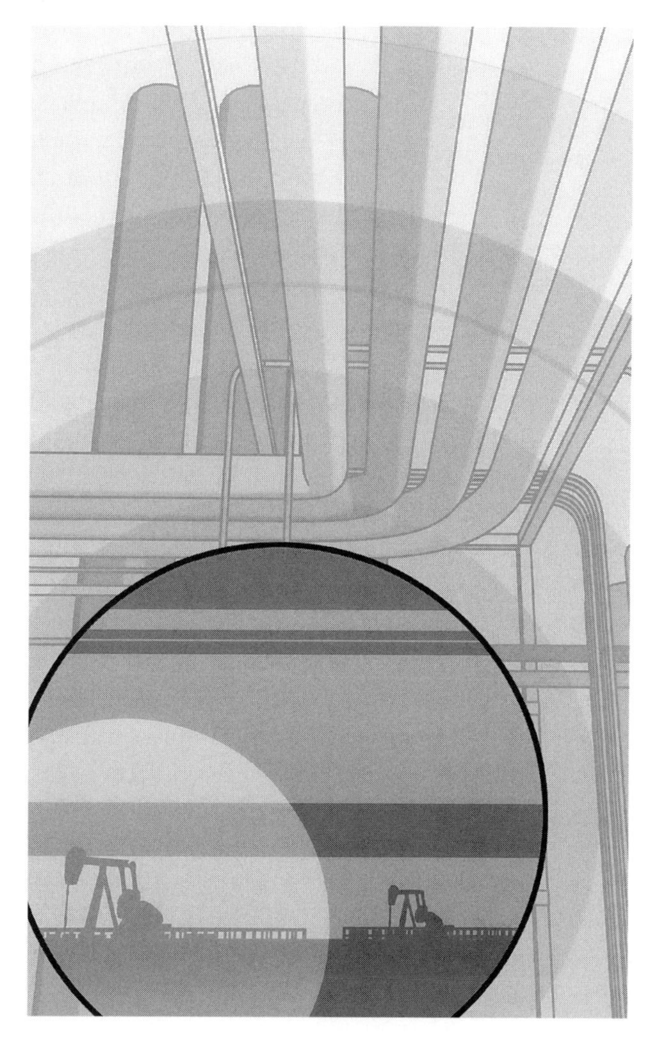

used, the policies and regulations designed to safeguard natural resources, and global energy needs.

After all, the world's people demand more petroleum every day. Population continues to increase, and the economies of some highly populous countries are expanding. For example, China became the world's second-largest consumer of oil in early 2004, when this nation's demand passed six million barrels a day. China's increasing oil consumption is considered a major reason that oil prices (not corrected for inflation) spiked in 2005.

FROM AGE TO AGE

Students should understand the history of energy resource use—and how one leading fuel historically has been replaced by another. The age of wood gave way to the age of coal, which in turn gave way to the age of oil. Around the world, the 20th century generally is considered the "Oil Age." How long will the Oil Age continue? The answer depends on when global oil production peaks and starts to decline.

The decline occurs when a finite nonrenewable resource such as oil cannot be produced in the amounts needed to meet demand. There is just not enough of the resource left to continue producing the amounts that are needed. Estimates of the peak of global oil production range from periods as early as 2003 to sometime around 2020. Regardless of which estimation is closest, you and your students are likely to live long past the end of the Oil Age.

Has global oil production already started to decline? Some people think so. The high price of oil in 2005 might partly reflect the inability of the world's oil producers to provide enough oil to meet rising demands—such as China's—for this vital energy resource. Increased oil prices are expected to follow the decline in global oil production. This development results in an oil sellers' market, at least for some time. When that happens, science and technology will be challenged to provide alternative and affordable new energy sources.

WHAT'S NEXT?

Usage of other energy resources will depend largely on the price of oil. As oil becomes more expensive, energy providers will work to replace it with less expensive alternatives. Options ranging from coal to solar power may become more economically viable.

After the decline in global oil production, anticipate outcomes such as these:

- Natural gas, which remains abundant and can be used in many ways, will replace many oil uses. Some say the world will enter the Natural Gas Age in the 21st century.

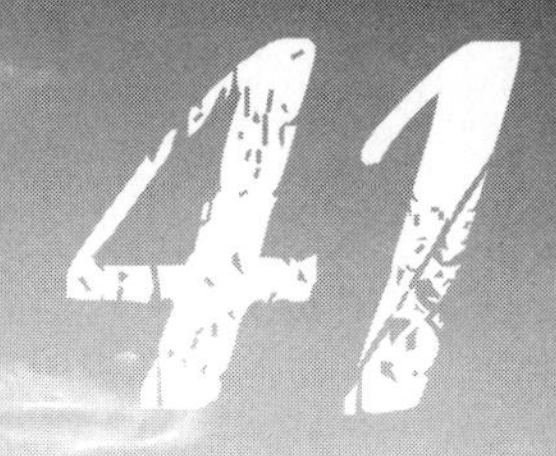

- Efforts will be made to expand the use of coal and nuclear energy to produce electricity.
- As gasoline becomes more expensive, alternative fuels and more efficient vehicles will be developed. Today's hybrid electric cars represent an example of this development. Hydrogen fuel cell technology eventually might replace internal combustion engines and the use of gasoline in vehicles. In fact, the Hydrogen Age may follow closely on the heels of the Natural Gas Age.
- Increased efforts to conserve energy—from all sources—almost certainly will be launched. "Green" technologies that involve energy conservation will become more profitable.

The viability of future energy alternatives will hinge at least partly on how well environmental issues are addressed. For example, expanding natural gas and coal use would depend on satisfactorily controlling atmospheric emissions including greenhouse gases. Increasing use of nuclear energy would depend on ensuring plant safety and developing environmentally sound ways of disposing of spent nuclear fuel and high-level radioactive waste. Also crucial will be the public's response to such developments, from nuclear reactors to wind farms.

In any case, new energy strategies will be necessary to maintain America's standard of living while supporting economic development in poorer regions around the world. Advances in science and technology can be expected to develop new, environmentally sound energy sources, but people's choices can either hinder or support these developments. A national energy policy has been debated in the U.S. Congress for years. A comprehensive and forward-looking policy—covering the broad scope of energy issues from conservation to new energy sources—will be vital as the Oil Age fades into history.

Chapter 42

AN EARTH-SYSTEM APPROACH TO UNDERSTANDING THE *DEEPWATER HORIZON* OIL SPILL

By Edward Robeck

The *Deepwater Horizon* explosion on April 20, 2010, and the subsequent release of oil into the Gulf of Mexico created an ecological disaster of immense proportions. The estimates of the amounts of oil, whether for the amount released per day or the total amount of oil disgorged from the well, call on numbers so large they defy the capacity of most students' comprehension. Similarly, the complex interactions among water, weather, living things, and other factors, as well as the extreme conditions found a mile under the water's surface, make it difficult to make sense of the effects of the incident. One way to help students comprehend this

disaster in meaningful ways is to adopt a *systems-thinking* approach by which the various effects and interactions can be mapped coherently. A systems-thinking approach will help students grasp the impacts of the oil spill and can also empower them to ask questions and focus on meaningful ways to learn more about this and other environmental incidents. A basic four-part systems diagram (Figure 1) can be used to facilitate students' use of an Earth-system model to analyze environmental events, depict their ideas about the impacts of the events, and extend their thinking, leading to new questions and avenues of inquiry.

SYSTEMS THINKING AND THE OIL SPILL

In general, a systems-thinking approach is characterized by holistic attention to the dynamics of interacting parts of a system. It is often contrasted with a *reductionist* approach that, in the extreme, separates a whole into parts that are then studied in isolation from each other. Consider, for example, understanding a natural environment as a group of distinct plant and animal types that can be named and counted. While there is much to be learned from this (reductionist) census, a more comprehensive understanding comes about from considering an ecosystem as sets of interacting elements that include living things, nonliving factors, energy, and the principles by which those elements affect and are affected by each other (e.g., thermodynamics).

Systems thinking has become an important tool among STEM professionals as a way of approaching inquiry using interdisciplinary perspectives. For example, oceanographers working to understand the *Deepwater Horizon* spill take into account much more than the characteristics of the oil and of ocean water, considering also the gases in the water, seabed topography, and the features of living things in the water column (e.g., whether or not they are able to swim away from toxins) (see *http://journalofcosmology.com/ClimateChange115.html*). Information about these characteristics and features is drawn from chemistry, geology, biology, mathematics, and many other disciplines as STEM professionals work together to produce models of what does or might occur. A systems-based approach facilitates this sharing by providing a framework for integrating the information across disciplines.

In Figure 1, the overall Earth system is understood as being composed of four interacting subsystems: the atmosphere, biosphere, geosphere, and hydrosphere. As the names imply, these are made up of the gases that surround Earth (atmosphere), the living things on Earth (biosphere), the rocks and minerals that make up the geologic structure of Earth (geosphere), and the water on Earth in all its forms (hydrosphere). These four subsystems could be further divided (e.g., some depictions divide the hydrosphere further to separate out the cryosphere, made up of the frozen water on Earth), and for

specific purposes other subsystems might be included (e.g., the exosphere, which includes energy and material in outer space). Each of the four main Earth subsystems can itself be seen as a system composed of even smaller subsystems (e.g., the hydrosphere is a system that includes river systems, aquifers, and ocean systems). For most purposes, however, the four systems in Figure 1 work well as a framework for analysis.

Figure 1

The Earth system is made up of four interacting subsystems, each of which can also be considered a system, with its own smaller subsystems.

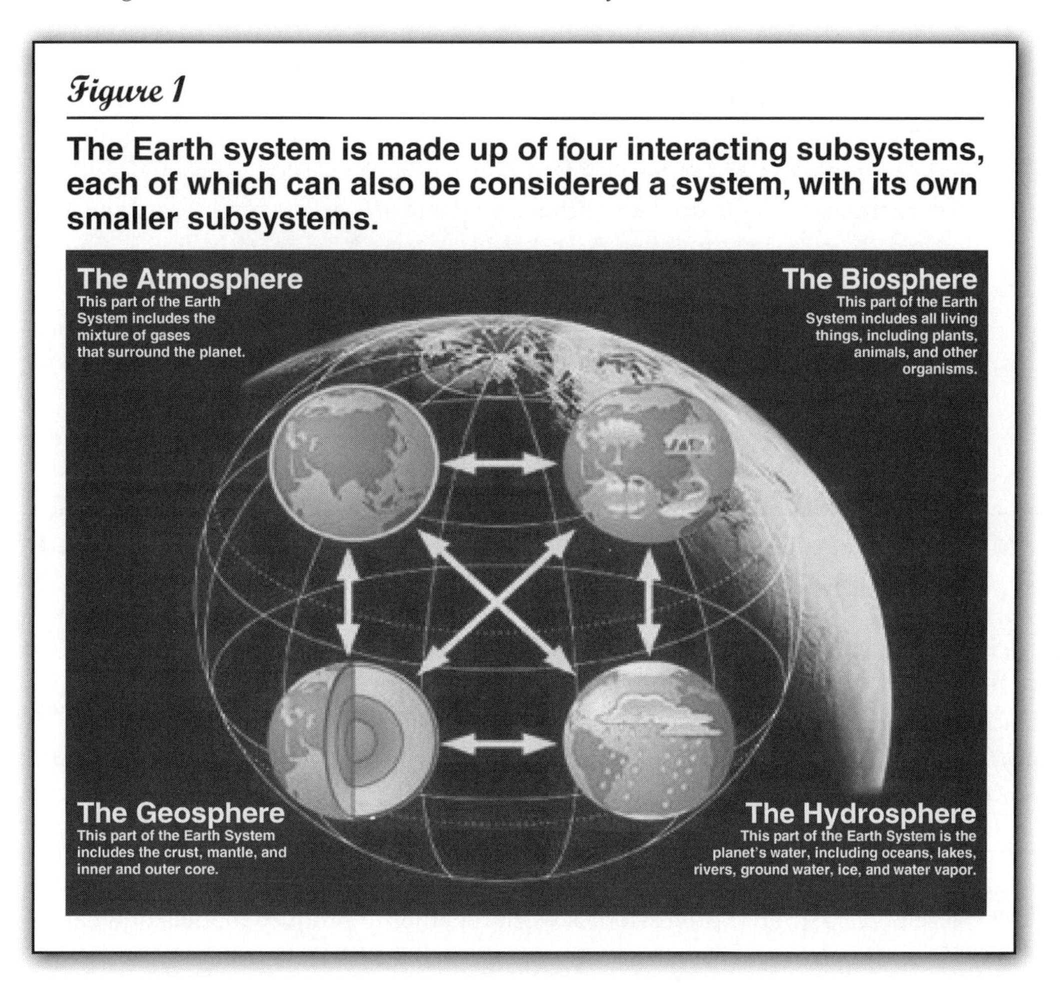

Within a systems-thinking perspective, attention may be limited to particular systems for the purposes of study, with tentative boundaries set that make sense for the questions being asked. One scientist might for a time define a system that is very small, such as a tide pool, while another might set boundaries that are very large, such as when studying the geologic history of an ocean basin. An important attribute of the *Deepwater Horizon* spill is that the system has to be studied using a range of boundary domains, from very small to very large. Whereas the boundaries for some questions might focus attention on a section of the Gulf, the affected area of concern and study for other questions includes the entire Gulf region, while other questions have led STEM professionals to consider the possibility of oil being taken up by the Loop Current, passed into the Gulf Stream, and then carried into the Atlantic Ocean. No matter what boundaries are chosen, the four Earth systems are helpful for sorting out the interactions that are taking place.

From the foundation provided by a basic mapping using the four Earth systems (Figure 2, items 1a–c), students can be led to ask questions that will enhance their understanding of the interactions among the systems, and of specific environment-related incidents. In a recent discussion in which eighth-grade students shared what they knew about the *Deepwater Horizon* oil spill, for example, students used a systems-based analysis as a way to capture and extend their reasoning about the oil spill. After a brief introduction to the Earth-system model, students were given a chart divided into four rectangles, each labeled with the name of one of the Earth systems (Figure 3). Working individually at first and then in groups, students labeled the diagram to

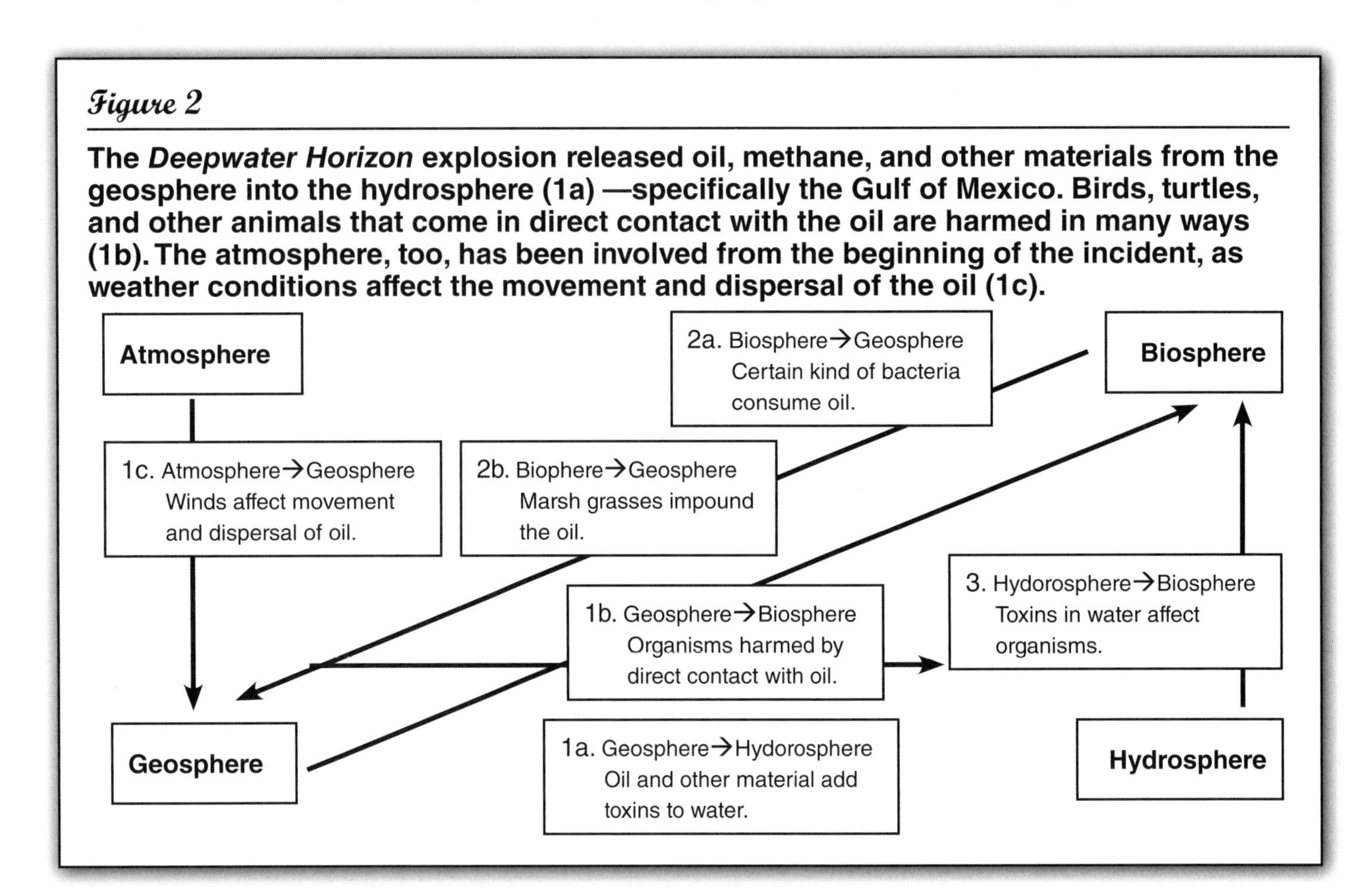

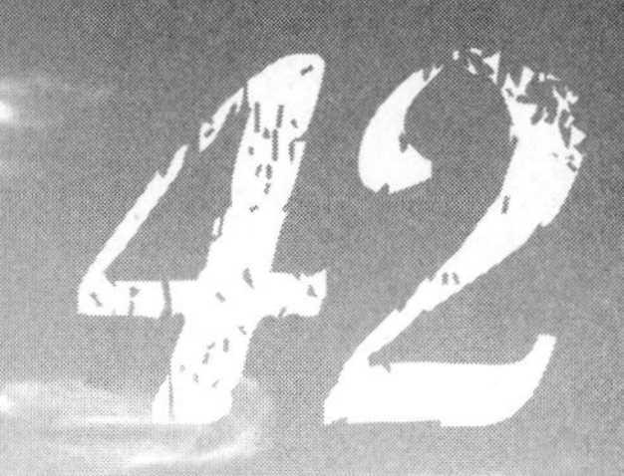

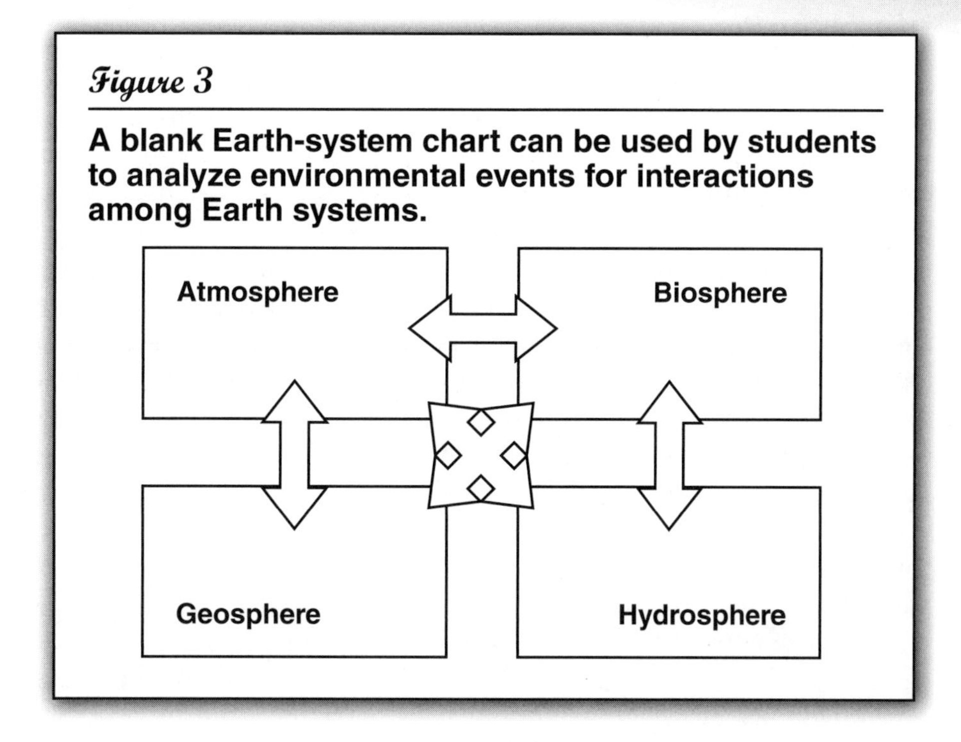

Figure 3

A blank Earth-system chart can be used by students to analyze environmental events for interactions among Earth systems.

create a schematic showing what they knew about the *Deepwater Horizon* incident. It was not long before they began to ask a range of questions implied by the Earth-system diagram. For example, students readily recognized that the oil (geosphere) affected living things (biosphere), since they had seen video of the efforts to clean oiled birds and relocate turtles along the coast. Going beyond those images, however, they began to ask questions such as whether the release of oil also affected other elements of the geosphere itself, including area beaches and the geology of the seabed from which the oil was flowing.

In this case, the Earth-system model was used as a tool to extend students' thinking as they communicated what they knew and considered other interactions for which they did not yet have examples. Along these lines, students recognized that they could not think of an interaction leading from the geosphere to the atmosphere. This led them to wonder about the impact on the air (atmosphere) that came about as a result of the oil (geosphere) that burned immediately after the initial explosion. Also, although they were unsure of whether the oil would evaporate directly into the air, they recognized that if the oil did evaporate, that process would be another way the geosphere affects the atmosphere in this case. They also asked questions about whether the oil affected the evaporation rate of water, which some of them had heard some scientists expected (*www.nhc.noaa.gov/pdf/hurricanes_oil_factsheet.pdf*), and what the effects of such changes might be.

Using an Earth-system approach, teachers can encourage students toward varied levels of understanding of environmental incidents. It is important in this regard to point out to students that the Earth-system diagram shows that interactions among the Earth's systems go in both directions. Recognizing this bidirectional feature can lead students to question what interactions might go in the other direction from those that are the initial focus of attention. For example, while the fact that the oil affects living things is clear, the bidirectional nature of Earth-system interactions can lead students to ask if there are also ways that living things (biosphere) are affecting the oil (geosphere). This type of interaction would be seen, for example, in the action of bacteria that scientists believe may consume a significant portion of the spilled oil (Figure 2, item 2a). Another example is that when the oil reached vegetation along the coastal marshes, the vegetation slowed the oil's movement, impounding it in the marsh areas (Figure 2, item 2b). Using the Earth-system diagram, students are encouraged to seek out possible interactions that are not initially part of their thinking.

A systems approach can provide a way for educators to guide students to analyze specific events, such as those events reported in the news or events depicted through images of the effects of the spill (Figure 4), providing a way to trace, understand, and raise additional questions about complex events as the effects of those events impact several Earth-system interactions. In some cases, one interaction can affect multiple systems. For example, as toxins are released into the water (hydrosphere), living things (biosphere) that live in the water are harmed (Figure 2, items 1a and 3). Similarly, marsh grasses impound the oil (biosphere → geosphere), slowing its return to open water (biosphere → hydrosphere). At this point, the oil entering the marsh could actually appear to be a good thing.

Yet the interactions don't end there. Many interactions ripple through several systems having multiple effects, and a full appreciation of the effects of interactions is gained by thoroughly tracking them across the Earth systems. In the short term, while it is true that the sea grasses will impound the oil and remove it from the open water column as it is held in the marsh areas, the toxic oil is expected to affect living things that use the marshes as nurseries (geosphere → biosphere). Also, much of the marsh grass may be killed. Long-term effects also emerge as this mapping of interactions proceeds. Since the marsh vegetation mitigates the effects of storm surges, absorbing much of the energy from the waves as they pass toward land, fewer marsh grasses mean that in the long term storm surges will have a greater impact on land and on the marsh itself. Using a systems approach to trace these effects among systems helps to make more comprehensible the complex interactions that are predicted to occur or are discovered as analysis of the event continues. It soon

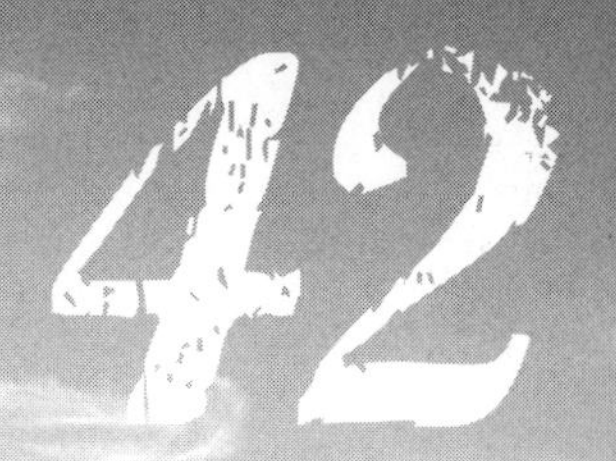

becomes apparent that a change in one system has impacts on each of the other systems, and that those impacts have still other effects as the interactions reverberate throughout the Earth system. The richly enmeshed nature of Earth's environments becomes quickly evident.

Figure 4

Emulsified oil remains on and pooled below vegetation in Pass a Loutre, Louisiana, following a storm the previous week.

When students capture their thinking using Earth-system diagrams, their understanding can be assessed in a variety of ways. For instance, a teacher can consider the extent to which students are able to express single-link interactions and then go beyond those to more complex interactions. Do students depict bidirectional interactions? Do their maps trace interactions through more than one or two of the Earth systems? Do their maps depict both short- and long-term effects? To engage fully with the interactions they read and hear about in the news, students may need to produce multiple maps, each tracing a limited set of effects across sets of interactions. Students' initial maps are often quite messy looking, with lines, arrows, and terms extending in different directions as students sort out ideas about how systems interact. As they explain these maps, however, through writing or oral descriptions, the increased clarity and sophistication of students' understanding is apparent. Students gain the ability to effectively express intricate ideas.

At a general level, a systems-thinking approach involves tracing inputs of materials and energy through interactions and to their outputs across a series of interconnected interactions. The questions mentioned above raised by students regarding how the oil might affect evaporation illustrate these interactions of matter and energy. Input of oil (geosphere) onto the surface of the sea (hydrosphere) is thought to affect the output of energy and matter through evaporation from the sea, which is an input to the air (atmosphere). Yet systems thinking can go to a more detailed level, too, as other concepts are added to the analysis.

Questions of scale, for example, become very important in systems thinking as matter and energy interact in more or less discrete ways. At a regional scale, interactions can be mapped among large features of Earth (Figure 5).

Figure 5

A satellite image framing a boundary around a specific system—the Bird's Foot Delta region

The Mississippi River (hydrosphere) has carried and deposited sediments (geosphere) on which vegetation (biosphere—appearing in red) has grown. At this scale, one can see that the vegetation on this delta is threatened by the oil (geosphere—seen as silver streaks) as it is being moved by water currents (hydrosphere) and wind (atmosphere). If the vegetation on the delta is killed by the oil, how might the delta (geosphere) be affected? How might that change where the river (hydrosphere) goes as it reaches the Gulf? What effect might removing the marsh grasses have on oxygen levels (atmosphere) in the water?

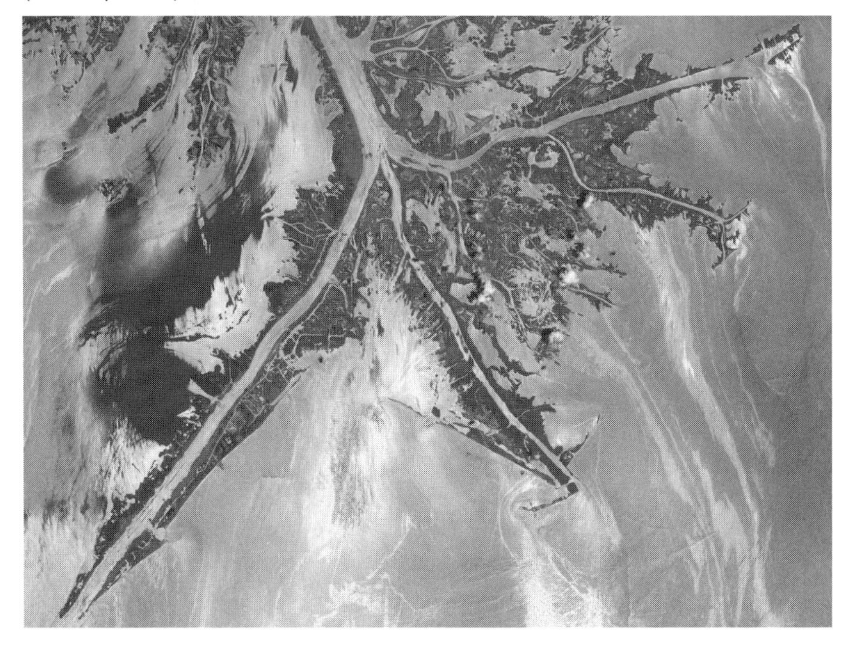

At the more common macro scale, students can see in photos how birds and other large animals are affected by the oil. Moving to a smaller scale, students begin to ask questions about how the oil might have effects on microscopic life, such as plankton. During summer months, these organisms reproduce in the shallow, nutrient-rich waters off the Gulf coast. As water currents and winds move oil into the shallow coastal areas, the toxins in the oil may kill plankton directly and also indirectly as the oil blocks sunlight from reaching the microscopic photosynthetic phytoplankton. Students can be encouraged to ask questions regarding which Earth systems might be affected by the decreased plankton concentrations.

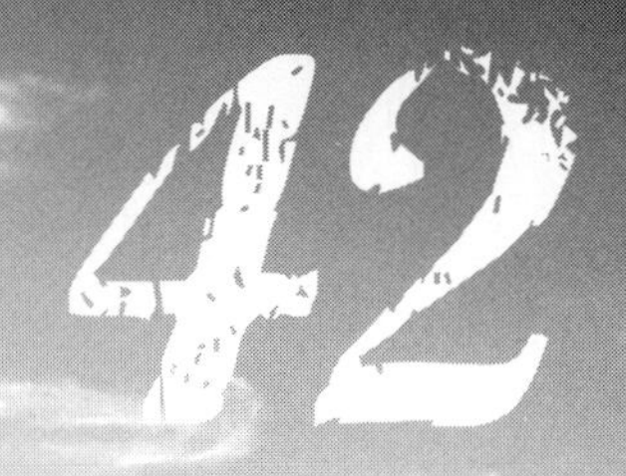

Moving to this scale also raises questions about the large "plumes" scientists have found in the Gulf that are made of tiny droplets, including methane and oil (see *http://news.sciencemag.org/scienceinsider/2010/06/huge-oil-plumes-confirmed-but-ef.html*). While these chemical droplets are far too small to be seen, their effect is potentially important if, as some predict, they deplete oxygen from the sea water (mostly, it is thought, as a result of bacteria that consume the chemicals). At a more global scale, the impacts the oil could have if it moves from the Gulf into the Atlantic are hard to anticipate, but sketching different possible system interactions helps students understand what may happen. As more is learned and more effects are recognized, a systems approach provides a coherent way for students to make sense of growing amounts of information, and the often conflicting predictions regarding effects.

CONCLUSION

A systems-thinking approach to the *Deepwater Horizon* incident and other real-world environmental events supports current initiatives in science education. The Earth Science Literacy Principles (available at *http://windows2universe.org/teacher_resources/main/frameworks/esl_main.html*) encourage the development of an Earth-system science perspective. Among the principles that the framework promotes is the idea that students need to understand that "the four major systems of Earth are the geosphere, hydrosphere, atmosphere, and biosphere" (Principle 3.1; Windows to the Universe 2009). These principles also encourage instruction through which students can come to appreciate that "humans significantly alter the Earth" (Big Idea 9; Windows to the Universe 2009).

Many other principles are also supported by the systems-thinking approach. As students consider the *Deepwater Horizon* incident and other environmental events using a systems perspective, they can identify those system interactions that humans have affected, and they can appreciate the dynamic consequences that those effects will have across all four of the Earth systems. Perhaps with this appreciation in mind, students will be able to more readily comprehend the potential effects of human activities, and the limits of human understanding of Earth's systems, so that they might be able to effectively respond to future disasters or, even better, stave off those disasters before they occur.

Acknowledgment

Special thanks to the eighth graders and staff at the Krieger Schechter Day School in Baltimore, Maryland, for their support of this project.

References

Smith, M., and J. Southard. 2001. *Investigating Earth systems: Water as a resource.* Armonk, NY: It's About Time.

Windows to the Universe. 2009. Earth science literacy framework. *http://windows2universe.org/teacher_resources/main/frameworks/esl_main.html.*

Chapter 43

CONVERTING SUNLIGHT INTO OTHER FORMS OF ENERGY

USING PHOTOVOLTAIC CELLS MADE FROM SILICON ALLOYS FOR SOLAR POWER

By Robert A. Lucking, Edwin P. Christmann, and Robin Spruce

The role of solar energy in a contemporary science curriculum springs from new concerns about how we power our homes and businesses and develop alternatives to declining fossil fuels. Given the many effective teaching materials on this topic available today, students will likely find the issues surrounding exploiting the Sun's energy to be intriguing and fun to explore.

BACKGROUND INFORMATION

Photovoltaic (PV) cells, also called solar cells, convert sunlight directly into electrical energy through the process of converting light (photons) into electricity (voltage), which is called the *PV effect*. The modern photovoltaic cell was developed in 1954, when scientists at Bell Laboratories discovered

that silicon (an element found in the sand of every beach) created an electric charge when exposed to sunlight.

Photovoltaic cells may one day surround our homes and the power devices we use in our day-to-day lives, and may free us, at least in part, from our dependence on oil. PV cells of the present generation are constructed by layering special materials called *semiconductors* into thin, flat sandwiches consisting of three distinct layers. The uppermost consists of *n-type silicon,* which releases negatively charged particles (electrons) when stimulated by light. The lower layer is composed of *p-type silicon* and develops a positive charge when struck by light. The middle layer is called the *junction layer*, and it acts as the medium for the flow of electrons.

Solar cells can be combined into modules that are known as solar panels, which today can provide sufficient power to run a small appliance; additionally, many such solar panels combined together create one larger system called a solar array. A typical home will use 10 to 20 solar panels combined as an array to potentially power the entire home. The panels can be mounted at a fixed angle facing south or they can be mounted on a tracking device that follows the Sun, allowing them to capture even more sunlight. Originally, solar panels were connected to banks of batteries, but the required purchase of the batteries led to additional expenses, and there were potential inefficiencies. Most of today's solar applications are tied to the larger power grid of the community, allowing individual solar purchasers to contribute to the greater good of the region whenever the homeowner's needs are met.

Older, traditional solar cells were made from silicon and affixed to a flat plate. Second-generation solar cells commonly found today are called *thin-film* solar cells because they are made from amorphous silicon or nonsilicon materials such as cadmium telluride. These thin-film solar cells use layers of semiconductor materials only a few micrometers thick. Because of their flexibility, these thin-film solar cells can double as rooftop shingles and tiles or the surface area for building facades.

Third-generation solar cells are now being made from a variety of new materials besides silicon, including solar inks using conventional printing press technologies, solar dyes, and conductive plastics. Some new solar cells use plastic lenses or mirrors to concentrate sunlight onto a very small piece of high-efficiency PV material. The PV material has become more sophisticated, and because so little is needed, these systems are becoming cost-effective for use by utilities and industry. However, because the lenses must be pointed at the Sun, the use of concentrating collectors is limited to the sunnier parts of the country. In short, solar panels are being made more cheaply, and new refinements, primarily in nanotechnologies, make them more powerful and more efficient in their use of the Sun's energy. Most experts believe that our

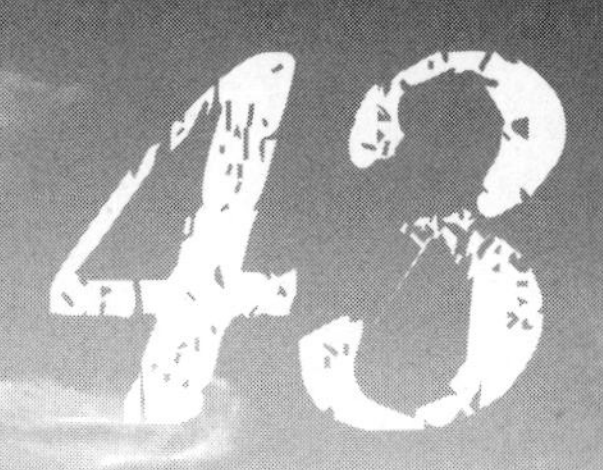

society is poised to make a significant shift in sources of energy brought to our homes and the rest of our daily lives.

HANDS-ON ACTIVITIES

Students are likely to question the practicality of such a shift and will want to explore considerations of the cost-effectiveness of solar options where they live. The best way of doing that is to begin by experimenting with actual solar units. Some websites suggest teachers find used solar cells that are being discarded from older applications and make their own panels for classroom experimentation; however, this is not a viable option in all parts of the country. Another option is to use a classroom solar energy kit offered by several companies today. Though the plastic pieces of a typical kit may at first seem toylike, the kit can be used to show many powerful concepts in dramatic fashion. For example, teachers can demonstrate that sunlight intensity is vitally important not just to the efficiency of the solar kit, but also in the classroom and to the larger world.

Using kits such as these will get students involved in conducting experiments and participating in firsthand explorations of these topics; however, students can also seek out supporting information through a variety of web materials. Students can see geographic variations in sunlight free of charge at the National Reviewable Energy Laboratory (NREL) website (*www.nrel.gov/csp/maps.html*), which includes a variety of maps showing solar intensity in the United States. Students in the Southwest will appreciate their good fortune on this measure when they see coverage in Maine, for example. Found on the Concentrating Solar Power Resource Maps page from the NREL website, students can find a link to the Solar Power Prospector Tool, which reveals sun radiation changes in the United States throughout the seasons. Another cost-free option is the Solar and Wind Energy Assessment website, a good source for information about solar energy concentration (*http://swera.unep.net*). Designed to supply information for concentrating solar collectors, this site provides a vast array of maps about solar intensity for locations in the United States and all over the world. The quality of the map images is outstanding, and students can examine data from such exotic locales as Ethiopia and Sri Lanka. By using resources like these, teachers can engage students to think critically about alternative energy. Questions for discussion could include which geographic regions are best suited to capitalize on the Sun's energy and what states or countries might reasonably focus on using other forms of alternative energy.

Using the solar kit, teachers can lead students to explorations in compounding and multiplying electrical energies and show them how to create a solar panel with customizable power outputs in one of two ways. A series connection can be made where the positive electrode from the cell is connected to the negative electrode of the next cell (as in the case of most flashlights). Alternatively, teachers can demonstrate a parallel connection by connecting the positive electrodes from the cell to the positive electrodes of the adjacent cell (as in the case of certain electronic devices). The point being made is that when two cells are wired together in a series, the voltage increases and the current remains constant, while when two cells are wired together in parallel, the current increases and the voltage remains constant.

UNDERSTANDING ELECTRICITY

For students to understand the deep structure of electricity, they will likely have to begin with some basic comparison to their known world and then move toward the physics of electrical impulses. Just as a water pump moves a certain number of water molecules and applies a certain amount of pressure to them, students will appreciate thinking of electricity as a special kind of water being pushed through a pipe in the form of a wire. In an electrical circuit, the number of electrons in motion is called the amperage or current, and is measured in amps. The "pressure" pushing the electrons along is called the voltage and is measured in volts. With knowledge of the amps and volts involved, it is possible to determine the amount of electricity consumed or produced, which is measured in watts. To determine more precisely the amount of electricity produced, the basic formula is W (watts) = V (volts) × A (amps), such as P (power) = I (current) × V (voltage).

COST-FREE WEB RESOURCES

Teachers looking to incorporate solar energy into their classrooms either as a short lesson or an entire unit will find myriad resources. Most provide step-by-step instructions and create opportunities for hands-on learning. By using the lessons as written, or perhaps by simply using them as inspiration, teachers can actively engage learners in experiential and authentic tasks. A good place to begin is with the U.S. Department of Energy's K–12 lesson plans and activities (*www1.eere.energy.gov/education/lessonplans*). Here, educators will find lesson plans related to solar energy including PV simulations, mini-rockets (powered by solar energy), and solar ovens. Also available from the Department of Energy is a complete list of all national laboratories that host teaching materials (*www.energy.gov/organization/labs-techcenters.htm*).

Another helpful source is the Florida Solar Energy Center (*www.fsec.ucf.edu/en/education*), which has developed solar energy units for grades K–12 and offers lesson plans to address many topics and every grade level. Some highlights are cooking using solar energy, photovoltaics, and fuel cells. Imagine a

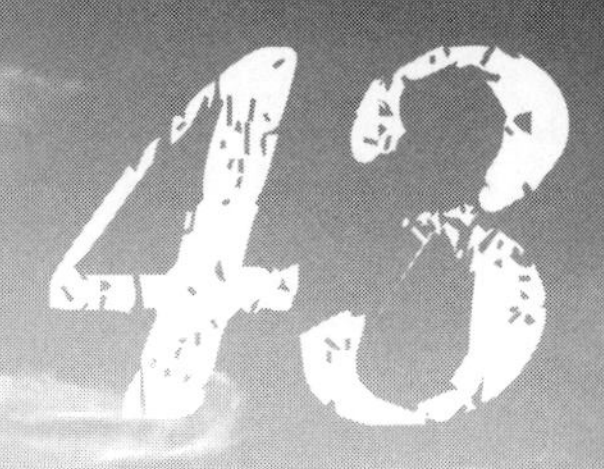

student's delight at firing off a rocket powered with the Sun's rays, or bringing home a batch of solar-baked cookies.

GreenLearning Canada (*www.re-energy.ca*) also shares excellent lessons that introduce students both to solar energy and solar electricity. Teacher lesson plans outline in detail materials instructors will need, as well as tools for assessing student progress. Our society is increasingly globalized and working with people not of our culture is a valuable life skill; learning about alternative energy use in another country will help U.S. students relate to their international peers.

Located in New Jersey, Global Learning Inc., has compiled many lessons and units for exploring solar energy. Some are specific to New Jersey, but all can be modified to fit into a variety of classrooms. Topics and lessons range from schools using solar energy, to ideas for science projects centered on tapping the energy of the Sun, to photovoltaics. To learn about any of these lessons, or to visit other sites recommended by Global Learning Inc., go to *www.globallearningnj.org/Solar1.htm*.

Lawrence Livermore National Laboratory (*www.llnl.gov*) offers continuing education opportunities to science educators interested in enriching their classroom teaching. One exciting option detailed at this site is the Department of Energy's Academies Creating Teacher Scientists. Teachers who are accepted to the program are eligible for course work over three summer sessions. During each eight-week session, teachers participate in hands-on experiments with science professionals working in the field. Teachers receive a stipend for each week worked and travel expenses, as well as funds to purchase equipment for their classrooms and to attend additional education opportunities such as conferences. There is no better way to improve student achievement and understanding in the classroom than through teachers' own education enrichment.

Another nationally supported resource can be found through the Argonne National Laboratory. While the laboratory itself is primarily a facility dedicated to nuclear energy research, its Division of Educational Programs provides teachers with curricular ideas for many different science topics (*www.newton.dep.anl.gov*). Lesson plan ideas are varied and span grade levels. Also on this site is "Ask a scientist," a portal that allows teachers and their students to correspond directly with practicing scientists; as every educator knows, real-world applications of school-based learning are a proven method for activating students' motivation.

Videos are another effective method for engaging students, and there are many options available. YouTube has hundreds of videos in its database relating to solar energy, including student-created research projects on solar energy use and how-to videos ranging from making a solar cell to installing

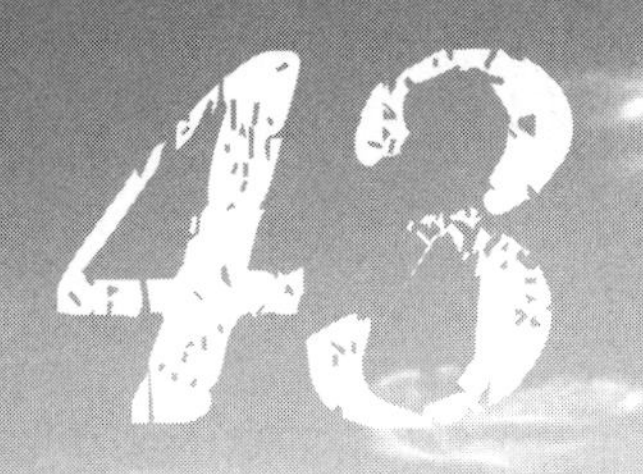

a solar system in a house. Some search terms that get good results are "solar energy for home," "solar energy facts," and "solar power." HowStuffWorks also has an informative video on solar energy (*http://videos.howstuffworks.com/multi-media-productions/1128-solar-energy-systems-video.htm*), and Green Energy TV has additional video and resources that are appropriate for the classroom (*http://greenenergytv.com*).

CONCLUSION

Experts, politicians, and large numbers of American citizens recognize that our present dependence on oil is a major obstacle to the progress of this nation; an examination of the potential of solar alternatives will likely be of considerable interest to young people. These matters extend beyond student interest, however, since they could contribute to a better future. The resources included here offer teachers many ways of engaging students that extend beyond lectures and involve young people in an active examination of potential new forms of energy.

Chapter 44
EVERYBODY TALKS ABOUT IT

By Steve Metz

"Everybody talks about the weather, but nobody does anything about it." The remark, generally attributed to Mark Twain, comes from a quote by Twain collaborator Charles D. Warner that appeared in the *Hartford Courant* on August 27, 1897. The comment undoubtedly got a laugh at the time. It was a favorite of my father, who often used it to describe various intractable problems, from traffic congestion to high taxes to underfunded schools.

The irony of Twain's joke, of course, is that we *were* doing something about the weather. In 1897 we were already 150 years into one of the most far-reaching experiments of all time. It goes something like this: Let's see what happens to our weather when we release massive amounts of carbon dioxide and other heat-trapping gases into the atmosphere. In 2007 we dumped about 30 gigatons—yes, *gigatons!*—of carbon dioxide into the atmosphere each year, largely from the burning of fossil fuels (IPCC 2007). Our worldwide petroleum habit currently consumes over 83 million barrels of oil *every day*, 24.1% of which occurs in the United States (BP 2007). What was a joke in 1897 is less of a laughing matter today. The results of this experiment are seen in news reports on a

daily basis: Global warming, disappearing glaciers, violent storms, rising sea levels, and other effects possibly related to changes in the atmosphere. The jury is still out on how much of all this can be attributed to anthropogenic causes, but there is little doubt that big changes are afoot on our planet.

Just in the past several years we seem to have reached a tipping point in the climate change discussion. Maybe it was The Intergovernmental Panel on Climate Change reports documenting changes in climate, or *An Inconvenient Truth,* the Academy Award–winning documentary film presented by Al Gore, or the Green Paper from the European Commission (COM 2007). Perhaps it was the April 2007 Supreme Court ruling that authorized the Environmental Protection Agency to regulate carbon dioxide emissions, or 2007's summer Live Earth concerts, reportedly watched by 2 billion people around the globe. Or maybe it was the striking visual images of melting glaciers and endangered polar bears. For whatever reason, climate change suddenly is big news.

In trying to educate my students about the exponential nature of our modification of the atmosphere, I like to point out that atmospheric CO_2 levels have increased dramatically, from about 280 parts per million (ppm) to almost 390 ppm, just since the beginning of the industrial revolution in the mid-1700s. I now think this might be a wrongheaded approach. Mention the year 1750 to an average teenager, and it will seem incredibly ancient—you might as well say "when dinosaurs roamed the Earth"—as all the past is generally relegated to the vast "dustbin of history." A better approach might be to discuss how two-thirds of this increase in atmospheric CO_2 has occurred during their parents' lifetimes (since 1960), and one-third just in their own short lives (since 1990). Even more striking: If current trends continue, the increase in CO_2 concentration in our students' lifetimes will *quadruple* the entire change that occurred before they were born. It is an unprecedented experiment in modifying global systems.

Chemistry students can investigate the reactions that produce CO_2 and other heat-trapping gases, study the solubility of CO_2 in Earth's water/oceans, or learn about the use of oxygen isotopes as proxies for long-term temperature measurement. In biology, students can study the precarious balance between respiration and photosynthesis, or consider the potential effects of climate change on delicate ecosystems. Physics classes can seamlessly incorporate weather and climate into energy investigations and discussions about the electromagnetic spectrum. In Earth science courses, understanding climate change is essential in comprehending glaciation, erosion, sea level changes, and other forces that have shaped—and continue to shape—our planet. Math classes can use rising CO_2 concentrations to illustrate exponential growth and history courses can discuss the effect of climate change on the rise and fall of civilizations.

An integrated, interdisciplinary investigation of climate change may perhaps inspire students to take action, on a personal or community level, to affect lifestyle and other changes that can help mitigate our negative impacts on the atmosphere. It is time that we prove Twain wrong, once and for all, by replacing his witticism with a new one for the 21st century: "Everybody talks about the weather, but now *we all* are doing something positive about it."

References

BP. 2007. Quantifying energy: BP statistical review of world energy. June 2007. *www.bp.com/liveassets/bp_internet/globalbp/STAGING/global_assets/downloads/S/statistical_review_of_world_energy_full_report_2007.pdf.*

Commission of the European Communities (COM). 2007. Adapting to climate change in Europe— options for EU action. Green paper from the Commission to the Council, the European Parliament, the European Economic, and Social Committee and the Committee of the Regions. *http://eurlex.europa.eu/LexUriServ/site/en/com/2007/com2007_0354en01.pdf.*

The Intergovernmental Panel on Climate Change (IPCC). 2007. Climate change 2007—The physical science basis. *www.ipcc.ch/pdf/assessment-report/ar4/wg1/ar4-wg1-frontmatter.pdf.*

Chapter 45

SCHOOL GREENHOUSE DESIGN TIPS

By James Biehle

Growing plants in a controlled environment such as a greenhouse can be a wonderful enhancement to any science program. Thoughtful planning, proper design, and faculty advocates who support the greenhouse as an important asset to their curriculums are necessary for a school greenhouse to be successful. This chapter discusses how to create a greenhouse at your school.

During the early planning stages of a project, before the budget is set and before the architect has located and designed the greenhouse, the faculty advocates need to identify how the greenhouse will be used in the science curriculum. Will large groups of students (e.g., a class of 24) be conducting activities

in the greenhouse at one time? Will the greenhouse be available to students at times other than normally scheduled class periods? Will individual or small group projects be carried out in the greenhouse in addition to activities involving the entire class? What type of climate should the greenhouse maintain? (The Missouri Botanical Garden, for example, has several greenhouse structures: one maintaining the climate of a tropical rain forest, another of a desert, another of the Mediterranean coast.) What types of plants will students grow and what will happen to the plants during summer vacations? Will other faculty members, student groups, and staff have access to and use of the greenhouse?

A greenhouse facility can be a large, freestanding facility in which students experiment with manipulating crop mutations and cross-pollination; a more modest space that is an integral part of the science department, surrounded by labs and classrooms; or simply an enlarged plant window that is extended out from the wall of a single lab/classroom to provide an area for living plants on a relatively small scale. Whatever the scale and location, design of the space is critical. Important design considerations include

- orientation,
- ventilation,
- cooling and heating,
- water supply and drainage,
- lighting,
- materials of construction and furnishings, and
- location.

ORIENTATION

Faculty must research the proper orientation and equipping of the greenhouse. In the northern hemisphere, the best orientation is due south; however, southeast and southwest orientations can also be acceptable. Greenhouses are usually constructed of glass or translucent extruded cellular polycarbonate and, therefore, allow a lot of daylight inside. This may be great for some school uses but too much light for others—many plants are sensitive to direct, intense sun that can burn leaves or cause plants to dry out quickly—which leads to the need to provide some form of shading.

Commercial greenhouses provide shading by means of shades of aluminum slats or mesh or ultraviolet-resistant fabric, which roll up and down on the outside of the structure like ordinary window shades. More elaborate and expensive options include motorized shades that can respond automatically to the amount of sunlight on a given day.

VENTILATION

Proper ventilation is crucial because without it the greenhouse can rapidly become unacceptably hot. Most greenhouses have vent panels in the roof and near the ground that can be manually opened and closed to allow fresh air in to mitigate heat buildup; the primary drawback to a manual operation is the chance that the person responsible will forget to close the vents either at night or on a cold day, or forget to open them the next morning. (For instance, I have lost plants in my greenhouse because I forgot to close the vent.) For a price, the vents can be automated and thermostatically controlled to open and close depending on the inside temperature of the greenhouse. An automatic ventilating fan system should also be considered with the fan at one end and an automated louver at the other, both controlled by a thermostat to bring in fresh air as needed.

COOLING AND HEATING

Evaporative cooling—the mechanical equivalent of the old cartoon of a person setting up a fan to blow across a block of ice—is a relatively inexpensive way of cooling a greenhouse while at the same time increasing the humidity level. Such a "swamp cooler" should be controlled by a thermostat.

In most climates in the United States, heating the greenhouse is also a necessity. While the space can rapidly heat up during a sunny day, it can quickly become cold enough to allow plants to freeze at night or even during the day when the weather outside is frosty. Small gas furnaces can serve the greenhouse space itself or the building heating system can be extended into the greenhouse. In a St. Louis–area school renovation, where a small lean-to greenhouse was installed on the roof of a heated space, the school's steam heating line was extended up through the floor and a small radiator with a thermostatically controlled steam valve kept the greenhouse warm on cold days. Faculty could also consider using solar heat reradiated from a *Trombe wall*—a massive wall facing the sun that absorbs the sun's heat during the day and reradiates it at night—as an energy-saving source that can be supplemented by a more traditional heat source.

WATER SUPPLY AND DRAINAGE

Another necessity, probably a no-brainer but occasionally overlooked, is water supply. Plants do generally need to be watered either manually or by an automated system. My small greenhouse includes a laundry sink with both hot and cold water and a threaded faucet to which I attach a hose and watering wand. Larger greenhouses should probably have one or two hose bibs centrally located to allow the connection of a hydroponic watering system or simply a hose. Also, when watering plants, the overflow needs someplace

to go, which requires one or more floor drains. The floor of the greenhouse should be constructed to slope toward these drains.

LIGHTING

Lighting should also be provided. Waterproof fixtures should be chosen, probably with fluorescent lamps; how the greenhouse will be used should determine how it is lit. Are grow lights needed to supplement the daylight? If the greenhouse is also an architectural amenity, should it be lit at night? Faculty should also consider the need for electrical receptacles within the greenhouse. Both the lights and the receptacles should be on ground fault circuit interrupter (GFCI) outlets.

MATERIALS OF CONSTRUCTION AND FURNISHINGS

The floor in a greenhouse should probably be bare, sealed concrete. Rubber mats can be placed in walkways between plant benches, allowing water to flow along the floor while minimizing the chance for a person to slip and fall. Plant benches should be either aluminum or structural plastic that is resistant to ultraviolet radiation. Freestanding plant benches will allow flexibility in arrangement as well as in periodic cleanup. A table or counter should be provided as a potting bench and tall cabinets for storage of materials and supplies. Such casework should not be made of wood or plastic laminate–covered particleboard as these materials will rapidly deteriorate in a moist environment. A door directly to the outdoors can often be helpful, particularly if the greenhouse is adjacent to an outdoor garden space.

LOCATION

Finally, faculty should make sure that the resulting greenhouse design works as a greenhouse—it is used and cared for, and not just a visual, remote addition to school grounds. Location is important. When properly planned and designed, a good working greenhouse or plant window can be both a science teaching resource as well as a wonderful architectural amenity.

Adapted with permission from Motz, L. L., J. T. Biehle, and S. S. West. 2007. *NSTA guide to school science facilities.* 2nd Ed. Arlington, VA: NSTA Press.

Chapter 46

CIRCUIT SAFETY

By Ken Roy

The study of electricity in general science or physics is fascinating for students. Observing electrons in motion, creating "lightning" in the laboratory, transferring electrons using static electricity, and watching hair stand on end due to volts from a Van de Graaff generator are often highlights for students. Unfortunately, a number of electrical dangers exist in the laboratory that are applicable to all types of science including biology, chemistry, physics, and Earth science. Electric shocks can be received from wall or bench receptacles, electrical switches, leads connecting equipment, extension cords, frayed wires, and more. One reason for safety concerns is the danger associated with exposure to an electrical current. Varying degrees of exposure can lead to paralysis, unconsciousness, or even death.

This chapter outlines the basic electrical dangers in an effort to show how the study and use of electricity can be fun *and* safe.

ELECTRICAL LAB HAZARDS: COME OUT WHEREVER YOU ARE

A lab has hidden electrical dangers that teachers and students must be aware of when working on science activities. The following list reviews potential dangers:

- *Batteries (dry cell):* Most science labs use dry cell batteries because they tend to be safe. Unless many of these batteries are linked together, it is not possible to secure an electric shock from the 1.5 V power source. Never try to recharge these batteries, however. Doing so is dangerous and could potentially result in an explosion.
- *Batteries (rechargeable):* The nice part about rechargeable batteries is the fact that they can be used many times. The down side is that the batteries can get hot if short circuited or charged with an incompatible charger. Make sure to take precaution when handling rechargeable batteries.
- *Batteries (car/auto):* Car batteries can be 12 V or more. They are dangerous for a number of reasons, especially because of their high amperage and containment of hydrogen gas. Such batteries might be used with caution in a high school science physics lab, but should not be used in lower-level labs.
- *Power supplies (low voltage):* Low-voltage power supplies usually provide a safe voltage up to about 12 V and cannot drive a current through a body to provide a shock. Low-voltage power supplies are a less expensive alternative to constantly replacing batteries for a high frequency of activities.
- *Power supplies (high voltage):* High-voltage power supplies can provide significant currents up to 150 mA and voltage between 40 and 500. The power supplies are inappropriate for use in most, if not all, high school laboratories.
- *Electrostatic machines:* Wimshurst machines and Van de Graaff generators are often found in high school laboratories. They can produce electrostatic charges at very high potentials and hundreds of thousands of volts but have relatively small amount of charge. These machines can be dangerous, however, and should only be operated by a knowledgeable adult doing demonstrations.
- *Wall receptacle:* Most labs have ground fault interrupt circuits (GFCI)-protected wall receptacles around 120 V. They also can be dangerous.

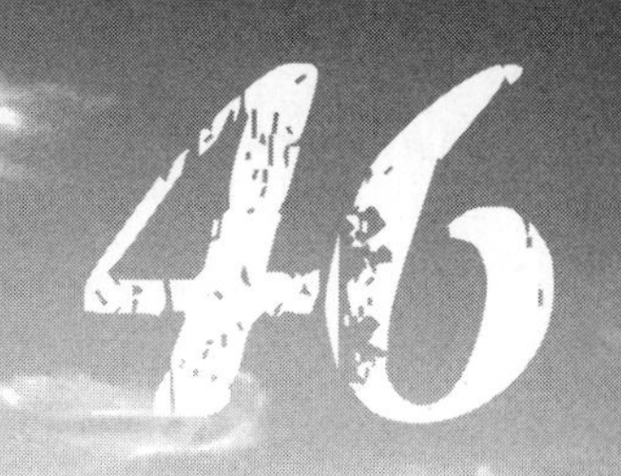

No safety device can protect the user against shocks. Never touch metal prongs in a wall receptacle even if it is GFCI protected.

PRUDENT ELECTRICAL PRACTICES

When using electricity, all laboratory occupants need to be trained in appropriate safety behavior. The following A–Z list provides a good starting point for training:

A. Make sure all laboratory occupants are aware of the appropriate use of electricity and dangers of misuse and abuse.

B. Know where the master switch is for electricity in the laboratory in case of an emergency and make sure signage points to the switch.

C. Before inspecting equipment, turn off power by unplugging the equipment or by shutting off the circuit breaker.

D. When using a circuit breaker, wear an insulated glove, turn your face away from the box, and then flip the switch. Arcing can occur!

E. To avoid making a closed circuit through the body when checking an operating circuit, keep one hand either in a pocket or behind your back.

F. Do not allow the cover plates of electrical receptacles or switches to be removed.

G. Prevent trip/fall hazards by placing wires away from places where people walk.

H. All conductive or metallic jewelry should be removed before working with electricity to help prevent shock.

I. Never pull the wire when unplugging cords. Always pull cords from the plug at the socket.

J. It pays to be water phobic when working around electricity. To help prevent shock, never use water or have wet hands when dealing with cords, plugs, and so on.

K. Use only GFI- or GFCI-protected circuits when working in the laboratory.

L. All tools, lab equipment, and so on should have nonconducting handles when working with electrical devices.

M. Use only double-insulated handheld equipment.

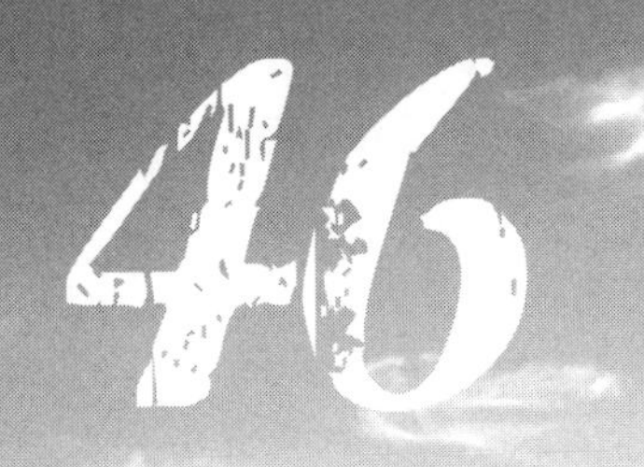

N. To prevent explosion or fire, never use or store highly flammable chemicals near electrical equipment.

O. Leads with 40 V or more should be shrouded or taped for shock protection.

P. Check circuits for proper grounding with respect to the power source.

Q. Access to electrical panels and disconnect switches must remain clear and unobstructed.

R. When using batteries, always inspect them first for cracks and leaking. Discard the batteries if any of these conditions occur.

S. The contents of batteries are corrosive and can be toxic or poisonous. Never open them.

T. When storing batteries, never allow the terminals to touch or short circuit.

U. Utility pipes such as water and gas are grounded. Do not touch an electrical circuit and utility pipe at the same time.

V. Never plug damaged electrical equipment into a wall receptacle. This includes frayed wires, missing ground pin, or bent plugs.

W. Never use extension cords in the lab. They are trip and potential electrical hazards.

X. For routine maintenance such as changing bulbs, make sure the device is unplugged before initiating the work. Review OSHA Lockout/Tagout Standard (29CFR 1910.147 and 1910.333) prior to working on any electrical device.

Y. Do not use water on an electrical fire. Only use ABC fire extinguishers.

Z. If a shock incident occurs, try to turn off the power via the master electricity disconnect in the lab. If that is not possible, use a dry insulating material (e.g., clothing or wood) to remove a victim from a live electrical circuit. Call the nurse immediately and begin CPR by a certified person.

DANGERS EXIST EVEN IN THE BIOLOGY LABORATORY

Numerous sources of electricity (and consequently potential dangers) exist in today's biology laboratories including microscopes with electrical plugs, autoclaves, incubators, microwaves, and electrophoresis units. An agarose

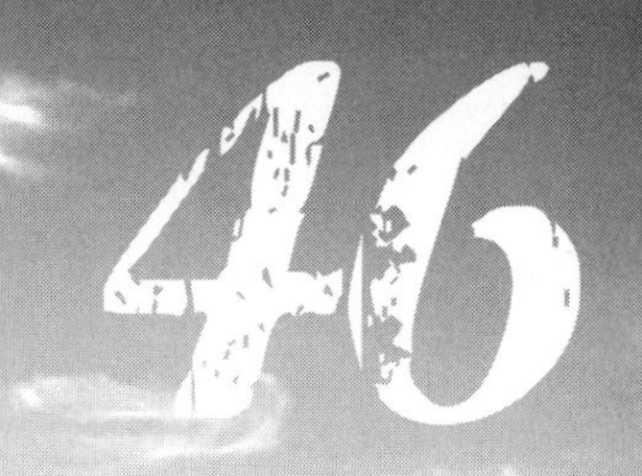

gel electrophoresis unit using around 100 V can cause a lethal shock with a current of 25 mA under certain conditions. Special circuit safety needs to be enforced with the use of this special equipment:

- Avoid unintentional grounding points and conductors such as sinks, jewelry, and pipes. Only nonconducting benches of wood or plastic are required.
- A thin film of moisture such as sweat serves as a good conductor of electricity. Think before you touch any part of the apparatus.
- Never run electrophoresis equipment while unattended.
- Never touch any cooling apparatus connected to the gel. The contents of the tubing can carry current.
- Should the electrophoresis buffer leak or spill, stop immediately and clean up the bench.
- Turn off the main power supply switch and wait 15 seconds (the voltage may not be completely discharged from internal capacitors) before making any disconnection or connections. After use, turn the power supply off before disconnecting both leads from the power supply.
- Always post appropriate warning signs on the power supply and buffer tanks.

FINAL WORD

The study of electricity can be fun. It is imperative, however, that all laboratory participants—students, teachers, and visitors—are prepared to deal with electricity safely for a rewarding and exciting experience.

Chapter 47

BATTERY SAFETY BASICS

By Ken Roy

Batteries commonly used in flashlights and other household devices produce hydrogen gas as a product of zinc electrode corrosion. The amount of gas produced is affected by the batteries' design and charge rate. Dangerous levels of hydrogen gas can be released if battery types are mixed, batteries are damaged, batteries are of different ages, or batteries are inserted incorrectly. The hydrogen gas can cause the battery compartment to rupture, allowing the hydrogen gas to mix with oxygen. This mixture of gases can cause a powerful explosion in the presence of a spark or excessive heat. However, when used properly, batteries are quite safe for classroom use. To reduce the risks associated with batteries, consider the following guidelines:

1. Use caution when linking dry-cell batteries. Doing so increases voltage when done in series, and amperage when done in parallel.
2. Use caution when working with rechargeable batteries. They can get very hot if they short circuit or are recharged with an incompatible charger.
3. Keep car batteries out of the lab.
4. Never mix different brands of batteries.
5. Never mix new and old batteries. The newer batteries can charge the older batteries and effect a voltage reversal with violent action.
6. Purchase only manufacturer-recommended products and accessories. Beware of inexpensive substandard batteries that might not meet U.S. safety standards.
7. Never mutilate (i.e.., crush or puncture) batteries. Hazardous chemical leakage can occur.
8. Never store batteries in equipment for a long period of time. Doing so can cause chemical leakage.
9. Never try to recharge batteries that weren't designed to be rechargeable.
10. Never overdischarge batteries. When they no longer can operate equipment, remove them.
11. Never use excessive force to install batteries or remove them from equipment. This can cause equipment damage, battery damage, and personal injury.
12. Never get batteries wet or use them wet. Discard batteries that are swelling or leaking—these are signs of corrosion and other potential safety issues.
13. Do not keep used batteries and remember to dispose of them properly.

STUDENT CODE OF BATTERY CONDUCT

When using batteries, students should be trained on appropriate safety behavior. The following is a simple list to get teachers started:

- Be aware of the hazards associated with using batteries, e.g., shock, corrosive chemicals, and explosive potential.
- When using batteries, always inspect them first for cracks, leaking, etc. Discard if any of these conditions occur.

- Never open a battery. The contents are corrosive and can be toxic or poisonous.
- When storing batteries, never allow the terminals to touch the terminals of other batteries.
- Be water phobic when working around batteries. Never use water or have wet hands when dealing with them.
- Always make sure the same style, size, voltage, amperage, and brand of batteries are used together.
- Handle batteries with care.
- Follow specific directions provided by the teacher for using batteries as a power source.
- Dispose of batteries properly.

On the web

Exploding flashlights: Are they a serious threat to worker safety?: *www.cdc.gov/niosh/fact0002.html*

Chapter 48

SUN SAFETY

THE STATS

By Ken Roy

According to the Health Physics Society (HPS) and the American Cancer Society (ACS), one in five persons will be diagnosed with skin cancer during their lifetime. Each year, more than one million new basal cell or squamous cell skin cancers are diagnosed in the United States alone. In addition, the most serious form of skin cancer—malignant melanoma—will be diagnosed in over 50,000 persons. Annually, approximately 10,000 people die of skin cancer, including over 7,500 from melanoma. It is worth noting that approximately 65–90% of melanomas are caused by ultraviolet (UV) radiation (CDC 2002).

What do these statistics have to do with high school science teachers? A number of professional associations, including the HPS, strongly encourage schools to develop skin cancer policies and take an active role in educating students about UV exposure. Science teachers can help protect themselves and their students by providing information on skin cancer, its causes, and prevention strategies. This is not just a summer issue, but one that affects us year-round!

BACKGROUND CHECK

There are several risk factors known to influence the development of skin cancer (CDC 2002):

- Excessive UV exposure: Sources of UV radiation, such as sunlight, sunlamps, and tanning beds, increase the risk for all major forms of skin cancer. Artificial sources of UV are carcinogenic and should be avoided.
- Childhood and adolescent UV exposure: Those with a history of more than one sunburn in childhood or adolescence have a higher risk of developing melanoma.

- Skin color and ethnicity: Darkly pigmented persons develop skin cancer at lower rates than lightly pigmented persons. Those with light hair or skin that freckles easily have a higher risk of developing skin cancer.
- Moles: A measurable predictor of melanoma is having large numbers or unusual types of moles.
- Family history: The risk of melanoma can increase to eight times the normal level depending on the number of affected relatives.
- Age: The older the individual, the more exponential the increase in the incidence of skin cancer.
- Environmental factors: Latitudes closer to the equator increase the UV radiation exposure level. Other environmental factors include cloud coverage, materials that reflect the Sun (e.g., water and sand), time of day outside, spring and summer exposure, and extent of ozone depletion.

SCHOOLS AND PREVENTION

With their background knowledge and understanding of physics and biology, science teachers can help educate students about risk factors and can play leadership roles in helping develop skin cancer prevention guidelines. The CDC's National Center for Chronic Disease Prevention and Health Promotion offers guidelines for skin cancer prevention efforts in schools (CDC 2002). Schools can offer education and skill-building activities to reinforce healthful behaviors in science courses or curricula, or in a school health program. These efforts can include developing polices to reduce UV exposure (e.g., use of sunscreen at school); providing information about the knowledge, attitudes, and behavioral skills needed to prevent skin cancer; providing professional development programs on sun sense for school employees; and involving family members in skin cancer prevention efforts. Science teachers should also promote sun-safety practices, such as the use of hats, long-sleeve shirts, and UV protective sunglasses, on field trips or on-site activities. The following prevention strategies should be periodically reviewed with students and supported by teachers (ACS 2009):

- Avoid the Sun between 10 a.m. and 4 p.m., especially for long periods of time.
- Look for shade, especially in the middle of the day when the Sun's rays are strongest. Practice the shadow rule: If your shadow is shorter than you, the Sun's rays are at their strongest.

- Cover up with protective clothing to guard as much skin as possible when you are out in the Sun. Choose comfortable clothes made of tightly woven fabrics that you cannot see through when held up to a light.
- Use sunscreen and lip balm with a sun protection factor (SPF) of 15 or higher.
- Cover your head with a wide-brimmed hat, shading your face, ears, and neck.
- Wear sunglasses with 99–100% UV absorption to provide optimal protection for the eyes.
- Follow these practices to protect your skin even on cloudy or overcast days, as UV rays travel through clouds.

FINAL THOUGHTS

With their knowledge base of biology and physics concepts, science teachers can have a major impact on fellow school employees and students by fostering good sun sense. Your students may not thank you now, but they will be grateful for their skin's health in the future!

Acknowledgment

Special thanks to dermatologist Andrew V. Atton in Glastonbury, Connecticut, for his professional review and contributions to this chapter.

References

American Cancer Society (ACS). 2009. Skin cancer facts. *www.cancer.org/Cancer/CancerCauses/SunandUVExposure/skin-cancer-facts*

Centers for Disease Control and Prevention (CDC). 2002. Guidelines for school programs to prevent skin cancer. Morbidity and Mortality Weekly Report 51 (RR04): 1–16.

INDEX

Page numbers printed in **boldface** type refer to figures or tables; those followed by n refer to footnotes.

INDEX

Q

R

S